Humid tropical
geomorphology

Humid tropical geomorphology

A study of the geomorphological processes and landforms in warm humid climates

A. Faniran and L. K. Jeje

Longman   London   Lagos   New York

Longman Group Limited
Longman House, Burnt Mill, Harlow,
Essex CM20 2JE, England
and Associated Companies
throughout the World

Longman Nigeria Limited
Ikeja, Ibadan, Owerri,
Zaria and representatives
throughout Nigeria

Published in the United States of America
by Longman Inc., New York

First published 1983

British Library Cataloguing in Publication Data

Faniran, A.
  Humid tropical geomorphology.
  1. Geomorphology
  I. Title   II. Jeje, L. K.
  551.4   GB401.5

  ISBN 0-582-64346-5
  ISBN 0-582-64351-1 Pbk

Library of Congress Cataloging in Publication Data

Faniran, A.
  Humid tropical geomorphology.

  Bibliography: p.
  Includes index.
  1. Geomorphology–Tropics. I. Jeje, L. K.
  II. Title.
GB446.F36 1983   551.4'0913   82-14896
ISBN 0-582-64346-5
ISBN 0-582-64351-1 (pbk.)

Printed in Great Britain at The Pitman Press, Bath

Contents

Preface

The humid tropical region appears currently to be in some ways at the frontier of scientific research. The reasons for this include the realization of late that models and ideas developed and applied in the humid temperate regions in particular are not always suitable for their hot and wet counterparts.

The incongruity applies to weather and climate, especially to meteorology, as well as to soils, vegetation and landforms. In each case, scientists, both indigenous and expatriates, have discovered 'unique' features which warrant special study. The result is the appearance of books and other forms of publications on the humid tropical regions.

In the area of geomorphology, several attempts have been made to highlight these 'unique' aspects, the boldest ones being Tricart and Callieux, *Landforms of the Humid Tropics, Forests and Savannas* and Thomas, *Tropical Geomorphology*. Moreover, several conferences and seminars have been held on the topic of humid tropical geomorphology, and their proceedings published (for example, J. Alexandre's *Dynamic Geomorphology in Tropical Regions*). However, the number and scope of these publications are limited, and, in fact, none has been found suitable as a major textbook for students in this region.

The present work is an attempt to provide in a single volume and in a systematic format the major contributions so far made in the field of tropical geomorphology. Since, perforce, emphasis in this kind of work should be on process, many chapters are so devoted. Even those on 'specific' landscapes (Section C) see these in relation to the operation of humid tropical processes and materials. Nevertheless, and because of the state of our knowledge, no conscious attempts have been made in the volume to create the impression of absolute uniqueness. Virtually all the forms and processes encountered in the humid tropical regions also occur in several other climatic regions: the difference lies, in certain cases, in the magnitude and intensity of these processes which create uniqueness in degree rather than in kind.

The book is written for geography and geology university students, especially those interested in geomorphology. Development agents, tourists, and others interested in the tropical region would also find it useful.

The authors acknowledge the keen interest and professional competence of the publishers as well as the cooperation of the clerical and technical staff of the two departments in which the co-authors work.

Adetoye Faniran

L. K. Jeje

List of tables

List of figures

List of Plates

SECTION A
Introduction

1 Geomorphology in the humid tropics

The term *geomorphology* in its simplest connotation means the science of landforms (from the Greek words *ge* (earth), *morphos* (shape); and *logos*, (a discourse). In practical terms, it involves the study of the form (shape) and development (process) of the earth's surface features with special reference to landforms. Geomorphology, especially in modern times, also seeks to investigate the nature of the intra- and inter-relationships of forms and processes. In this way, it has moved far away from the historically based definition of 'the science of the evolution of landforms'.

The history or beginning of geomorphology, like that of most sciences, can be traced essentially to the ancient Greeks, with notable additions by subsequent workers, especially the Romans and Arabs, who recognized many landforms and landscapes, as well as the physical and other processes of their formation. The discipline also benefited immensely from the work of mediaeval scientists such as Leonardo da Vinci (1452–1519), Agricola (Georg Bauer) (1494–1555), and Bernard Palissy (*c*. 1510 to *c*. 1589), all of whom recognized and studied fluvial processes of land formation and thereby laid to rest the biblical idea of the diastrophic origin of the earth and its surface features. The history of geomorphology as a discipline is already very well told in many books (see Chorley *et al.*, 1964) and so needs no reiteration here. Suffice it to say that it features a number of landmarks, including its birth with the Neptunists in the eighteenth century; the development of the uniformitarian concept of landform formation by Charles Lyell (1797–1875), among others; the formulation and systematization of the subject as a scientific discipline in its own right by W. M. Davis, otherwise known as the Davisian cyclical synthesis; the growth of climatic geomorphology, from the work of Powell, Gilbert, Dutton, von Richtofen, Passarge and Davis among others; and the deluge of quantitative analysis in the post-1950 era. This latest phase, starting with the work of Horton (1945), has helped to orientate geomorphology towards a more secure scientific footing through serious concern with measurement, vigorous statistical analysis of data, association with kindred sciences as well as with practical problems; and a conscious attempt to break with the tradition of narrow scholasticism which characterized much of the previous works.

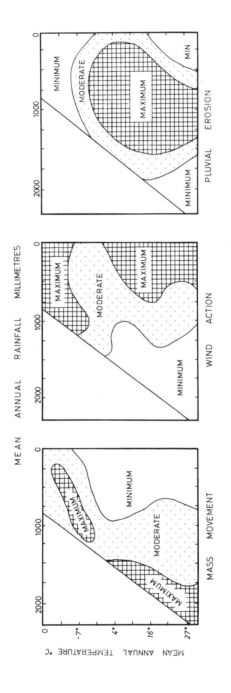

Fig. 1.1 Climate and morphogenetic processes (after Peltier, 1950)

A number of the landmarks enumerated above have of course developed into specific fields of study – namely, dynamic or process geomorphology, which is concerned with the study of processes such as weathering, pedogenesis, denudation and deposition; structural geomorphology, with emphasis on geological attributes such as endogenic dynamic processes and the effect of rocks and rock structures; and climatic, including climato-genetic, geomorphology, to name only a few. The present text is based on the principles and assumptions of climatic (climato-genetic) geomorphology; without necessarily offering it as an apologia, for that scheme, like most of the others, is yet to become firmly established.

According to the protagonists of the concept of climatic geomorphology, variations in climatic elements exert a controlling influence on all dynamic (geomorphological) processes and take precedence over all other components, including structural components. Consequently, every climatic zone has its own specific complex of dynamic processes which create distinguishable landforms and landform assemblages (Figure 1.1).

The origins of climatic geomorphology can be traced to the work of von Richtofen, who worked in China; Jessen, Walther and Thorbecke in Africa; and Sapper in Central America and Malaysia. These pioneers were followed by scholars like Büdel, Peltier, Cailleux, Tricart and others, whose works remain a legacy to the entire field of geomorphology.

The concept of climatic geomorphology has some central themes. The most important of these themes are that (a) landforms and landform processes differ significantly between different climatic areas; (b) that these differences are the result of areal variations in climatic elements or parameters; and (c) that climatic changes, including the most recent ones in Quaternary times, have not masked the evidence of climate – landform relationships where they exist. We discuss each of these themes briefly.

Distinctive climato-genetic landforms

The concept that a specific climate can produce a distinctive landform assemblage probably dates to Davis (1899) and his theory of landform evolution, a theory later highlighted by Cotton (1942), who described the climatically controlled savanna landscape. Climatic geomorphology postulates that different climates, by affecting processes, develop unique assemblages of landforms. These processes and their relationship with climate have been systematically analyzed with the view to defining morphogenetic regions on a world-wide basis (Peltier, 1950; Büdel, 1948, 1963; and Tricart and Cailleux, 1965). Thus Peltier (1950) defines climatic–morphogenetic regions using simplified climatic parameters such as mean annual rainfall and temperature with reference to their hypothetical effects on weathering and erosion processes. Among his resultant morpho-climatic regions, he recognizes a zone of intense chemical weathering (Selva) – the humid tropics. Further, the analysis by Peltier in 1962 suggests first-order distinctions between glaciated,

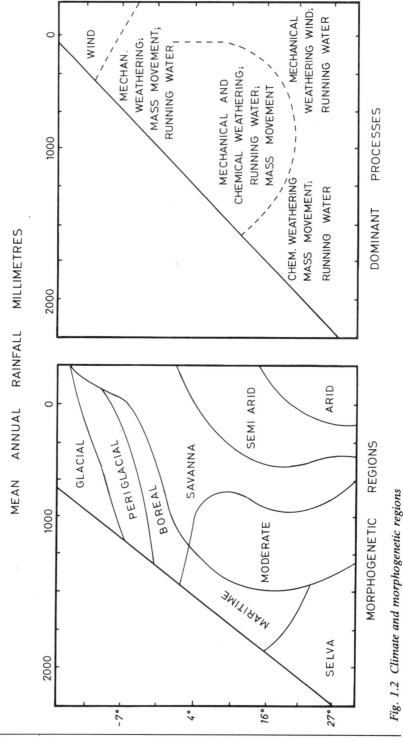

Fig. 1.2 Climate and morphogenetic regions

tropical and all other fluvial landscapes and second-order differences between desert, semi-desert and temperate fluvial landforms (Figure 1.2). Büdel (1948, 1963) and Tricart and Cailleux (1965), also using this synthetic approach, similarly identified among other morpho-climatic regions a humid tropical morphologic zone. Strakhov (1967) also recognizes the humid tropics as the zone of maximum efficiency of chemical weathering (see Figure 3.4).

Common to all these schemes is their over-generalization and subjectivity. In fact, no serious empirical evidence is yet adduced from morphometric analyses to substantiate the idea of distinctive climatogenetic landforms. The main reason for this situation is the material available for most morphometric work in geomorphology – the topographical map. Not only are these maps, especially those of the developing world, defective in the information portrayed (Eyles, 1966; Ebisemiju, 1976), but the best of them still fail to demonstrate genuine and recognizable differences in the landform of areas with different rainfall, temperature and even vegetational groups. It is only when one operates either at the broad, generalized level (see Peltier, 1950, 1962), or at the level of specific and peculiar landscape components, that the envisaged 'distinctive' climato-genetic landforms are brought out or identified. Also, the chances that the results obtained were fortuitous, given the coarseness of the measures used, cannot be ruled out; this type of work also might have suffered from the very common error of 'forcing' an answer to justify what is often referred to as 'an illusion of differences' in landform to be expected from observed differences in vegetation cover (Stoddart, 1969). In fact, Peltier's work is the best available example of the inductive approach to the recognition of climato-genetic regions (see Peltier, 1950).

This is not to deny relationship between landforms and the controlling factors, especially of climate and vegetation. In fact, a close study undertaken in Fiji (Viti Levu) shows clearly that there is finer topographic dissection of the drier, grass-covered leeward side than the forest-covered windward side; which information is often concealed in most aerial photographic prints as well as topographical maps. The implication here is that detailed study of forms and processes is necessary to recognize better and more convincing evidence of climatically controlled landforms and landform assemblage.

Other authors have approached the problem of defining morphological zones differently by relating sediment yield in rivers to climate. Thus Douglas (1969), in his study of sediment yields in rivers, and of volumes of sediments in dated depositional basins, shows that surface erosion in the humid tropics differs from that in other parts of the earth's surface. Also Fournier (1960) and Corbel (1964), analyzing erosion data for different climatic zones, demonstrate that climate affects sediment yield, although relief is an important factor (cf. Tables 1.1; 1.2).

Notwithstanding the recognition of the humid tropics as a morpho-climatic zone by these scholars, other researchers do not support the concept of a tropical zone within which some distinctive geomorphic processes and landform assemblages occur. King (1967) argues that hill-slopes all over the earth's surface undergo the same processes of parallel retreat, with differences in form related to structure or changing importance of certain slope elements. Fripiat and Herbillion (1971) state clearly that they 'do not believe that transformation sequences or synthesis mechanisms observed in tropical regions are essentially different from those occurring under other climatic conditions'. Stoddart (1969), reviewing the work of Langbein and Schumm (1958) and Schumm (1965), points out that since several factors besides temperature and rainfall control sediment yield – most especially man's influence, which may increase sediment yield by up to 100 per cent – the erection of a system of climatic regions based on sediment yield may be difficult to sustain.

Notwithstanding these objections, it is evident that certain landform assemblages can be attributed to present-day factors and processes and thus to climate, as in the case of periglacial and desert landforms. Although certain landforms, products of deep chemical weathering and subsequent erosion, such as tors, corestones, inselbergs, pediments, duricrust capped hills and duricrust pavements, are generally regarded as characteristic of the tropics, these landforms have also been observed in other climatic zones. Since there is evidence to show that climate has changed several times in all the present climatic zones through the Tertiary into the Quaternary, it is uncertain under which climate these landforms evolved in the extra-tropical areas. In any case, there is evidence to suggest that some of these landforms in extra-tropical areas predate the prevailing climates – for example, tors in Dartmoor, England, have their origin in the past under climatic regimes similar to what now obtain in the humid tropics (Linton, 1955). Jahn (1974) also concedes that the granitic tors in the Sudeten mountains in Poland owe their origin to deep chemical weathering in the late Tertiary. In fact, Bakker and Levelt (1964) and Bakker (1967) have used deep chemical weathering as evidence of tropical Tertiary climates in parts of Europe. Thus, to the extent that several landforms in the humid tropics have resulted largely from continuous deep chemical weathering and the erosional stripping of the regolith, these landforms can be regarded as unique to this environment.

However, in proffering such arguments one must be aware of other problems, especially that of equifinality; that is, that very few forms have unambiguous features indicative of origin, as similar forms can result from the operation of markedly different processes. Thus what needs to be done and has been attempted in this volume, in the spirit of present-day geomorphology, is to study process–form relationships at varying levels including the extent and nature of climatic controls on the forms and processes of landforms.

Climatic control on geomorphic processes and forms

The major sources of energy for geomorphological processes, particularly the exogenic types, relate to climatic variables – namely, atmospheric and soil moisture and temperature. These processes, including weathering, erosion and deposition, are later discussed in separate chapters (chapters 3, 4 and 5 respectively); only brief comments are made here on weathering and erosion in an attempt to illustrate the general theme of this book.

The general relationship between rate and nature of weathering and atmospheric and soil climate is best illustrated by the zonal concept of weathering and soils, originally enunciated by Russian scientists, including Dokuchaiev and Glinka (cf. Rose, 1962). At the broadest level, distinction is made between the mechanical processes dominant in periglacial and hot desert regions and the chemical/biological processes of the warm humid climates. Examples of detailed field studies of this broad generalization are few, but are growing with time. In South Africa, soil engineers have found a good empirical relationship between soil character and some climatic parameters – for example, potential January (warmest month) evaporation, expressed statistically as

$$N = 12 \frac{EJ}{Pa} , \ldots (1.1)$$

where EJ = Potential January evaporation,
and Pa = annual precipitation.

The isopleth $N = 5$ closely corresponds with the boundary between disintegration–product (mechanical weathering) soils ($N>5$) and decomposition–product (chemical weathering) soils ($N<5$). With $N>6$, hydrous mica is prominent in the weathered product; with N lying between 2 and 5, basic rocks weather to montmorillonite and acid rocks to kaolinite; with $N<2$, kaolinite is more common than montmorillonite; and with $N<1$, lateritization occurs (Figure 1.3). The distinction made here between clay minerals and type of weathering (mechanical/chemi-

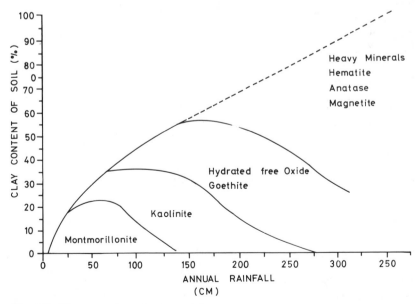

Fig. 1.3 Climate and weathering products

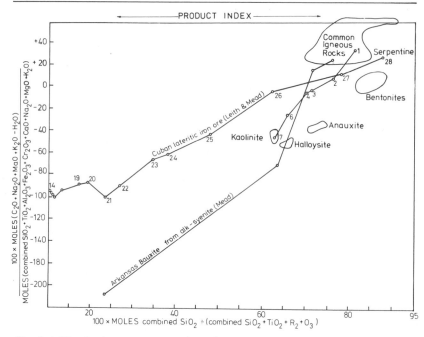

Fig. 1.4 Weathering processes and products

cal) is definitely of great geomorphological significance (cf. Tanada, 1951; Sherman and Uechara, 1956; Keller, 1957; cf. also Figure 1.4).

Similarly, in recent times, increasing attention has been given to the study of erosional processes and forms in different climatic regions, following Peltier's (1950) proposition of 9 morpho-climatic regions. Examples include Rapp (1960) and Jahn (1961) for periglacial environments and Rougerie (1960) and Douglas (1967, 1968) for humid tropical environments. Others, especially those which attempt to

TABLE 1.1 Total world erosion (m^3 × 10^6) per annum

	Arid	Normal	Humid	Total
Warm				
Equatorial	1.0	300.0	52.5	353.5
Inter-tropical	3.7	225.0	20.0	248.7
Extra-tropical	19.0	360.0	80.0	459.0
Total				
Temperate	175.0	885.0	152.5	1 061.0
Cold		1 050.0	325.0	1 550.0
Extra-polar	770.0	550.0	90.0	710.0
Polar	77.5	350.0	75.0	502.5
Total	147.5	900.0	165.0	1 212.5
Total unglaciated	346.2	2 835.0	642.5	3 823.7
Glaciated lands	550.0	4 000.0	200.0	4 750.0

Source: Corbel, 1964, pp. 410–11.

TABLE 1.2 Rates of regional erosion (m²/km²/yr)

Climate	Arid under 200 mm		Normal 200–1 500 mm		Humid Over 1 500 mm	
	Mountain	Plain	Mountain	Plain	Mountain	Plain
Hot 15°N–15°S	1.0	0.5	25	10	30	15
Tropical 15°–23.5°N and S	1.0	0.5	30	15	40	20
Extra-tropical Temperature over 15°C	4	1	100	20	100	30
Temperate Temperature 0–13°C	50	10	100	30	150	40
Cold Temperature under 0°C	50	15	100	30	180	–
Polar	50	15	100	30	150	–
Glaciated, polar	50		1 000		2 000	
Glaciated non-polar	–		–	–	2 000	

Source: Corbel, 1964, p. 407

summarize available information at different times, include Corbel (1959, 1964), Fournier (1960) and Strakhov (1967).

The conclusions drawn by the various workers, although contradictory in certain cases, point in at least two directions. First, they hint that the nature and rate of both weathering and erosion processes vary in different climatic regimes, which differences can be studied both experimentally in the laboratory and empirically in the field. Secondly, the available information is not sufficient for the establishment of laws (predictive statements), which can make sense only after a lot more detailed work has been carried out (Douglas, 1966, 1968a, b). It is the view of the present authors that many more of such studies will need to come from the humid tropical regions where the operative processes are still not well known because of the paucity of useful data.

Climatic change and climato-genetic landforms

That climates have changed over geologic time is now firmly established in the scientific literature. Some of these changes occur over a fairly long period, whereas others occur in short and quick successions. The

Quaternary period appears to illustrate the latter type of climatic change, during which over a period of about 2 million years as many as between 8 and 10 fluctuations have been recorded in some places.

One important implication of short-lived climatic fluctuations is that equilibrium landforms will not have time to develop, which by extension means that the zonation of landform processes and forms will be very unlikely. It is possible that this is what obtains over a large part of the world, and so makes the establishment of zonal landforms so difficult. If climates have changed continuously in the last 50 million years, and rather more rapidly in the last 2 million years, any attempt to establish a firm correlation between present-day processes and the landforms on which they operate is certain to be a fruitless exercise.

However, apart from zonal soils and climax vegetations, many workers have described regional (zonal) landscapes, either on the basis of broad, generalized characteristics or because of local differences (cf. Birot, 1960; Tricart, 1972; Thomas, 1974). In the light of these accounts the idea of short-lived variations will need to be modified. It may be that these short-lived climatic changes (mostly during the Quaternary) did not alter the existing zonal landforms, which still persist, or that these are exceptions to the general rule of short-lived climates.

The humid tropics appear to be one of these exceptions. It is not so much that climates did not change in the humid tropics; rather, these changes appear to have affected mostly the marginal areas, while those which probably affected the entire region appear to be mostly the short-term changes. It may therefore be concluded that climatic changes, especially the most recent ones (the Quaternary), have been less important as agents of landscape formation in the humid tropics, where warm humid conditions have existed for many millions of years. In fact, humid tropical conditions are known to have extended to present-day hot deserts, humid temperate, and even cooler regions where deep-weathering profiles among other diagnostic features have been described (cf. Dury, 1971).

The significance of structural stability

Another important variable in the evolution of zonal landforms is structural stability, described in the literature generally as diastrophism (Thomas, 1974, p. 1). Even if climates remain stable for a long time, structural instability is likely to impose the same type of constraint on the development of the type of equilibrium landscapes implied by climatic geomorphology as short-lived climatic variations. This is particularly true of areas where the rates of diastrophism vary temporally and spatially, a situation likely to occur over wide areas such as those of climatic zones or regions.

Again, in the case of the humid tropical areas, a large part is known to be underlain by old and relatively stable landmasses (see chapter 2), the tectonically mobile zones being restricted to Indonesia, Central America and the Caribbean. It is therefore to be expected that, given a relatively stable geology and climate, geomorphological processes must have operated for long enough time and over a wide enough area to produce distinctive forms in at least parts of the humid tropics. A few studies are reported below on this general theme.

Progress in humid tropical geomorphology

Two books have appeared in recent times with themes similar to that of the present work. These are Tricart's *The Landforms of the Humid Tropics, Forests and Savannas* (1972) and Thomas's *Tropical Geomorphology* (1974). Tricart argues very strongly, like Peltier (1950), Birot (1960), Büdel (1963) and Strakhov (1967) among other protagonists of climatic gemorphology before him (particularly those of the French school of geomorphology), for a distinctive, tropical morphogenetic zone. Thomas is, however, more cautious, and in fact has been strongly influenced by the arguments of Stoddart (1969a, b), among others, to the effect that we do not have enough information yet for a dogmatic position. Thus, whereas Thomas described processes (chemical weathering, mass movement and surface flow) and forms or terrains (domed inselbergs, pediments, weathering profiles and products, including laterites, and planation surfaces) which are mostly characteristic of, if not exclusive to, humid tropical regions, he declined to set these within 'a specific range of bioclimatic conditions' (1974, p. 1). Nevertheless, Thomas has contributed, perhaps more than any other geomorphologist, to the development of humid tropical geomorphology, having published more than ten major works on this theme, based on a very varied experience in West Africa (particularly Nigeria), Central Africa, Australia and South-east Asia.

Thomas's contributions lie essentially in his study of the processes and forms associated with deep weathering, a phenomenon well-known to scientists for more than half a century before him (cf. Falconer, 1911; Woolnough, 1918, 1927; Wayland, 1934). The idea being advanced is that the theme of deep weathering is central to humid tropical geomorphology. Not only is the process relatively unique to the humid tropical envornment, it is also important in soil formation and plant life, as well as in influencing the operation of the major geomorphological processes and the resulting landforms (Dury, 1971; Faniran, 1970, 1974).

The chapters that follow are arranged in a fashion similar to that of Thomas (1974), but in a rather more systematic and comprehensive manner. In Chapter 2, the humid tropical environment is described in order to complete this introductory background setting for the major topics described in sections B, C and D of the book. Section B is devoted to a description and discussion of the geomorphological processes operating in the humid tropics: weathering (Chapter 3); erosion Chapter 4); and deposition (Chapter 5). As is to be expected, forms resulting from the operation of these processes at the local level are described. However, the character and examples of humid tropical landscapes are described in

Section C (chapters 6 to 12). The general appraisal of the landscape is presented in chapter 6, while specific examples are described in chapters 7 (rivers and river patterns), 8 (duricrusts), 9 (slopes), 10 (residuals), 11 (tropical karsts), and 12 (tropical coasts). Chapter 13 relates the processes and forms to the important issue of resource and the satisfaction of human wants.

References and further reading

Bakker, J. P. (1967) *Weathering of granites in different climates*, in *L'Evolution des Versants*, P. Macar (ed.), Congress Collegium, L'Université, Liège, **40**, 5168.

Bakker, J. P. and **Levelt, T. W. M.** (1964) 'An inquiry into the problem of polyclimatic development of peneplains and pediments (etchplains) in Europe during the Senonian and Tertiary period', *Publ. Serv. Carte Geol. Luxembourg*, **14**, 27–75.

Birot, P. (1960) *The Cycle of Erosion in Different Climes*, trans. C. I. Jackson and K. M. Clayton, Batsford, London.

Büdel, J. (1948) 'Das System der Klimatischen Geomorphologie (Beilrage zur Geomorphologie der Klimazonen und Vorzeit-Klimate V)' *Verhandlugen Deutscher Geographentag*, München, **27**, 65–100.

Büdel, J. (1963) 'Klima-genetische Geomorphologie', *Geographische Rundschau*, **15**, 269–85.

Chorley, R. J., Dunn, A. J. and Beckinsal, R. P. (1974) *The History of the Study of Landforms*, 2nd edn., Methuen, London.

Corbel, J. (1959) 'Vitesse de l'érosion', *Zeits Geomorph.* N.F., **3**, 1–28.

Corbel, J. (1964) 'L'érosion terrestre, étude quantitative', *Ann. Géogr.*, **73**, 385–412.

Cotton, C. A. (1942) *Climatic Accidents in Landscape Making*, Whitcombe and Tombs, Wellington, New Zealand.

Davis, W. M. (1899) 'The geographical cycle', *Geogr. J.*, **14**, 481–584.

Douglas, I. (1967) 'Natural and man-made erosion in the humid tropics of Australia, Malaysia and Singapore', *Publ. Int. Ass. Hydrol. Scient.*, **75**, 17–29.

Douglas, I. (1968) 'Erosion of granite terrains under tropical rain forest in Australia, Malaysia and Singapore', in *Symposium on River Morphology*, Bern.

Douglas, I. (1969) 'The efficiency of humid tropical denudation system', *Trans. Inst. Br. Geog.*, **46**, 1–16.

Dury, G. H. (1971) 'Relict deep weathering and duricrusting in relation to the palaeoenvironments of middle latitudes', *Geogr. J.*, **137**, 511–22.

Ebisemiju, S. F. (1976) 'The structure of the interrelationships of drainage basin characteristics', Ph.D. thesis, University of Ibadan.

Eyles, R. J. (1968) 'Stream net ratios in West Malaysia', *Bull. Geog. Soc. Amer.*, **79**, 701–12.

Falconer, J. D. (1911) *The Geology and Geography of Northern Nigeria*, Macmillan, London.

Faniran, A. (1970) 'Landform examples from Nigeria, No. 2: the deep weathering (duricrust) profile', *Nig. Geog. J.*, **13**, (1), 87–8.

Faniran, A. (1974) 'The extent, profile and significance of deep weathering in Nigeria', *J. Trop. Geog.*, **38**(1), 19–30.

Fournier, F. (1960) *Climat et Erosion: la relation entre l'érosion du sol par l'eau et* *les précipitations atmosphériques*, Presses Univ. de France, Paris.

Fripiat, J. J. and Herbillon, A. J. (1971) 'Formation and transformation of clay minerals in tropical soils', in *Soils and Tropical Weathering, Proceedings Bandung Symposium*, UNESCO, 1969, 15–24.

Horton, R. E. (1946) 'Erosional development of streams and their drainage basins', *Bull. Geol. Soc. Amer.*, **56**, 275–370.

Jahns, A. (1961) *Quantitative Analysis of some Periglacial Processes in Spitzbergen*, Warsaw Univ.

Jahns, A. (1975) 'Granite tors in the Sudeten mountains', in *Progress in Gemorphology*, E. H. Brown and R. S. Waters (eds), Inst. Br. Geog., p. 7.

Keller, W. D. (1957) *The Principles of Chemical Weathering*, Lucas, Columbia, Mo.

King, L. C. (1967) *The Morphology of the Earth*, 2nd edn., Oliver & Boyd, Edinburgh.

Langbein, W. B. and Schumm, S. A. (1958) 'Yield of sediment in relation to mean annual precipitations', *Trans. Am. Geophys. Union*, **39**, 1076–84.

Linton, D. L. (1955) 'The problem of tors', *Geogr. J.*, **121**, 470–87.

Peltier, L. C. (1950) '*The geographic cycle in periglacial regions as it is related to climatic geomorphology*', *Ann. Ass. Am. Geog.*, **40**, 214–36.

Peltier, L. C. (1962) 'Areal sampling for terrain analysis', *Prof. Geogr.*, **14** (2), 24–8.

Rapp, A. (1960) 'Recent development of mountain slopes in Karkenvagge and surroundings, northern Scandinavia', *Geogr. Annal.*, **42**, 65–200.

Rose, A. A. (1962) *Soil Science*, Israel Program for Scientific Trans. Jerusalem.

Rougerie, G. (1960) 'Le faconnement actuel des modèles en Côte d'Ivoire forestière', *Mem. L'Instit. francais d'Afrique Noire*, Dakar, **58**.

Schumm, S. A. (1965) 'Quaternary paleohydrology', in *The Quaternary of the United States*, Wright, H. E. Jr. and Frey, D. G. (eds), Princeton University Press, 783–94.

Sherman, G. D. and Uechara, G. (1956) 'The weathering of divine basalt and its pedogenic significance', *Proc. Soil Sci. Soc. Amer.*, **20**, 337–40.

Stoddart, D. T. (1969a) 'Climatic geomorphology', in *Water, Earth and Man*, R. J. Chorley (ed.), Methuen, pp. 473–85.

Stoddart, D. R. (1969b) 'Climatic geomorphology', in *Progress in Geography*, Edward Arnold, London, pp. 160–222.

Strakhov, N. M. (1967) *Principles of Lithogenesis*, 1, Oliver & Boyd, Edinburgh.

Tanada, T. (1951) 'Certain properties of the inorganic colloidal fraction of Hawaiian soils', *J. Soil Sci.*, **2**, 83–96.

Tricart, J. and Cailleux, A. (1965) *Introduction à la Géomorphologie Climatique*, Paris, S.E.D.E.S.

Thomas, M. F. (1974) *Tropical Geomorphology: a study of weathering and landform development in warm climates*, Macmillan, London.

Tricart, J. (1972) *The Landforms of the Humid Tropics, Forests and Savannas*, trans. C. J. K. de Jonge, Longman, London.

Wayland, E. J. (1934) 'Peneplains and some other erosional platforms', *Bull. Geol. Surv. Uganda, Ann. Report*, Notes 1, **74**, 366.

Woolnough, W. G. (1918) 'Physiographic significance of laterite in Western Australia', *Geol. Mag.*, N.S., **6** (5), 385–93.

Woolnough, W. G. (1927) 'The duricrust of Australia', *J. Proc. Roy. Soc. N.S.W.*, **61**, 25–53.

2 The humid tropical environment

The humid tropics: definition and boundary

The tropical region proper refers to the area between the Tropic of Cancer and Tropic of Capricorn – that is, areas between latitudes $23\frac{1}{2}°$N and $23\frac{1}{2}°$S; it covers about 40 per cent of the earth's total surface area, or about 36 per cent of its land area. However, as defined here, this is not equivalent to the humid tropics, as the area so defined encompasses semi-arid and arid zones.

The humid tropics has been delimited by various authors using different criteria. Kuchler (Fosberg *et al.*, 1961), using vegetation criteria, defines the humid tropics as areas with relative absence or paucity of xeromorphic features in the vegetation, such areas being characterized by tropical rainforest, semi-deciduous, deciduous forest and savannas. Fosberg (1961), in agreeing with this definition, stresses that the area is covered by mesophytic vegetation which shows no great predominance of xeromorphic characteristics or putative adaptations to reduced transpiration. However, apart from being indirectly based on climate, the map of the humid tropics by Kuchler covers semi-arid zones which by no stretch of imagination can be regarded as humid tropics.

Koppen (1936), Garnier (1957, 1961), Hodder (1957), Gourou (1961) and Tricart (1972), among others, use climatic criteria in delimiting the humid tropics. All of them agree that this area falls under winterless climates with the mean temperature of the coolest month at 18°C. This, however, will exclude the highlands, where the mean temperature of the coolest months are less than 18°C, but which yet experience no winter. Reducing the temperatures of such areas to sea-level temperatures readily facilitates their inclusion (Nieuwolt 1977). Although these authors believe that total annual rainfall constitutes a valid criterion for defining the humid tropics, there is no consensus on the amount and distribution of the rainfall. Gourou (1961) limits the total annual rainfall to 400 mm, whereas Tricart (1972) limits it to 750–800 mm. Nieuwold (1977) accepts Köppen's (1936) values of 450–600 mm.

The definition based on certain values of these climatic parameters are unsatisfactory to Garnier (1958, 1961) for various reasons, the most important of which is that it fails to resolve the problem of humid tropicality as a climatic term. Garnier (1958) thus conceptualizes humid tropicality as obtaining when, over a given period such as a week, a month or a season, the moisture supply is such that actual evapotranspiration equals potential evapotranspiration, and soil moisture is not depleted below a certain critical value. Based on his observations on the climatic elements in Nigeria, Garnier established that humid tropical conditions could be identified by reference to a climatic situation in which actual evapotranspiration equals potential evapotranspiration; mean minimum temperatures are 18°C or higher; mean vapour pressure is at least 20 mb. As these represent a relatively dynamic condition, Garnier indicates that they cannot be satisfied throughout the year, and that thus in any particular place, a month in which the conditions identified are satisfied can be designated a 'humid tropical' month. Thus the climate of different stations can be identified according to the number of such months in the year when all these conditions are satisfied. However, since the general application of this scheme may be difficult in identifying the climate of the various stations in the tropics, Garnier (Fosberg *et al*. 1961) defined the humid tropics as being characterized by (a) mean monthly temperature for at least eight months of the year of at least 20°C; (b) vapour pressure and relative humidity for at least six months of the year averaging at least 20 mb and 65 per cent respectively; and (c) mean annual rainfall totalling at least 1 000 mm and for six months precipitation of at least 75 mm per month.

The cardinal factor in Garnier's delimitation of the humid tropics is the balance between precipitation and evapotranspiration, which other authors have conceptualized and used in different ways. Some authors have devised precise indices relating precipitation to potential evapotranspiration – for example, the aridity index.

Martonne's (1926) annual aridity index is $P/T + 10$, while his monthly index is $12 P/T + 10$. Gaussen's index is $P/2T$ and Birot's (1959) is $P/4T$ where

$$P = \text{average precipitation in mm}$$
$$T = \text{mean temperature in °C.}$$

Birot empirically derives his own formula on the bases that the column of water evaporated in a month under a moderately humid atmosphere, and from a soil prevented from drying out, is about 100 mm for a temperature of 25°C. Under such a situation, rainfall higher than 100 mm is required for the water to infiltrate into the soil and contribute to soil water reserves. With an average temperature of 20°C, Birot's formula gives 80 mm as the limit of the humid month. Thus, based on Birot, humid tropical conditions could be identified with reference to a climatic situation in which the mean of the minimum temperatures is 20°C or higher; actual evaporation equals potential evapotranspiration with rainfall of at least 80 mm for each month. These conditions have to be satisfied for the whole year in any place before it can be described as humid tropical. This definition can only apply to areas under Koppen's Af climate.

Types of humid tropics

Most authors distinguish between different types of humid tropical areas. Whereas Fosberg, Garnier and Kuchler (1961) recognize two types of humid tropics (constantly humid and seasonally humid), Köppen (1936) establishes three types: Af – tropical rainforest climate; Am – tropical rainforest monsoon type; and Aw – tropical savanna. Tricart (1972) recognizes four types; constantly humid, constantly humid with a very short dry period, seasonally humid with a long dry period. The recognition of the above types of humid tropical areas is not as easy as delimiting the areal coverage, which has been done using either one or a combination of several parameters by different authors. Based on the mean sea-level temperature of 18°C for the coolest month and on rainfall, Köppen (1936) puts the northern boundary at about latitude 13°N in Africa except in East Africa where Somalia is excluded, latitude 20°N in Central America but reaching southern Florida, latitude 23½°N in India and latitude 18°N in South-east Asia. He puts the southern boundary at between latitude 18° and 23½°S in South America, 10°S in Central Africa extending to 23½°S in the coastal area of Mozambique, latitude 20°S in Queensland and latitude 15°S in northern Australia. All the mountain ranges in these areas are excluded.

However, this delimitation does not coincide with the 'geomorphological humid tropics' of Tricart, who observes that whereas the boundaries of the humid tropics as delimited by Köppen are fairly accurate where a zone of arid and semi-arid land separates it from the mid latitudes, the boundaries are not so clear where the humid tropics merge gradually into the mid latitudes, as in southern China up to latitude 30°N and in the Atlantic Brazil, in the Rio Grande do Sul region, as far as latitude 30°S, all of them on the eastern sides of the continents. In such places, the physiognomic characteristics of the vegetation are more or less the same as in the tropical rainforest except that there are differences in terms of species composition. Such places also exhibit the typical weathering pattern and profiles characteristic of the humid tropics (Ruxton and Berry, 1957) and have the typical reddish lateritic tropical soils. Tampa and New Orleans at latitudes 28°N and 30°N respectively in the USA typify conditions in such marginal zones. The mean of the lowest monthly rainfall in Tampa is about 40 mm, with rainfall in form of thunderstorms for about one quarter of the year. In New Orleans, it is about 90 mm. The mean of the lowest monthly temperature in Tampa is 16°C, 12°C in New Orleans, while the highest average temperature for the hottest month in Tampa is 27°C, and in New Orleans it is 30°C. As both places rarely experience frost, biological and chemical actions are continuous all the year round.

Tricart (1972) regards these marginal areas as sub-humid tropics, a classification invalidated by the concept of humid tropicality as defined by Garnier. Such areas can be regarded as part of the humid tropics because the landforms and the geomorphological processes, especially the weathering pattern in such areas, are more or less the same as in other areas of the humid tropics. Thus for the purpose of this work, areas regarded by Tricart as sub-humid tropics will be classified along with the rest of the humid tropics. Adejuwon (1979) was perhaps mindful of this problem in commenting that 'while changes from tropical to non-tropical patterns are noticed at latitudes 20°–25° in the western sides of the continents, tropical conditions are experienced beyond latitude 30°N on

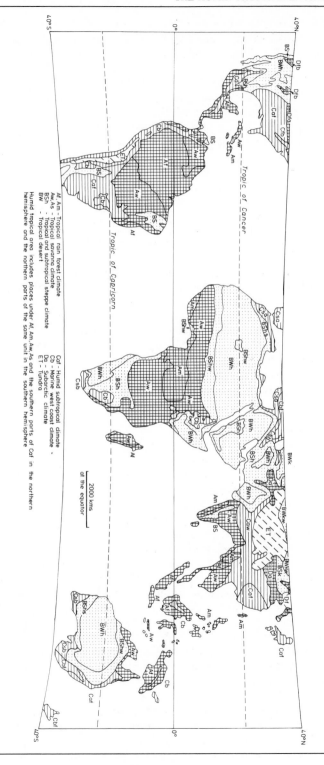

Fig. 2.1 Koppen's climate classes between latitudes 40°N and 40°S

Af, Am - Tropical rain forest climate
Aw, As - Tropical savanna climate
BSh - Tropical and subtropical steppe climate
BW - Tropical desert

Caf - Humid subtropical climate
Cb - Marine west coast climate -
Dø - Subarctic climate
E T - Tundra

Humid tropical area includes places under Af, Am, Aw, As and the southern parts of Caf in the northern hemisphere and the northern parts of the same unit in the southern hemisphere

the eastern side'. Based on the above considerations, areas under humid tropical conditions in each continent are shown in Figure 2.1. However, it must be emphasized that the boundary of the humid tropics will fluctuate greatly as climatic conditions vary from year to year, as was evidenced during the recent drought (1971–1974) in the sahelian zones of West Africa when the northern boundary of the humid tropics probably shifted as far south as latitude 12°N. Thus the boundaries cannot be regarded as immutable. Nevertheless, for geomorphological purposes, the area delimited is sufficiently characteristic in terms of the operating processes, if not in those of forms.

Climatic parameters and their effects

Temperature and rainfall parameters are generally regarded as the most important parameters of the natural environment; others, like relative humidity and evapotranspiration, are to a large extent closely related to them and so can be subsumed under the former parameters in so far as they affect the geomorphology of the humid tropical regions. The intensity of these climatic parameters play very important roles in determining the nature of the geomorphological processes and the character of the resultant landforms in the humid tropics. A sketch outline of the characteristics of these parameters and their role in the geomorphology of the humid tropics is contained in what follows. These characteristics are, however, further discussed in subsequent chapters.

Temperature

The major factor of temperature is insolation, which is most abundant in the humid tropics for a number of reasons.

At any point between these latitudes, the sun is overhead at midday twice in the year, during which time the angle of incidence of solar radiation at the equator is about 78°, in which case a high amount of solar radiation is intercepted within the tropics all the year round. Around the equator, the sun is almost always overhead so that the area records the highest incidence of solar radiation. Consequently, between the tropics, the amount of solar radiation intercepted at the earth's surface is quite high, calculated at about 220 000 cal/cm²/year (Adejuwon, 1979a). Lockwood puts the figures at 150 kilo cal/cm²/year in South-east Asia, and between 120 to 140 kilo cal/cm²/year in equatorial Africa. Not only is the annual interception of solar radiation very high, but the seasonal variation is relatively low because of the little variation in the length of daytime throughout the year.

The very important characteristic of the humid tropical environment recognized earlier as the high temperature prevalent all the year round with the coolest months having an average temperature of about 18°C is therefore understandable.

THE HUMID TROPICAL ENVIRONMENT

The means of the highest monthly temperatures vary from station to station, averaging about 28°C. The hottest places are in the drier marginal zones of the Aw and Am climates. Zungeru in Nigeria has a mean maximum temperature of 37°C in the dry season. Most stations in India and mainland South-east Asia also experience mean maximum temperatures in excess of 37.8°C during the few months preceding the monsoons. Darwin and Cairns in northern Australia also experience mean maximum temperature of up to 32.3°C in the months preceding the monsoons.

Table 2.1 shows the temperature pattern in some of the stations in the humid tropics.

Another important characteristic of the humid tropics is the relatively low seasonal and diurnal range; the nearer a station is to the equator, the less the seasonal and diurnal variations and vice versa.

Around the equator, actual mean temperatures are about 25°C, and both seasonal and diurnal range are less than 5°C. Recorded seasonal temperature range for Belém and Manaus, both in Brazil, are 1.0°C and 1.7°C respectively; for Quito in Ecuador, 1.0°C; for Entebbe in Uganda, 1.7°C; for both Colombo in Sri Lanka and Singapore in Malaysia, 2.0°C; for Freetown in Sierra Leone, 2.5°C; for Libreville in Gabon, 3.1°C; for Lagos in Nigeria, 3.2°C; for Accra in Ghana, 3.6°C; and for Nairobi in Kenya, 3.9°C. On islands along the equator, seasonal variations are still less due to the moderating influence of the sea. Thus Seychelles has 2°C, the Solomon Islands, 1.5°C, the Gilbert Islands, 0.2°C and Djarkata in Java, 1.0°C.

Further away from the equator, Bangkok in Thailand has 4.7°C; Darwin in northern Australia, 5.2°C; Zungeru in Nigeria, 5.0°C; both Rangoon in Burma and Rio de Janeiro in Brazil, 5.6°C; Cairns in Australia, 6.1°C; Durban in Natal, 7.0°C; Beira in Mozambique, 7.2°C; Kambwe in Zambia, 9.0°C; Khartoum in Sudan, 10.0°C; Calcutta in India, 10.8°C; Mandalay in Burma, 10.9°C; and Patna in India, 12.0°C. Seasonal temperature variations are also lower on islands than on mainlands; thus Suva in the Fiji Islands has only 3.6°C.

Diurnal temperature range is again lowest near the equator, where it hardly exceeds 3.0°C. In Nigeria, Lagos records diurnal range of between 1°C and 3°C during the rainy season, increasing to 5.6°C in the dry season, while Kaduna records mean diurnal range of between 5.6°C and 8.3°C in the rainy season, increasing to 11.1°C to 13.9°C during the dry season.

Apart from the above, it is to be noted that stations on the same latitude in the same continent may not necessarily have the same temperature, as they may be influenced by different types of currents. Thus Luanda, on the coast of Angola on latitude 8° 59′S influenced by the cool Benguela current, has a seasonal temperature range of 16.4°C, whereas Tamatave in Madagascar on latitude 18° 2′S influenced by the warm Agulhas Current has a seasonal range of 5.6°C.

Continentality is another influence on both seasonal and diurnal range. The seasonal temperature ranges for Freetown in Sierra Leone and Mongalla in the Sudan, both on about latitude 9°N, are 2.5°C and 6.0°C respectively. The diurnal range for Lagos in Nigeria on latitude

TABLE 2.1 Temperature in the tropical zone (in °C)

Locality	Latitude	Mean annual	Mean of hottest month	Mean of coldest month	Average extremes Max.	Min.	Recorded extremes Max.	Min.
Apia, Samoa	13° 48'S	25.8	25.1	15.1	32.7	18.3	35.5	16.1
Singapore, Malaya	1° 18'N	27.2	27.2	26.4	33.8	20.5	26.1	18.8
Sandakan, N. Borneo	5° 49'N	27.4	28.1	26.5	34.4	21.6	37.2	21.1
Mazaruni Station, Guyana	6° 50'N	26.0	27.2	25.1	—	—	—	—
Manos, Brazil	3° 8'S	27.2	28.2	26.5	36.6	21.1	38.3	18.8
Yangambi, Zaire	0° 45'N	24.5	25.2	23.9	—	—	—	—
Eala, Zaire	0° 3'N	25.6	26.3	24.6	—	—	37.2	15.2
Dula, Cameroons	4° 3'N	25.4	26.9	23.8	32.7	20.0	35.0	18.3
Lagos, Nigeria	6° 27'N	26.9	28.5	25.3	35.0	19.4	40.0	15.5
Sokoto, Nigeria	13° 2'N	28.0	32.7	23.7	43.3	10.0	46.1	3.8
Rangoon, Burma	16° 47'N	27.3	30.6	24.8	38.3	15.0	41.6	12.7
Mandalay, Burma	21° 59'N	27.6	32.2	21.3	43.3	10.5	44.4	7.2
Bambesa, Zaire	3° 26'N	24.4	25.1	23.5	—	—	—	—
Masindi, Uganda (1 146 m)	1° 35'N	21.8	22.6	20.7	—	—	36.7	9.4
Cairns, Queensland	16° 55'S	24.8	27.2	21.0	—	—	—	—
Rio de Janeiro, Brazil	22° 54'S	22.7	25.6	20.0	35.5	13.8	38.8	10.0

Source: after P. W. Richards, 1957

TABLE 2.2 Mean monthly soil temperature (University of Ibadan) in °C

Depth	5 cm				10 cm				20 cm		30 cm		60 cm	120 cm	Gross
Time	07	10	16	18	07	10	16	18	10	16	10	16	10	10	Temp.
January	24.8	25.7	32.2	30.7	26.4	24.7	30.9	28.9	27.8	28.9	28.8	29.1	28.1	28.8	19.4
February	26.2	27.9	35.4	33.4	28.2	28.3	33.0	32.8	29.7	31.2	30.4	29.9	28.5	29.6	21.7
March	26.3	28.4	35.5	33.3	28.4	28.8	33.6	33.4	29.9	32.1	30.8	30.3	29.9	30.3	20.6
April	25.8	28.2	33.8	31.8	29.7	26.4	30.1	32.3	29.2	31.4	29.8	30.3	29.1	29.8	20.8
May	25.6	28.1	32.8	31.2	27.2	28.2	31.8	31.8	38.9	30.7	29.4	30.0	29.1	29.8	20.8
June	24.9	27.1	30.1	29.6	26.6	27.2	30.2	30.2	30.7	29.2	28.7	28.5	28.1	29.3	20.8
July	24.1	25.4	29.1	28.1	25.7	26.1	28.7	29.3	27.1	28.2	27.4	27.1	26.8	28.5	20.8
August	23.9	22.8	25.9	25.2	23.3	23.3	26.1	24.4	26.6	27.6	26.7	25.7	25.0	25.4	19.7
September	22.2	23.4	29.3	29.3	28.4	25.7	25.9	27.2	29.1	26.9	28.1	27.2	26.7	27.6	20.6
October	22.7	24.4	28.4	27.4	22.4	22.8	28.0	25.8	27.5	23.9	27.9	27.9	27.2	28.2	21.1
November	23.3	27.2	30.1	28.4	23.2	25.9	27.0	24.4	31.0	30.2	28.8	29.1	28.2	28.7	20.6
December	23.1	26.4	30.2	28.7	27.0	27.4	29.0	28.3	28.2	29.8	28.9	28.5	28.4	28.9	17.2
X-Year	24.4	26.3	31.2	29.7	26.2	26.3	29.6	29.3	28.6	29.7	28.7	28.6	27.8	28.8	20.4

6° 20'N averages 4.0°C, whereas for Kaduna on latitude 10° 30'N it averages 10°C.

Temperature patterns under different vegetations and on exposed surfaces, which are of more relevance to geomorphological processes, are unfortunately not well known. However, from the little available data in Southern Nigeria, Evans (1939) has shown that temperature on the ground floor of the rainforest (Shasha Forest Reserve) is generally less than air temperature and is characterized by a lower range. The seasonal range is also low, with a mean maximum of 26.8°C in the dry season and a mean minimum of 25.1°C in the wet season. In the savannas, where the grass affords little protection to the surface, ground temperatures are higher, with a mean minimum of 40°C and a mean maximum of 50°C – that is a range of 10°C. Diurnal figure is higher at about 30°C or more, due to rapid heat loss from the ground surface at night and the high incident radiation maintained in the top soil during the day.

At present little is known about the soil temperature pattern due to paucity of data, as only very few meteorological stations include equipment to obtain soil temperatures in any systematic manner. For instance, in the whole of Southern Nigeria only the University of Ibadan Station has reliable soil temperature data. Table 2.2 summarizes the available data from this source.

Due to the high ground surface temperatures, soil temperatures in the savanna are higher than in the forest. Also, both diurnal and seasonal ranges are high. Even then, these still depend on the density of vegetal cover. Thus as shown in Table 2.3, temperature ranges increase steadily from the Guinea Sudan to the sahel Sudan.

TABLE 2.3 Annual range of soil temperature in different vegetation zones in Nigeria

Station	Latitude	Annual tempera- ture range at 30 cm	Annual tempera- ture range at 120 cm	Vegetation
Makurdi	7° 45'N	5.3°C	3.2°C	Guinea Savanna
Ilorin	8° 30'N	5.5°C	3.2°C	Guinea Savanna
Kaduna	10° 30'N	5.6°C	3.4°C	Sudan Savanna
Zaria	11° 30'N	6.7°C	4.1°C	Sudan Savanna
Maiduguri	11° 50'N	8.6°C	5.4°C	Sahel Savanna
Kano	12° 00'N	8.8°C	6.3°C	Sahel Savanna

Source: Ojo, 1977, p. 152

Temperature and geomorphological processes

In the humid tropics, as in other parts of the world, temperature in the form of heat plays both direct and indirect geomorphological roles as follows:

1   High solar radiation incident on exposed rock surfaces directly effects thermal weathering of such rocks through the processes of flaking, spalling and boulder cleaving (Ollier, 1975). These features are ubiquitous on rock outcrops all over the hot regions of the world. Whereas in the more humid areas such weathering may involve moisture swelling of exposed rock surfaces, there is little doubt that in the savanna areas, the extremely high diurnal and average annual temperature range may be significant in this aspect of weathering. According to Vischer (1945), diurnal temperature of about 50°C experienced on exposed surfaces in the drier parts of the humid tropics can subject such rock surfaces to considerable strain from expansion and contraction.

2   High temperature is particularly important in the process of chemical weathering of rocks. Although the intensity and depth of rock weathering depend on water, temperature exerts very strong influence in conformity with Van't Hoff's law, which states that given the same amount of moisture, an increase of 10°C in temperature doubles or trebles rates of chemical reaction. All things being equal, therefore, chemical weathering would be expected to attain its greatest intensity and depth in parts of the humid tropics (Strakhov, 1967).

3   Heat, by effecting biochemical reactions in plants, is of vital importance to certain geomorphological processes, especially where moisture is not a constraint to plants' vascular activities. As already emphasized, equatorial areas offering optimum thermal and hygric conditions for plant growth experience a greater amount of above-ground vascular plant activities than other parts of the humid tropics. The net values of biomass production in the forests of Southern Nigeria range from 3 000–4 000 gm/m²/yr, while values for the savanna zones range from 1 000–3 000 gm/m²/yr (Adejuwon, 1979b). There is considerable production of litter, particularly in the rainforest area. Similarly, the rate of decomposition and mineralization is very fast as a result of high temperature and high pH. Little humus is found on the ground except where the absence of oxygen slows down decomposition.

In the pre-decomposition and mineralization stages, litter on the ground lessens splash erosion and absorbs water which, in the case of partially decayed organic matter, may amount to three to four times its dry weight. When fully decomposed as humus, it increases infiltration into the soil at the expense of overland flow (Wilkinson and Aina, 1976).

Similarly, in the savannas where the daytime temperature of the topsoil may be as high as 50°C, litters are rapidly decomposed with practically no humus formed. Humus is known to cement soil aggregates and to increase soils' resistance to erosion. Therefore Williams' (1969) finding in northern Queensland that, given the same amount of rainfall, splash erosion in the savanna is about twenty times that of the rainforest, is easy to understand.

4   Temperature is also important in determining the rate of solubility of certain elements in the soils. According to Ollier (1975), silica appears more soluble at higher temperatures, a phenomenon responsible for the desiliconization of soils in the humid tropics. Although analyses of the silica content of rivers in the humid tropics (Douglas, 1969; Thomas, 1974) do not support this assertion, the

evidence of desiliconized clay minerals in freely drained tropical soils remains to be explained otherwise.

5   Finally, the high seasonal range of soil temperatures to depths of 60 cm is perhaps one of the most significant factors in the upward movement of water and dissolved elements (Thomas, 1974, p. 63) in the weathering profiles of the drier savanna zones.

Moisture

Rainfall

The characteristics of rain-producing airmasses are similar all over the humid tropics. These areas are charactierized by cT airmasses. The former in Africa originate between latitudes 20°N and 40°N over the Sahara, and the drier interior of South Africa; in Asia they originate over the Tibetan massifs north of the Tropic of Cancer; in Australia, over the interior deserts. The origin of mT airmasses is over the oceans. Both airmasses are warm, but mT are humid whereas cT are dry – except in Asia, where they are initially dry and cold only to become warm and humid on crossing the warm ocean.

These airmasses are separated by a zone variously called Inter Tropical Convergence Zone, or Inter Tropical Front or Inter Tropical Discontinuity (ITD). This zone, although not as sharply defined as the Polar Front, is still easily recognizable as it is characterized by vigorous convectional activity with air currents rising to 6000–10 000 m, leading to the formation of altocumulus and cumulonimbus clouds which yield violent thunderstorms. The 'front' tends to follow the apparent movement of the sun but with its accompanying belt of high temperatures reaching different latitudes in different continents (Figure 2.2).

The amount of rainfall in a given region is influenced by many factors, including relief, wind direction relative to coastal orientation and distance from the ocean. For instance, where humid airmasses moving across a continent are forced to rise, instability results which is often accompanied by heavy rains. For example, Debunscha at the foot of the Cameroon mountain records an average annual rainfall of 1 017 cm, while Cherrapunji at the foot of the Assam mountain ranges has an average annual rainfall of 1 161 cm. Other factors, such as local overheating by insolation, especially of bare surfaces in areas close to the equator, and artificial heating in towns (heat islands), can stimulate rapid vertical air movements and convectional activities.

Rainfalls are often accompanied by thunderstorms which are associated with squall lines often associated with the disturbances in the Inter Tropical Convergence Zone. These storms occur locally over areas with diameters of 10–16 km (Nieuwolt, 1977); but they contribute up to 90 per cent of the total rainfall in the savanna zones.

Three patterns of rainfall distribution distinguished in the humid tropics are as follows (Figure 2.3):

1 Areas between latitudes 5°N and 5°S with rain throughout the year, but marked by a single maximum. Such areas include the interior of the

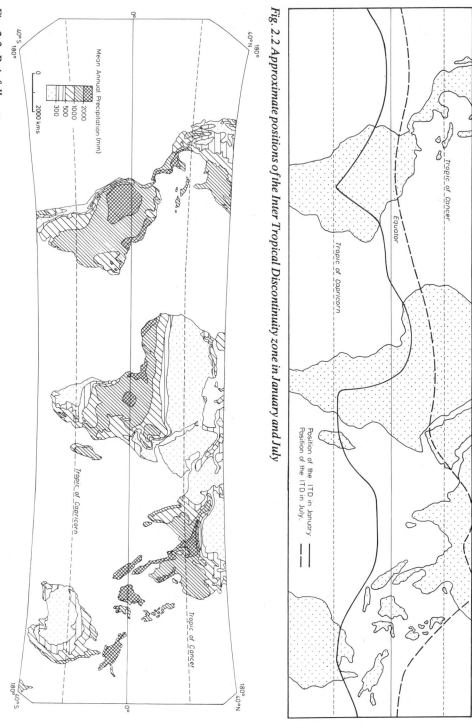

Fig. 2.2 Approximate positions of the Inter Tropical Discontinuity zone in January and July

Position of the ITD in January
Position of the ITD in July

Fig. 2.3 Rainfall patterns

Mean Annual Precipitation (mm)

2000
1000
500
300

0   2000 kms

Congo and Amazon basins, the islands along the equator and the eastern coast of Malagasy. Mean annual rainfall in this zone varies from 100 cm to about 300 cm but could be higher as a result of local orographic effects.

2 Areas between latitudes 5° and 10° north and south of the equator with two short, dry seasons in either July or August and November to February north of the equator and June to July, and December to January south of the equator. Such areas marked by double maxima include the coastal areas of Nigeria, East and Central Africa around the Congo basin and the interior of South America. Mean annual rainfall in this zone varies from about 70 cm in Nairobi to over 300 cm in Akassa, Nigeria. It is even higher in certain areas due to orographic effects.

3 Areas with a short rainy season marked by a single maximum and a long dry season lasting from October to March or April or May north of the equator, or from April to October or November south of the equator. These include the rest of the humid tropics in Central and South America, East Africa, west Malagasy and the monsoon areas of West Africa, India, South-east Asia and northern Australia. Mean annual rainfall varies from about 80 cm in Central India to above 1 000 cm in the foothills of the Assam ranges.

Beckinsale (1957) puts the value of the mean daily precipitation in the humid tropics at between 12 mm and 19 mm. Records from different stations, however, show the values to have a closer range nearer the equator and a wider range in the savanna. For instance, at Akure in Nigeria (latitude 7° 20′N) it varies between 7 mm and 22 mm. Table 2.4 shows the annual mean values for locations scattered all over the tropics. These values are highest in areas subject to monsoon rainfall especially at hill foot stations where up to 200 mm of rain may occur in a day.

TABLE 2.4 Annual mean rainfall per rain day (days with over 1 mm rain)

Station	Mean rainfall (in mm)	Station	Mean rainfall (in mm)
Quito, Ecuador	8.5	Entebbe, Uganda	12.4
Georgetown, Guyana	13.3	Bombay, India	22.4
San Salvador, El Salvador	16.1	Calcutta, India	15.5
San Juan, Puerto Rico	10.1	Rangoon, Burma	20.9
Accra, Ghana	14.4	Djarkata, Indonesia	13.5
Ondo, Nigeria	13.0	Hong Kong	21.2

The nature of individual rainfall varies. In Southern Nigeria, at the onset and the end of the rainy season, the rains are in the form of heavy showers of short duration, starting with a maximum intensity of up to 75 mm/hour or more, and declining in intensity after about 30 minutes, when the storm gives way to a steady downpour with an intensity of about 12.5 mm/hour and less. Garnier (1967) discovered that at Ibadan between 1952 and 1955, for individual rains yielding over 12 mm, 85 per cent of the rain falls during the first 30 minutes of the storm, whose average duration is 2 hours. A maximum of 112 mm/hour was achieved once during the period. Rainfall intensity is normally greater in drier areas, where about 90 per cent of the rains are in the form of thunderstorms. It is also greater in the monsoonal areas of South-east Asia where, according to Dobby (1973), rainfall intensities of 50 mm/hour are common.

Rainfall distribution is characterized by high variability with periods of heavy rains alternating with those of sub-average rains. Records from Southern Nigeria show that a very wet year may be followed by two drier years. On an annual basis, the relative variability index is about 15 per cent in the perennially wet zones, increasing to between 15 and 20 per cent in the seasonally wet and dry zones, between 20 and 25 per cent in regions with about three to four months of dry season, and to about 40 per cent in areas with longer dry seasons. In most places, the driest months have the highest variability index, whereas months with the highest rainfall figures experience the lowest variability.

Relative humidity

In most parts of the humid tropics, humidity is near saturation point at night, but decreases rapidly during the day. Seasonal variation is equally marked, being highest in the wet season and lowest in the dry season. Relative humidity values are at their maximum near the equator all the year round. Also, relative humidity is constantly high at high elevations. At about 1 000 m along the equator, it is close to saturation most of the time. For West Africa, Ojo (1977) has shown that relative humidity varies between 60 and 80 per cent (15.00 GMT) and 100 per cent (6.00 GMT) during the wet season and between 50 and 80 per cent during the dry season in places south of latitude 9°N. Values decline with increasing latitude, so that along latitude 20°N, relative humidity is less than 10 per cent both morning and night.

Evapotranspiration

Evapotranspiration in the humid tropics depends on the relative humidity and the capacity of the air to absorb water vapour. Other conditions conducive to evapotranspiration losses include intensive solar radiation and turbulence by convectional currents.

Values of potential evapotranspiration vary between 1 000 and 1 500 mm per annum in areas between the equator and latitudes 5° north and south of the equator, except on highlands where they are lower owing to high relative humidity, heavy cloud cover and the consequent decrease in the amount of net radiation available for potential evapotranspiration. As from latitudes 8°N and S, values rise sharply, reaching about 2 000 mm at approximately latitudes 15°N and S. Values of actual evapotranspiration at 1 100 mm more or less coincide with those of the potential evapotranspiration in the thick rainforests of the Amazon and Congo basins; elsewhere, they are considerably less (Franzle, 1979). Barry (1969) puts the values at about 300 mm in the savanna zones. Mean annual water surplus in Nigeria is from over 1 000 mm along the coast to 500 mm along latitude 9°N (Garnier, 1957); north of this latitude, there is a mean annual deficit of over 500 mm.

Moisture and geomorphological processes

Moisture in the atmosphere, especially in form of rain, is of direct and indirect relevance to the operation of most geomorphological processes all over the earth's surface. It has direct influence on the rate and amount of denudation on exposed surfaces, as well as – together with temperature – the rate of chemical reactions in rocks. It determines the rate of biochemical reactions in plants, which in turn serve as a constraint on certain geomorphological processes such as soil erosion, while enhancing others like chemical weathering.

The most obvious geomorphological effect is the detachment of particles from soil clods by raindrop impact, their movement by splashing and the subsequent removal of the detached and splashed particles by running water, which also detaches and removes more soil particles. (See chapter 4, where splash distances among other things are shown to vary depending on the type of soil, nature of sediment, land use and the degree of soil compaction.) Distances reported vary from 20 cm for coarse particles to 92 cm for fines from cultivated soil and 137 cm for fines on compacted soil (Smith and Wischmeier, 1962).

Because of paucity of data, the actual amount and rates of rain-splash erosion in the humid tropics are still unknown, and in any case there is the difficulty of making a distinction between the amount of sediment detached by raindrop impact and that detached by runoff. However, there is little doubt that, except for surfaces completely blanketed by dense vegetation and litter, splash erosion is ubiquitous, especially at the onset of rains in the forest areas, and for most of the season in the savanna areas, where more than a third of the thunderstorms are of high intensities (Stocking and Elwell, 1973; Jeje, 1977).

Water is the most important reactant in chemical weathering (see chapter 3). Although total rainfall is not an adequate indicator of the importance of precipitation in weathering, there is no doubt that high rainfall, together with other climatic factors like high temperatures and high relative humidity, are jointly important in chemical weathering. In areas of continuous precipitation where total rainfall is greater than the potential evapotranspiration, weathering is likely to be continuous, aided by a constant removal of certain elements in solution which are washed down into the deeper layers of the soil and removed in the drainage water. The ferralitic and ferruginous soils of the humid regions are, for instance, more leached of sodium, magnesium and calcium in comparison with

28

those of the arid regions (Ollier, 1975). The sand and silt fractions of these soils also contain more quartz than soils in the more arid areas of the tropics, an indication of a thorough weathering of the less stable minerals. In drier areas, weathering is pronounced in the rainy season, followed by upward movement of water, drying out of soil and the crystallization of salts near the surface.

Finally, rainfall plays a dominant role in mass movement (see chapter 4). It is of the utmost importance in slow movements like soil creep. During periods of excessive rainfall, soils stabilized on steep slopes by plant roots can absorb a large amount of water, thus setting up severe hydrostatic pore pressure which increases the shearing stress of these soils. With further saturation, the soil and the underlying regolith (the products of rock weathering, in the form of a mantle of loose rock and soil fragments), become increasingly soggy and can start to assume the characteristics of a viscous fluid. This, especially on steep slopes, will further increase the shear stress in the slurry material until a stage is reached when the regolith becomes unstable, suffers mass failure and moves downslope (Jeje, 1979b). Such is the severity of these processes in certain parts of the humid tropics – for example, Java and New Guinea (Simonett, 1967) – that some authors believe them to constitute an integral part of landscape evolutionary cycle in the area. Thus Wentworth (1943) and White (1949) went as far as postulating a regolith stripping cycle for the humid tropical area, involving deep chemical weathering of local rocks and subsequent stripping of the weathered debris from slopes by landslides and other rapid mass movement processes.

Vegetation

Based on the characteristics of vegetation in the humid tropics, two distinct plant communities can be recognized. They are the forests and the savannas (Figure 2.4).

Forests are favoured where rainfalls are high and evenly distributed throughout the year; where high rainfalls compensate for short dry seasons, generally of less than five months, or where water is available locally despite longer dry seasons – for instance, along river courses (gallery forest). As these forests are described elsewhere (see Richards, 1952), it is unnecessary to describe them in detail. They can be classified into three groups: the tall evergreen forest; the dry forest; and the secondary forest. The first type, which Franzle (1979) observed in the Amazon basin to consist of hierarchically arranged, highly organized complexes comprising a large number of species and ecotypes, is characterized by tall, buttressed emergents, a tall slender understorey intertwined by lianas, abundance of epiphytes and sparse undergrowth. The dry forests of South-east Asia are not exactly the same as the cerradao of Brazil, because the former contains mostly teak (Tectona grandis) and bamboo (Dendrocalamus gigantea), but both are marked by deciduousness and are characterized by a shrub layer. Secondary forests resulting from anthropogenic interference are characterized by a thick

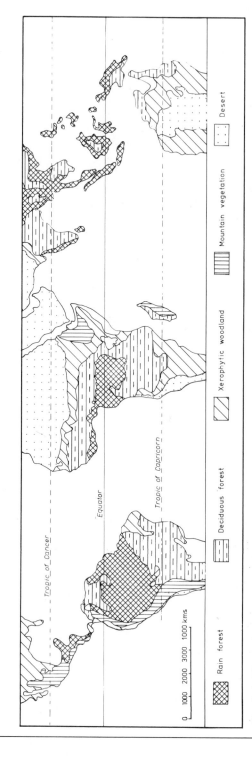

Fig. 2.4 Vegetation patterns in the tropics

tangle of light-loving plant species and are usually denser than the lower storeys of the primary rainforest.

Savannas are non-climax vegetation, being edaphically or anthropologically induced. In West Africa, they seem to have resulted from annual burning which destroyed fire-tender trees, and impoverished the soil. The savannas are characterized by grasses growing in tufts in which trees, 2 m to 5 m high, occasionally occur. However, in certain cases, the trees may constitute the dominant element, as in savanna woodland. In the Sudan savanna of West Africa and the Campos Limpos of South America, bare soils are exposed for a substantial part of the year in between the grass. This is also true of the Campo Cerrado and the Sahel, which are largely comprised of shrubs which, during a considerable part of the year, are leafless, leaving large areas of bare soil exposed to insolation and rain splash.

Vegetation and geomorphological processes

Apart from its obvious role in the physical weathering of jointed rocks through the wedge action of roots growing along joint planes, vegetation shields ground surfaces from rain-splash erosion and surface wash; anchors and retains regolith on slopes through the action of roots; improves soil structure through the supply of organic matter; and contributes organic acids to the chemical weathering process (Young, 1972).

Protection of the ground surface from erosion relates to the interception of rainfall by the forest canopy. The amount of water intercepted by vegetation depends on the types of plants, their stage of growth, their location and density. It is also a function of the quantity, duration and intensity of precipitation. Kirkby (1965) puts canopy interception in temperate environment at the first 10 mm of rainfall and 20 per cent of the subsequent rainfall in any one storm. However, studies by Freise (1936) in the Hylea forest of Brazil show that 20 per cent of rainfall evaporates from the canopy, 34 per cent drops to the ground, 29 per cent forms treetrunk runoff, while the remaining 18 per cent disappears in the bark through evaporation and absorption. Mohr and Van Baren (1954) put the interception values in the thick forest of Indonesia at 20–30 per cent. Recent studies in the rainforest have shown that the quantity of water penetrating the stand increases with the duration and intensity of the rainfall, but the relative amount of interception is inversely proportional to the amount of precipitation. Thus Jackson (1971) found that in tropical forests interception values of up to 70 per cent for 1 mm of rainfall decreases steadily to 13.5 per cent for 40 mm.

Given these interception values, theoretically the protection afforded the surface from erosion will increase with the vegetation density. Forest communities are therefore expected to shield the ground surface more effectively than the savanna communities. However, it is well known that emergents of the A-layer are up to 40 m and above, while trees of the B-layer are often higher than 15 m. At such heights, water drops from leaves approach the terminal velocity of raindrops, so that exposed soils are liable to rain splash and wash erosion. Thus primary rainforest characterized by a sparse ground or field layer (E) and with a very thin layer of litter 1–3 cm thick (Richards, 1952, p. 217), which may even be absent, is liable to more rain splash and wash erosion than other types of forest (Jeje, 1977; Oyegun, 1980; see also chapter 4). Miniature earth-pillars resulting from these processes have been reported under this forest (Ruxton, 1967; Faniran and Areola, 1974). The more open deciduous forest, apart from having a denser ground layer due to more sunlight reaching the ground, is also characterized by a thick layer of litter, resulting from defoliation in the dry season, which lessens rain splash and wash erosion. Generally, as a result of a high infiltration rate, surface wash is minimal in both forests. Thus suspended sediment yield varies from 60–100 tons/km²/yr on lowlands, rising to 200 tons/km²/hr on high lands (Fournier, 1960): soluble materials are transported in the soil so that solution loss is an important geomorphological process. Secondary forests offer better protection from splash erosion and surface wash than both types of forest.

The drier savannas, characterized by tufted grasses growing as individual clumps, leave wide expanses of bare surfaces especially during the dry season. The *cerrado* of Brazil are so sparse that they perennially leave large, bare surfaces in between individual shrubs. Thus grasslands all over the humid tropics are liable to severe rain splash and wash erosion especially at the onset of the rains, with the severity of erosion decreasing as vegetation becomes thicker. Protection from erosion is least where the ground is covered by thorn scrubs, as in the sahel of West Africa, which constitutes one of the zones of most severe fluvial erosion in the world, with denudation rates of between 1 000 and 2 000 tons/km²/yr (Fournier, 1960).

Organic acids are important in the leaching of certain elements and thus in the process of chemical weathering. This is especially the case in the humid tropics, where the amount of litter supplied by the forest is considerable, and where the rate of decomposition and mineralization is also equally quick to release fulvic and humic acids. Tricart (1972) contends that the latter dominate in the humid temperate zone, while the former dominate in the humid tropics. Laboratory experiments by Ong *et al.* (1970) have shown that organic acids comprise of macro ions which can readily associate with any mineral element to form organic complexes. For elements such as copper, iron, zinc, lead and so on the amount of elements mobilized increases with increasing concentrations of organic acid and with increasing values of pH. Organic acids breakdown or pepticize these elements into colloid. More than any other factor, the constant availability of organic acids in the rainforest zones may account for the more intense and deeper pattern of weathering than in the other areas characterized by sparsity of litter production.

As a result of this continuous vigorous chemical action on most rocks, many areas in the humid tropics, especially in areas with a high amount of annual rainfall, are characterized by thick weathered rock debris. Most slopes, except those exceeding certain threshold angles (about 30°) on acid igneous and metamorphic rocks, and covered by regolith and rock rubble whose thickness varies inversely with slope steepness. In fact, slopes exceeding 30° even in heavily fissured rocks like quartzites and quartz-schists and on heavily jointed basic rocks are covered by thin regolith and rock rubble. These slope materials are potentially unstable because of the effect of gravity and the frictional stresses.

Finally, in the humid tropics, almost all surfaces covered by soil and regolith are colonized by vegetation. Except for imperceptible movements by debris creep, regolith on the steep slopes is stabilized and kept in place by the closely knitted plant roots. However, as indicated earlier, rapid mass movements occur frequently on steep slopes, especially where there is a prolonged period of heavy rainfall.

Geological environment

The humid tropical areas lie astride complex geological structures which can be classified into the Precambrian shields, sedimentation basins, fold mountain systems and the volcanics. Because much of the tropical areas are relatively little known geologically, the brief description of the broad geological structures that follows may represent an oversimplification of essentially complex structures (Figure 2.5).

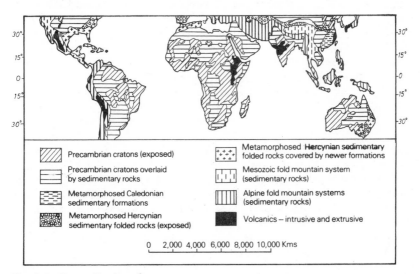

Fig. 2.5 Generalized geology

The Precambrian shields

These are the stable blocks or cratons, comprising the nuclei of all the continents to which mountain systems of various ages have been 'welded', and whose surfaces are masked in several places by sedimentary formations.

Precambrian rocks cover most of tropical Africa except for the

33

depositional basins of West Africa, Central and East Africa and Kalahari areas. In South America, the shield covering about 50 per cent of the tropical areas occurs as the Guyana, São Luiz, central Brazil, São Francisco and Uruguay–Tandila cratons, separated by several basins. In Central America, the shield occurs in north-western Sonora in Mexico. In tropical Australia, it covers most of the western part of the continent, outcropping mainly in the Pilbara area, and underlying younger sediments in the Rum Jungle area to the north-west, and in most of the Northern Territory. In Asia, the entire subcontinent of India – the Deccan shield – is a craton. The Sunda platform underlying eastern Sumatra, Malaysia and Borneo is also an ancient craton. The northern extension of the Australian craton into southern New Guinea is described as the Sahul shield by Dobby (1973).

The origin and composition of these rocks vary, but the consensus of opinion points to the fact that they originated as sedimentary rocks (Haughton, 1963). In West Africa, the oldest (Dahomeyan) formation comprises mainly of para-rocks: ferromagnesian, and argillo–arenaceous sediments metamorphosed into mica schists, gneisses, quartzites and migmatites; and granulites and schists (Haughton, 1963). The Birriman system is comprised of metamorphosed lavas, pyroclastic, hypabyssal basic intrusives, phyllites, and greywackes highly folded and intruded by granites. The Tarkwaian, Togo and Falemian formations comprise of quartzites, phyllites, schists, metamorphosed limestone and dolomite, while the Rockwell river and Buem series comprise of arenaceous, calcareous and argillaceous sediments with intercalated lavas – basalt, andesite, trachyte and tuffs. The Younger granites comprise of acidic masifs which intruded the meta sediments in places like the Jos plateau in Nigeria. This particular intrusion was preceded by massive extrusion of rhyolitic lavas – quartz porphyries and felsites. The composition of the cratons in the other continents hardly differs from the above.

Apart from common composition, the cratons have a nearly identical complex geological history. The following is a simplified geological history of most of them. The oldest parts of the cratons date back to over 3 000 m.y.; cf. the Pibara system in Australia at 3 050 m.y. (Richard, 1975), the Dahomeyan in West Africa at about 3 400 m.y. (Holmes and Cahen, 1957), and the Guyana craton in Brazil at about 3 400 m.y. (Harrington, 1975). Not only are these cratons about the same age, but in almost all of them about six to seven orogenic cycles have been recognized. These cycles, marked by intensive folding, magmatization and formation of high-grade metamorphics and the intrusion of igneous rocks in the African craton, include the Dahomeyan 3 200–3 400 m.y., the Birriman and Bulawayan 2 800–3 100 m.y., the Shamwaian 2 650 m.y., the Limpopo 2 000 m.y., the Karagwe 1 040 m.y., the Goodhouse 875–900 m.y. and the Katanga 620 m.y. (Holmes and Cahen, 1957). Although Furon (1963) recognized only four orogenic phases, Haughton (1963) confirms the observations of Holmes and Cahen by recognizing seven distinct cycles in the West African Precambrian. Six to seven orogenic cycles have also been recognized in the cratons of Brazil by Harrington (1975). These comprise the Gurian 2 700 m.y., the Trans-Amazonian 1 700–2 200 m.y., the Espinhação

1 300–18 00 m.y., the Roraima 1 550–1650 m.y., the Uruaguyan 900–1 300 m.y., the Brazilian 700–900 m.y. and the Pan American cycle 600–640 m.y. The same number of orogenic cycles has been recognized in the Precambrian of Australia (Rickard, 1975).

These cratons were affected by severe epeirogenic activities involving warping, tilting and uplifting in the Palaeozoic, especially in association with the orogenic movements in adjacent geosynclines. Such movements led to the formation of basins on the cratons which have since served as foci of sedimentation. The cratons also experienced sagging, warping, faulting and uplifting in the Mesozoic, especially between the Jurassic and early Cretaceous possibly in association with the dismemberment and drifting apart of Gondwanaland and the formation of the rift-valley systems in East Africa and along the lower Niger and Benue in West Africa (King, 1962). The deformations accompanying these movements resulted in the evolution of several basins on the cratons.

The Tertiary period witnessed several epeirogenic activities in the adjacent fold mountain system that affected the cratons, which were uplifted in most places to form well-marked plateau, as in African, Indian, Guyanan and Brazilian plateaux. The movements were also accompanied by widespread outpouring of lava, as in India and Ethiopia.

Sedimentation basins

These basins can be classified into three groups: (1) basins originating as uneliminated Precambrian geosynclines – for example, the Indo-Gangetic basin in India, Amazon, Orinoco and Parana-Paraquay basins in South America; (2) basins dating to the Precambrian period and overlying the cratons – for example, Carpentaria, Canning, Fitzroy, Ord, Bonaparte, Daly and Gorgina in northern Australia; and (3) basins formed as a result of crantonic warping, sagging and faulting in the Palaeozoic–Mesozoic period – for example, Parnaiba and São Francisco in Brazil; Senegal–Mauritania, middle Niger, Chad–Benue, Congo, Lake Victoria, Rift valley and Kalahari basins in Africa; Godavari–Kristna, Villupuram and Cauvery basins in India; Lower Irrawaddy and Sitang in Burma, Chao Phraya and lower Mekong basins in Thailand and Cambodia; basins in the Basin and Range Province, Central Mesa and Balsas Mexcala in Mexico. Both the first and second groups were downwarped during the Palaeozoic and Mesozoic.

The older basins are filled with sediments dating to the Precambrian. The oldest rocks overlying the craton in the Carpentaria basin, comprising very thick unmetamorphosed sequences of sedimentary rocks, have been dated at 1 400–1 800 m.y. (Brown, 1975). All the older basins include glacial deposits like tillites, varves and erratics dating to the Carboniferous in South America and to the Permian in north Australia in their basal formations. Between the Triassic and Jurassic, these basins were filled with terrestrial and marine sediments, the former reflecting the different prevalent climatic regimes. Some of these basins experienced widespread lava flow. The Parana basin was covered by basalt up to 1 000 m thick and about 1.2 million km², the occurrence of which Bigarella (1975) puts at Rhaetic, and which Harrington (1975) ascribes to the Cretaceous.

The depositional sequence in the Congo basin since the upper Cretaceous probably depicts the geological and gemorphological processes in these basins. The oldest deposits comprise basal conglomerate, grits and lateritic rubble which, according to Cahen (1954), indicated a warm, humid phase. In early Tertiary, peripheral craton areas were uplifted, with corresponding sagging and deformation of the basin. This uplift was also experienced in Brazil and Australia. The effect was a widespread erosion of the craton and sedimentation of the basins. In the Congo basin, these early Tertiary sediments are composed largely of sands, soft sandstones, silt, silicified clays and limestone which were both marine and terrestrial. From the latter, Cahen (1954) deduced three alternating arid and wet phases. During the mid-Tertiary, the cratons and adjacent mountain chains were again involved in uplift movements with corresponding downwarping of the basins. In the case of the Congo basin, the uplift occurred along all its periphery except to the north. The present-day basin was established at the end of the Pliocene by further uplift along its periphery and the formation of a lake at the centre, which became the local base level (Haughton, 1963). These movements resulted in a great thickness of terrestrial sediments in this and the other basins.

Since the Pleistocene, these basins have been filled with series of deposits related to different climatic regimes. In the basins close to arid zones, lacustrine/fluvial and aeolian deposits alternate, reflecting wet and arid phases, and in mountainous regions, the basins are filled with sediments reflecting the alternation of cold and warmer climates.

The South-east Asian and the South American basins are still subject to continuous sedimentation, as the mountain chains adjacent to them are still undergoing active uplift and severe fluvial erosion.

The fold mountain systems

These include the Andean Cordillera in South America, the Sierra Madre Oriental in Mexico, the southern extension of the Himalayan system in Burma and Thailand (Ponnyadung, Arakan Yoma and Pegu Yoma in Burma and Bilauk Taung in Thailand), the island mountain systems of Indonesia and New Guinea and the northern part of the eastern highlands in Queensland, Australia. The older Hercynian systems in Queensland are generally less than 2 000 m; the younger systems in Indonesia are higher than 2 200 m, attaining summit altitudes of about 5 500 m in New Guinea; while in the Andean Cordillera, summit altitudes exceeding 6 000 m are common.

These mountain systems, composed of complexly folded sedimentary rocks into which series of massive batholiths were intruded, share certain common characteristics. The Tasman geosyncline, the forerunner of the fold mountain system in eastern Australia, developed in the early Palaeozoic, more or less at the same time as the Andean geosycline west of the Brazilian cratons and the Qaxacan geosyncline in Mexico. In all, about three to four distinct orogenic phases occurred in these mountain systems during the Palaeozoic. Severe folding and granitic intrusion occurred in the Tasman geosycline between the Cambrian and Permian period (Rickard, 1975). The Andean geosyncline was repeatedly affected by folding and granitic intrusions between late Devonian and early Permian times (Harrington, 1975). The Qaxacan geosyncline in Mexico also experienced repeated severe folding in the Ordovician and early Permian times (Czerna, 1975). In the Tertiary the Andean Cordillera witnessed four distinct orogenic movements, each consisting of several phases of folding and granitic intrusions, which led to the evolution of the present complex mountain system. Most of these Cainozoic fold mountain systems, especially the Andean Cordillera, have been under continuous epeirogenic movements since the Pleistocene, as evidenced by the strong earthquakes and volcanic eruptions frequently affecting them. This continuous uplift has led to severe erosion of the mountain sides and the continuous sedimentation of the intermontane basins and footslopes.

Volcanics

These comprise volcanic plateaux underlain by rhyolites, dacite, basalt, and trachyte several hundred metres thick, and volcanic mountains occurring in association with the fold mountain systems in Indonesia, Mexico and South America, and independent of any mountain system in East Africa and Central America. In the latter, individual volcanic mountains form a chain composed of five segments extending for 1 400 km, while in East Africa, Kilimanjaro and Kenya, at 5 896 m and 5 200 m respectively, are the best known.

The volcanic plateaux in Mexico and Nicaragua were formed as a result of intensive vulcanicity towards the end of the Oligocene through the Miocene (Czerna, 1975). The volcanoes occurred mainly in association with the Tertiary orogenic cycles in the fold mountain systems, and also following the intense epeirogenic activities in the Pleistocene. They also form along subduction zones associated with crustal movements along the southern continuation of the San Andreas fault in Mexico (Czerna, 1975) and the Guatemala trench in Central America (Moody, 1975). Volcanic plateaux in Nigeria, especially the Biu plateau, was also formed as a result of several fissure eruptions in the mid-Tertiary.

Most of the cratons which cover the largest areas of the humid tropics have been stable since the Tertiary, and in most places the long periods of crustal stability have coincided with periods of alternating climatic regimes. Moreover, where mostly wet climatic phases prevailed, crustal stability together with the climatic and vegetal parameters already described have been instrumental in deep chemical decomposition of the rocks, severe leaching of the soils and widespread lateritization and silcrete formation in drier environments. These are considered in detail later in this book.

Conclusion

One important feature of the above picture of the humid tropical environment is that of an insufficient data base. It has not been easy to delimit unambiguously and characterize the region successfully, which for a long time has acted as a hindrance to scientists. It is only very recently that scientists have started to establish and maintain experiments aimed at unravelling the mystery of the humid tropical environment, and then only when indigenous scientists join the staff of various universities and research institutes. The information provided by these scientists is summarized in what follows.

References and further reading

Adejuwon, J. O. (1979a) *An Introduction to the Geography of the Tropics*, Nelson, London.

Adejuwon, J. O. (1979b) 'Potential primary biological productivity: its implications for irrigation in Nigeria', Unpub. mss. Dept. of Geography, University of Ife, Ile-Ife.

Barry, R. G. (1969) 'The world hydrological cycle', in *Water, Earth and Man*, R. J. Chorley (ed.), Methuen, London.

Beckinsale, R. P. (1957) 'The nature of tropical rainfall', *Tropical Agriculture*, **34**(2), 76–98.

Bigarella, J. J. (1975) 'Brazil' in *The Encyclopedia of World Regional Geology, Part I, Western Hemisphere*, R. W. Fairbridge (ed.), J. Wiley, London.

Brown, D. A. (1975) 'Australia' in R. W. Fairbridge (ed.), *op. cit.*

Birot, P. (1959) *Géographie Physique Générale de la Zone Intertropicale*, Paris, C.D.U.

Cahen, L. (1954) *Géologie du Congo Belge*, Liège.

Czerna, Z. de (1975) in R. W. Fairbridge (ed.), *op. cit.*

Dobby, E. H. G. (1973) *South East Asia*, 11th edn., Hodder, London.

Douglas, I. (1969) 'The efficiency of humid tropical denudation systems', *Trans. Inst. Br. Geog.*, **40**, 1–16.

Evans, G. C. (1939) 'Ecological studies on the rainforest of southern Nigeria II, the atmospheric environmental conditions', *J. Ecology*, **27**, 437–82.

Faniran, A. and Areola, O. (1974) 'Landform examples from Nigeria: a gully', *Nig. Geogr. J.*, **17**(1), 57–60.

Fosberg, F. R., Garnier, B. J. and Kuchler, A. W. (1961) 'Delimitation of the humid tropics', *Geogr. Rev.*, **51**(3) 333–47.

Fournier, F. (1960) *Climat et Erosion: la relation entre l'érosion du sol par l'eau et les précipitations atmosphériques*. Presses, Univ., Paris.

Fournier, F. (1962) *Carte du Danger d'Erosion en Afrique au Sud du Sahara*, C.E.E.–C.C.T.A. Presses Univ., Paris.

Franzle, O. (1979) 'The water balance of the tropical rainforest of Amazonia and the effects of human impact', *Applied Sciences and Development*, **13**, 88–117.

Freise, F. (1936) 'Das Binnenklima van Urwaldern in sub-tropischen Brasilien', *Petermans Mitt.*, **82**, 301–7.

Furon, R. (1963) *Geology of Africa*, Oliver and Boyd, Edinburgh.

Garnier, B. J. (1953) 'Some comments on the rainfall at University College, Ibadan during 1952', *Research Note No. 3*, Department of Geography, University of Ibadan.

Garnier, B. J. (1957) 'The climatic aspects of irrigation development in Nigeria', *Advancement of Science*, **13** (52), 351–4.

Garnier, B. J. (1958) 'Some comments on defining the humid tropics', *Research Note No. 11*, Department of Geography, University of Ibadan.

Gourou, P. (1961) *The Tropical World* (5th edn. 1980), Longman, London.

Harrington, H. J. (1975) 'South America', in R. W. Fairbridge (ed.) *op. cit.*

Haughton, S. H. (1963) *The Stratigraphic History of Africa South of the Sahara*, Oliver & Boyd, Edinburgh.

Hodder, B. W. (1957) 'A note on delimiting the humid tropics: the case of Nigeria in West Africa', *Research Note No. 10*, Department of Geography, University of Ibadan.

Holmes, A. and Cahen, L. (1957) *Géochronologie Africaine 1956, Resultats acquis au 1er juillet 1956*, Mem. Ac. R. Sc. Col. Brussels 5, Fasc. 1.

Jackson, I. J. (1971) 'Problems of throughfall and interception assessment under tropical forest', *J. of Hydrology*, **12**, 234–54.

Jeje, L. K. (1977) 'Some soil erosion factors and losses from experimental plots in the University farm, University of Ife, Ile-Ife, Nigeria', *Nigerian Geog. J.*, **20** (1), 59–69.

Jeje, L. K. (1979a) 'Erosion under rubber plantations in Araromi Rubber Estate, Ondo State', Paper presented at the Departmental Seminar, April 1979, Dept. of Geography, University of Ife, Ile-Ife.

Jeje, L. K. (1979b) 'Debris flow on a ridge near Ile-Ife, Western Nigeria', *J. of Mining and Geol.*, **16**, 1.

King, L. C. (1962) *The Morphology of the Earth*, Oliver & Boyd, Edinburgh.

Kirkby, M. J. (1969) 'Infiltration, throughflow and overload flow', in *Water, Earth and Man*, R. J. Chorley (ed.), Methuen, London.

Köppen, W. (1936) 'Das geographische System der Klimate', in *Handbuch der Klimatologie*, **1**, C. Berlin, Gebr. Borntrager.

Lockwood, J. G. (1974) *The Physical Geography of the Tropics: an introduction*, Oxford Univ. Press.

Mohr, E. C. J. and Van Baren, F. A. (1954) *Tropical Soils*, Interscience, London.

Moody, J. D. (1975) 'Central American – regional review', in R. W. Fairbridge (ed.), *op. cit.*

Nieuwolt, S. (1977) *Tropical Climatology: an introduction to the climates of the low latitudes*, J. Wiley, London.

Martonne, E. De (1946) 'Géographie zonale: la zone tropicale', *Annales de Géographie*, **55**, 1–18.

Ojo, O. (1977) *The Climates of West Africa*, Heinemann, London and Ibadan.

Ollier, C. D. (1975) *Weathering*, 2nd edn., Longman, London.

Ong, H. L., Swanson, V. E. and Bisquo, R. E. (1970) 'Natural organic acids as agents of chemical weathering', *U.S. Geol. Survey Prof. Paper*. 700-c.

Oyegun, R. O. (1980) 'The effects of tropical rainfall on sediment yield from different landuse surfaces in sub-urban Ibadan', Ph.D. thesis, University of Ibadan.

Richards, P. W. (1952) *The Tropical Rainforest*, Cambridge Univ. Press.

Rickard, M. J. (1975) 'Australasia – regional review', in R. W. Fairbridge (ed.), *op. cit.*

Ruxton, B. P. (1967) 'Slopewash under mature primary rainforest in northern Papua', in *Landform Studies from Australia and New Guinea*, Jennings, J. N. and Mabbutt, J. A. (eds.), Cambridge Univ. Press., pp. 85–94.

Ruxton, B. P. and Berry, L. (1957) 'Weathering of granite and associated features in Hong Kong', *Bull. Geol. Soc. America*, **68**, 1263–92.

Simonett, D. S. (1967) 'Landslide distribution and earthquake in Bewani and Torricelli mountains, New Guinea', in *Landform Studies from Australia and*

New Guinea, Jennings, J. N. and Mabbutt, J. A. (eds), Cambridge Univ. Press, pp. 64–84.

Smith, D. D. and Wischmeier, W. H. (1962) 'Rainfall erosion', *Advances in Agronomy*, **14**, 109–48.

Stocking, M. A. and Elwell, H. A. (1973) 'Prediction of subtropical storm soil losses from field plot studies', *Agric. Meteorology*, **12**, 193–201.

Strakhov, N. M. (1967) *Principles of Lithogenesis*, **1**, trans. J. P. Fitzsimmons, Oliver & Boyd, Edinburgh.

Thomas, M. F. (1974) *Tropical Geomorphology: a study of weathering and landform development in warm climates*, Macmillan, London.

Tricart, J. (1972) *The Landforms of the Humid Tropics, Forests and Savannas*, trans. by C. J. K. de Jonge, Longman, London.

Vischer, S. S. (1945) 'Climatic maps of geological interest', *Bull. Geol. Soc. Amer.*, **56**, 713–36.

Wentworth, C. K. (1943) 'Soil avalanches in Oahu, Hawaii', *Bull. Geol. Soc. Amer.*, **54**, 53–64.

White, S. E. (1949) 'Processes of erosion on steep slopes of Oahu, Hawaii', *Amer. J. of Science*, **247**, 168–86, 248, 508–15.

Wilkinson, G. E. and Aina, P. O. (1976) 'Infiltration of water into two Nigerian soils under secondary forest and subsequent arable cropping', *Geoderma*, **15**, 51–9.

Williams, M. A. J. (1969) 'Prediction of rainsplash erosion in the seasonally wet tropics', *Nature*, **222** (No. 5159), 763–5.

Young, A. (1972) *Slopes*, Oliver & Boyd, Edinburgh.

Geomorphological processes in the humid tropics

The forces that create and modify landforms anywhere on the earth's surface are basically of two types:

1 the endogenic or tectonic agents emanating from the earth's interior, and producing earthquakes, folded and faulted structures, volcanoes and so on; and
2 the exogenic or external gradational agents, which reflect in the main climatic and related influences.

Although the first set of forces are very important in their own right and, in fact, influence the second type, they are not, however, of interest here. The point has been stressed earlier that in the area of interest in this work, geological structure in the form of dominant tectonic forces is rare over large areas and so can be held constant. Endogenic forces are not common over large parts of the humid tropics, which consist essentially of stable blocks or cratons made up of old basement complex structures. Moreover, tectonic forces are not affected by climatic parameters, and so do not fit into the general theme of this work.

Accordingly, emphasis is placed on the gradational or denudational processes which effect the breakdown of the basic geology of rocks; erosion and transportation of the weathered or loose surface material; and finally the deposition of the transported material. All these processes operate at or near the earth's surface to produce the land forms which geomorphologists, among other earth scientists, study. The operations of these forces – weathering, erosion and transportation (removal), and deposition – is universal not only in the sense that they occur all over the earth's surface, including deserts and ice-capped regions; but also because the processes are common to all the exogenic agents of water, wind, ice and, to some extent, gravity, some of which bear little or no relation to climate. At the same time, however, the nature, rate and end result of the process vary spatially, a thesis which is explored and exemplified for the humid tropics in the three chapters in this section.

3   Rock weathering in the humid tropics

One of the most fundamental phases of the geomorphological scene at the earth's surface is the breakdown, by the processes of disintegration and/or decomposition, of the rocks at or near the earth's surface. This is so for various reasons. First, it is the initial phase of the chain of events which result in the modification of earth's surface features by the various gradational agents. Secondly, the products of rock weathering, the regolith, is a crucial prerequisite for the effective operation of most of the gradational or denudational agents. This is because the regolith, although varying in depth and constitution, provides the loose materials of suitable sizes for removal by mass wastage, running water, wind and glacial

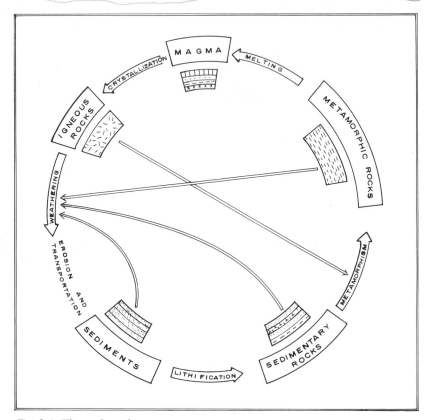

Fig. 3.1 The rock cycle

erosional processes. Although fresh (intact) bedrock is frequently exposed on hillsides and cliff faces, as well as in some deserts and ice-worn arctic regions, yet an estimated 90 per cent of the earth's surface is mantled by regolith and other unconsolidated materials, on which the denudational agents operate. In fact, it is difficult to imagine any detachment, transportation and deposition of rock fragments on fresh, unweathered bedrock.

The process of rock weathering is also an important – indeed, indispensable – component of the rock cycle. Ollier (1979) considers weathering as a convenient way to begin the study of the geological (or rock) cycle, a process which facilitates the production of material for the processes of sedimentation, consolidation, diagenesis and metamorphism among others (Figure 3.1). The rock cycle is a very important phenomenon which, operating within the larger hydrologic cycle, of which it is only a phase, distributes the materials of the earth's crust, helps to preserve the delicate balance between the processes of mountain building and destruction, and consequently maintains the stable equilibrium between landmasses and ocean basins (Faniran and Ojo, 1980).

Finally, the process of rock weathering has important implications for natural resources (Figures 3.2 and 3.3). Apart from helping to aggregate and concentrate ore minerals such as iron ore, bauxite, tin ore,

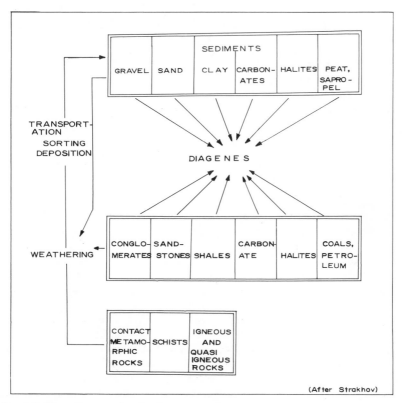

Fig. 3.2 The minor cycle of matter on earth

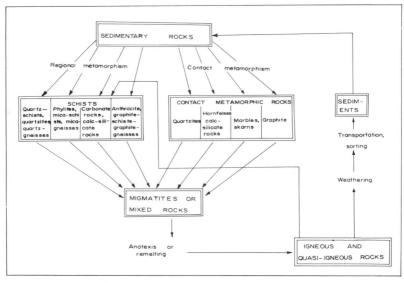

Fig. 3.3 The major cycle of matter on earth

opal and so on (Faniran, 1971*a*), the regolith also facilitates the ingress of water into the overburden and so transforms the ordinarily impermeable and impervious granites and gneisses into sources of abundant groundwater reserves (Faniran and Omorinbola, 1980; Chapter 13).

The last process – that is, deep weathering – has special relevance in the humid tropics, where the process of deep weathering is not only peculiar (see Chapter 1) but also particularly suitable for the desired effects. This process, which is discussed further below, is the most intense form of weathering and operates at the maximum within the area demarcated as the humid tropics.

Definitions, type and scope

The term *rock weathering* applies to the physical and chemical alteration of the rocks exposed at the earth's surface. This idea of alteration has, of course, been expressed in several ways in the literature (cf. Polynov, 1937; Reiche, 1950; Keller, 1957; Ollier, 1975; Faniran and Areola, 1978; Faniran and Ojo, 1980). Of all these, perhaps the most comprehensive viewpoint is that contained in Ollier, as follows:

> Weathering is the breakdown and alteration of materials near the earth's surface to products that are more in equilibrium with newly imposed physico chemical conditions. Many rocks were originally formed at high temperature, high pressure and in the absence of air and water, and a large part of weathering is, in fact, a response to low temperatures, low pressures, and the presence of air and water. Many kinds of alterations are possible, however, and even a material that is

an end-product of weathering under one set of conditions may become the raw material of a weathering process if conditions change. (p. 1)

A similar idea is implied in the process of metamorphism in geology, by which process rocks change in response to increased temperature and pressure. The difference is that whereas metamorphism involves rearrangement and restructuring of rock grains and minerals, in most cases making them more compact and resistant, weathering involves the opposite process of breakdown and (in most cases) loss as well as formation of new (secondary) substances.

A number of terms are often used in the study of rock weathering, particularly by Russian scientists, which require some explanation. Among these are the weathering crust, supergenesis, deep weathering, pedogenesis, diagenesis, katagenesis, halogenesis, hydrogenesis, biogenesis and technogenesis.

The *weathering crust* refers to the part of the lithosphere which is dominated by the weathering process. The 'crust' is formed in the zone of weathering, and its composition and structure are entirely determined by the interaction of the lithosphere, the hydrosphere, the biosphere and the atmosphere, especially the troposphere (Lukashev, 1970; Faniran and Areola, 1978). It is more than the regolith, being a peculiar thermodynamic layer characterized, like the latter, by low temperature and pressure, but whose lower boundary extends further than the decomposed layer to the surface reached by oxygen penetration.

The most important contributions to the development of the concept of the weathering crust have been made by Russian scientists, most prominent among whom are Dokuchaiev, Glinka, Zemyatchauskii, Vernadskii, Fersman, Polynov, Vinogradov and Lukashev.

There are two closely related trends in this development. One is concerned with the regularities involved in the processes of weathering, migration and accumulation of chemical and other elements and compounds in the weathered zone; the formation of various types of weathering products; and the distribution of these materials over the earth's surface in relation to climatic, geochemical and other conditions and processes; the other investigates the ancient crust and its mineral resources, including the formative processes of these resources in the oxidation and supergenesis zone. In other words, the first trend concerns itself with current soil-forming processes – namely, the 'recent' weathering crust – while the second trend has developed into what is now being studied under the topic of deep weathering and duricrusts.

Supergenesis is a term related to 'weathering crust'. First introduced by Fersman, it is used for the specific geological and geochemical processes occurring in the outermost part of the earth's crust. These processes are the physical and geochemical processes of disintegration, decomposition, transportation and redeposition of mineral material in the upper crust as well as of mineral formation in the hydrosphere. Examples of processes embraced by this term are weathering, *pedogenesis* or soil formation, and *biogenesis* or mineral formation by the action of organisms, sedimentation and other phenomena related to the migration, transportation and accumulation of chemical elements and compounds within the weathering crust. *Diagenesis* describes the changes that take place in the composition and properties of sediments after their formation. *Katagenesis* is the alteration in mineral material at the boundary between dissimilar rocks. *Halogenesis* is the natural precipitation of salts and saline solutions and brines, hydrogenesis is the natural condensation of moisture in the air spaces of surficial rock or rock material, and *technogenesis* the processes caused by man's technological activities, such as engineering, mining, and agriculture.

However, the most significant of these terms in the context of the present volume is *deep weathering* (see Faniran, 1968, 1970a, 1974). Although the concept of the weathering crust refers to the entire mantle of weathered material overlying fresh rock wherever it occurs, deep weathering is often restricted to the type of rock weathering which takes place under a set combination of environmental conditions, including climatic, biotic, geomorphic, geologic, site and chronologic factors (Thomas, 1974). Deep weathering is very prominent in the humid tropics and so takes the lion's share of the discussion in this chapter. Other types of weathering are discussed in relation to it, being the most intense form of rock weathering by natural processes so far encountered at the earth's surface.

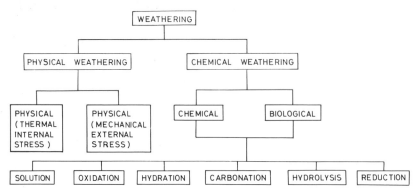

Fig. 3.4 Types of rock weathering

Broadly speaking, weathering processes are of two types – the mechanical or physical type, and the chemical type (Figure 3.4). Mechanical weathering is described in terms of rock disintegration, implying that pieces of rock are segregated without alteration. In contrast, chemical weathering, or rock decomposition, involves the breakdown of the chemical structure of the mineral grains that make a rock.

The distinction between mechanical and chemical weathering is, however, becoming an artificial one. Not only are both processes governed by the same principles of re-establishing equilibrium under changing environmental conditions, but also both occur together in many places, while large-scale mechanical fracturing of rocks aids the penetration of air, water and organic acid, the main agents of chemical weathering. Nevertheless, the two types are usually described separately (cf. Bloom, 1969; Ollier, 1975; Butzer, 1976; Faniran and Areola, 1978). These descriptions are not repeated here, but interested readers are referred to the above sources among others. Aspects which are important in the humid tropics are reviewed in what follows with particular emphasis on the nature, factors and products of deep chemical weathering.

The nature of deep chemical weathering

As observed above, rock weathering results from a change in the physical and/or chemical environment of the original rock. In the case of chemical weathering the change involves the chemical environment and obeys the chemical equilibrium laws, also called Le Chatelier's Principle, which in effect relates to the state of stability of matter. The stability state in a system is that when the phases of the system do not undergo any further change with the passage of time. This means that chemical reactions or changes will continue until equilibrium is reached, both with the environment, and with the other elements in the system. Changes set in again if components are added and/or removed from the system, and stop when a new equilibrium is reached. Two examples illustrate this point; as follows:

(1) Potash feldspar + water → kaolinite + silica + potash

$$2KAlSi_3O_8 + 3H_2O \rightarrow Al_2Si_2O_5(OH)_4 + 4SiO_2 + 2KOH$$

Potash feldspar + water → illite + silica + potash

$$3KAlSi_3O_8 + 2H_2O \; KAl_2(Al,Si_3)O_{10}(OH)_2 + 6SiO_4 + 2KOH$$

In (1) the alteration of potash feldspar to kaolinite involves the loss of potash in solution; in (2) some potash is retained, and the product is illite.

The various types of chemical reactions involved in chemical weathering – oxidation, carbonation, hydrolysis, hydration, chelation, and so on are all well known (cf. Reiche, 1950, Loughnan, 1969; Ollier, 1975); what is not so well known is the nature of the operation of these processes in different climates. General observation suggests that climate plays an important role in determining the kinds and intensities of weathering processes. Climate controls weathering directly, through the temperature and precipitation of a region, and indirectly, through the vegetation cover.

Perhaps the best illustration of the above viewpoint is found in the humid tropics, where the best examples of extreme (deep) chemical weathering are found in the form of certain tropical regoliths with rich concentrations of iron and aluminium oxides that are widely mined as ores (Faniran, 1971a). The extreme weathering situations in these environments can be illustrated as follows:

$$Al_2Si_2O_5(OH)_4 + 5H_2O \rightarrow Al_2O_3 \cdot 3H_2O + 2H_4SiO_4$$

Kaolinite + water → hydrated + silicic
aluminium   acid in
oxide   solution

Here, the end product of the reaction in (1) (hydrolysis) above is further acted upon to produce another end product, consisting essentially of stable elements and compounds found in many tropical soils or regoliths and variously called laterites, ferricretes, bauxites and so on (Dury, 1969; Faniran, 1970b).

Alongside the above case for a unique humid-tropical (deep) weathering environment are some recent works which tend to argue against the concept of a tropical zone within which an exclusive group of distinctive weathering processes occurs (Thomas, 1974, p. 13). Fripiat and Herbillon (1971), in particular, do not believe that the transformation sequences or the synthesis mechanisms of weathering differ in different climates.

The opponents of the concept of a distinct tropical weathering environment support their position by such observations as deeply weathered profiles and duricrusts located in extra-tropical regions (Dury, 1971), and argue against the idea of these materials being relict in extra-tropical areas. They also point out the variable depths of weathering in present-day humid tropical regions. Nevertheless, the preponderance of laterites and related products of deep weathering, as well as the great depths of weathering measured in humid tropical regions (cf. Ruxton and Berry, 1957; Faniran, 1974; cf. also Figure 3.5) are sufficiently challenging to warrant further investigation and analysis. Examples of this kind of investigation include Jenny (1941), Sherman (1952), Keller (1957) and Pedro (1968).

Jenny studied the influence of temperature on weathering, and shows that the amounts of clay in soils increase with annual temperature, thus partly explaining the characteristically high clay content of most tropical regoliths. Similarly, Sherman (1952) demonstrates that different clay minerals tend to form under different soil moisture (rainfall) conditions. It is significant to note in this last figure that kaolinite and goethite occur within the rainfall figures of the humid tropics, and montmorillonite at the dry marginal areas. Gibbsite is another common mineral in tropical regoliths, and has been found in abundance in Malaysia, especially where rainfall rises from 2 000 to 3 500 mm per annum leading to heavy leaching (West and Dumbleton, 1970). Finally, within the humid tropics itself, Pedro (1968) distinguishes three weathering environments, based on mean annual precipitation. These are:

1 zone of bisiallitization, with the formation of 2:1 lattice (montmorillonite) clays (rainfall <500 mm p.a.);

2 zone of monosiallitization (kaolinization) rainfall between 500 and 1 500 mm p.a.); and

3 zone of allitization, in which gibbsite occurs along with kaolinite (rainfall > 1 500 mm p.a.).

The zones suggested by Sanches Furtado (1968) follow a similar pattern.

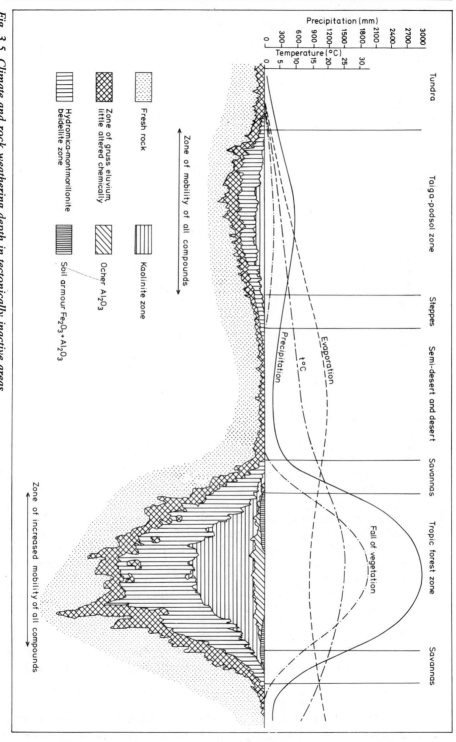

Fig. 3.5 Climate and rock weathering depth in tectonically inactive areas

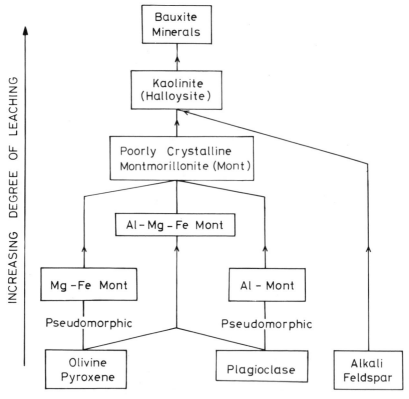

Fig. 3.6 Sequence of weathering

It is of interest to compare the above pattern with the sequence shown in Figure 3.6. In both cases some idea of the extent or intensity of weathering is given, resulting in the concentration of the most stable minerals in the regolith or saprolite. The importance of the leaching process is also indicated. The leaching process itself is, of course, a function of both the solubility of the elements and compounds themselves and the weathering environment, especially in relation to the pH and the Eh of the solution (see Table 3.1). The local water-table and water-balance situations are also important. All these have been ably discussed by Loughnan (1969) and Thomas (1974) among others. Overall, under normal environmental conditions, the least mobile cations – namely, aluminium oxides, iron oxides and silica, in that order – are the only ones which survive the deep weathering process, and so constitute the main components of the end product of deep weathering.

Unfortunately, deep weathering itself has not been studied in sufficient detail, so that only a few general statements can be made:

1 weathering reaches its most advanced stages, leaving only the most stable substances such as the clays;
2 all the highly mobile and most of the intermediate (medium) mobile substances, sometimes including silica, are leached out of the profile;

TABLE 3.1 Weathering products in relation to some specific environments

Environment	pH	Eh	Behaviour of metallic ions	Mineralogy
Nonleaching, hot. Rainfall 0–12 in.	Alkaline	Oxidizing	Some loss of alkalies. Iron present in ferric state.	Partly decomposed parent minerals. Illite, chlorite, montmorillonite, an mixed layered clay minerals. Hematite, carbonates and salts.
Nonleaching below the water table	Alkaline to neural	Reducing	Some loss of alkalies. Iron present in ferrous state.	Partly decomposed parent minerals. Illite, chlorite, montomorillonite, and mixed layered clay minerals. Siderite and pyrite. Organic matter present.
Moderate leaching temperate climate. Rainfall 25–50 in.	Acid	Oxidizing to reducing	Loss of alkalies and alkaline earths. Some loss of silica. Concentration of alumina, ferric iron, and titania.	Kaolinite with or without degraded illite. Some hematite present. Organic matter generally present.
Intense leaching, hot. Rainfall 50 in.	Acid	Oxidizing	Loss of alkalies, alkaline earths, and silica. Concentration of alumina, ferric oxide, and titania.	Hematite, goethite, gibbsite, and boehimite with some kaolinite. Organic matter absent or sparse.
Intense leaching, cool. Rainfall 50 in.	Very acid	Reducing	Loss of alkalies, alkaline earths, and much silica, and titania retained.	Kaolinite, possibly with some gibbsite or degraded illite. Organic matter abundant.

3   the process involves both the downward and the upward movement of the sesquioxides and their precipitation and demobilization within the profile (this is another way of saying that the deep-weathering process can be thought of as combining the pedological processes of leaching and sesquioxide concentration as well as the capillary groundwater and sedimentary processes in geology);

4   under favourable conditions, especially of rock type, rocks weather directly into gibbsite or bauxite (cf. Figure 3.6); however, the more common process is that leading to the formation of a less homogeneous material, such as the laterite or fersiallitic duricrusts consisting of considerable and comparable quantities of silica and the sesquioxides of iron and aluminium – in other words, the formation of virtually all types of duricrusts, particularly those associated with typical deep-weathering profiles, can be linked to the deep-weathering process.

Nevertheless, these views have to be seen in relation to the following limitations, among others:

1   the remote, harsh and debilitating tropical environment that has affected the course of research, which not only started rather late, compared with espeically the temperate lands, but has also limited the number, scope and rigour of empirical and laboratory (experimental) work;

2   differences in microclimate, drainage, parent material and vegetation among others are certain to produce spatial variation in the rate and nature of the deep-weathering process (Douglas, 1980);

3   much confusion and controversy still exist in the literature, especially in relation to such basic issues as the factors, processes, terminology and nomenclatures of deep weathering, as shown in what follows.

The factors of deep weathering

From what has been said so far in this chapter, two important notions are combined in the concept of deep weathering. These are depth (of weathering) in the spatial sense – that is, the thickness of the regolith, and the extent of change (alteration) in the original rock affected by the chemical reactions. The concept of deep weathering thus involves chemical weathering of rocks to considerable depths – namely, thick regoliths, as well as an advanced degree of change in the original rock (see Figure 3.7). The end product is usually characterized by low weathering-product indices (WPI) expressed as ratio (Reiche, 1943, 1950; Thomas, 1974), while the clay content (another common expression of the degree of weathering) is considerable and is made up mostly of stable components.

For a long time, deep weathering was studied separately from the related topics of, say, bauxitization and lateritization. It was only in the 1920s that Woolnough brought these related ideas together in his study

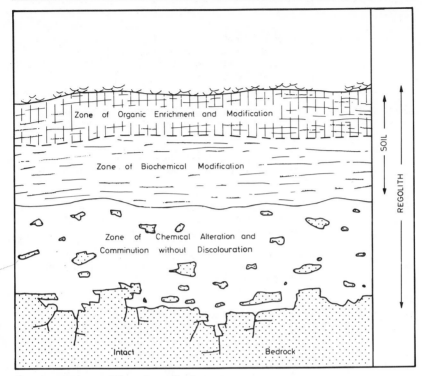

Fig. 3.7 Soil, regolith and weathering

of 'The duricrusts of Australia' (Woolnough, 1927). Since then, many workers have elaborated this process–form relationship in which duricrusts (including laterites, ferricretes, bauxites and so on) are described as having formed largely by the process of deep-weathering, aspects of which include lateritization, bauxitization, ferralitization among others (Faniran, 1968, 1969a and b, 1974; Faniran and Areola, 1978).

The above argument is relevant to the extent that the factors described in the literature in connection with these other processes (cf. Harder, 1952; De Swardt, 1964; Prider, 1966; Thomas, 1974; Ollier, 1975) are also applicable in the case of deep weathering (cf. Table 3.2).

TABLE 3.2 Major factors affecting depths of weathering

Climatic factors	Temperature: high temperatures increase the rate of endothermic chemical reactions.
	Precipitation: high precipitation increases the availability of the principal reagent in weathering process: water.
Biotic factors	Vegetation cover: a dense forest canopy protects the surface from wash processes and provides organic acids

which are capable of mobilizing certain rock minerals, especially iron by chelation. Conversely, open vegetation of the savanna type favours immobilization of iron and favours surface runoff.

Geomorphic factors	Landsurface stability: weathering penetration is favoured by a low rate of surface denudation prevailing on gentle slopes. Age of landsurface: prolonged stability (persistence of ancient landsurfaces) allows deep profiles to develop.
Site factors	Free drainage: elevated sites promote downward movement and frequent renewal of groundwater essential to rapid decomposition of rocks. Reception zones: sites experiencing convergent runoff may have increased water supply, but this may combine with poor site drainage.
Geologic factors	Rock type: presence of minerals particularly susceptible to alteration increases rate of weathering penetration and may promote early disaggregation of rock. Rock texture: rock texture affects behaviour under weathering attack. Crystalline rocks of coarse texture disaggregate more rapidly than fine-textured rocks, which may undergo more rapid alteration. Texture in sedimentary rocks affects permeability and rate of weathering penetration. Rock fissility: faults, joints and fractured grain boundaries promote weathering penetration, especially in crystalline rocks. Hydrothermal alteration: rocks previously subject to varying forms of hydrothermal activity may become additionally susceptible to groundwater weathering on exposure.
Chronologic factors	Climatic change: variations of climate and vegetation with time alter the balance of weathering and erosion. Pluvial conditions in the arid zone during the Tertiary and Pleistocene have led to the presence of relict deep weathering. Tectonic change: variations of crustal stability affect landsurface stability and the period available for weathering penetration.

Whereas Thomas (1974) lists the factors affecting depths of weathering generally (Table 3.2), Harder (1952) and De Swardt (1964), among others, list those relevant to the process of lateritization as follows:

1 Presence of rocks with easily soluble minerals, yielding residues rich in alumina and iron;
2 effective rock porosity and permeability facilitating easy access as well as circulation of water (solutions) – the free circulation ensures the removal of dissolved matter, so as not to encourage the establishment of equilibrium conditions in saturated solutions;
3 normal to abundant rainfall with either a seasonal regimen or at least breaks in between;
4 Abundant vegetation and other biotic components, including bacteria; organic acids in particular act as effective solution and precipitation agents;
5 tropical or warm temperatures which accelerate the rate of chemical reaction and promote the process of clay formation (see Figure 3.8);
6 low to moderate topographic relief, allowing free movement of the water table and minimization of removal processes; and
7 A long period of stable geological structure.

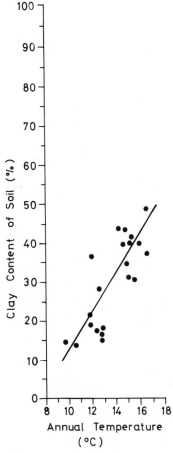

Fig. 3.8 *The clay content of soils derived from basic rocks in relation to mean annual temperature*

Of these factors, the most controversial is perhaps the topographic factor where opinion is divided among

(a) those who believe in the low-relief hypothesis (for instance, Woolnough, 1927; Harder, 1952; Ollier, 1974);

b) those who advocate the high-relief hypothesis (for example, Playford, 1954; Prider, 1966); and

c) those who do not subscribe to either view but argue that deep weathering and duricrusts can occur on any type of relief and slope (Hallsworth and Costin, 1953).

These three views are discussed in Faniran (1969*a*) among other places. The conclusion drawn there broadly agrees with Ollier's (1975, p. 127), as follows:

Many examples of deep weathering are found in areas of considerable relief. These may be relics of once much thicker layers of regolith, and it is possible that weathering occurred when the areas had much less relief than at present.

Similarly, Thomas's (1974, p. 15) observation on the factors of weathering as they operate in the humid tropics is very pertinent. It is stated as follows:

... it is clear that the inter-tropical zone is one within which many of these factors [see Table 3.2] operate to favour deep chemical decay of rocks. Large areas of high temperature, and moderate to high rainfall, promote the weathering processes.... Stable landsurfaces of considerable antiquity ... provide a geological and geomorphic basis favouring deep penetration of weathering, and thorough decomposition of rocks. Weathering processes are most favoured in the humid tropics, where surface denudation is not particularly rapid (Fournier 1962; Douglas, 1969).

In conclusion, there seems to be sufficient information in the literature to support the view that the various factors of weathering appear to assume their most favourable forms in the humid tropical regions, particularly the Af and Am climates. Their combined influence wanes in the marginal savanna and semi-desert regions, where weathering depths are less and the minerals weather mostly to intermediate states, except where they re relics of earlier, more humid periods (Thomas, 1974, p. 15; Figure 3.4).

The products of deep weathering

Ordinarily, the products of rock weathering, especially chemical weathering, are described as saprolite, regolith, gruss or weathered mantle. They consist of

1 an intimate mixture of rock and mineral fragments in all possible stages of decomposition;

2 organic substances; and

3 solution and colloidal suspension.

The mineral matter consists of clay minerals, which may be illite, montmorillonite or kaolinite, and hydrated sesquioxides of aluminium, iron, titanium and so on, mixed with dissolved and colloidal substances. There are also new substances of coagulated colloidal silica, alumina, ferric iron-oxide and reactions of these with dissolved materials – for example, phosphoric acid. Also included in this group are such deposits from indigenous solutions as amorphous silica of some desert hardpans, the calcium carbonate accumulation in arid soil profiles, and other products of base-exchange reactions.

However, the products of weathering that are of interest here are more limited, being restricted to those resulting from deep chemical weathering in humid tropical environments (cf. Table 3.1). They are distinguishable in at least two ways: in profile, and by their physical and mineralogical characteristics. In profile, the material is distinguishable by its considerable depth and/or its characteristic zonation or stratification (Faniran, 1968, 1974; Faniran and Areola, 1978). The physical and mineralogical characteristics of the material vary according to position in profile as well as with the type of the parent rock, process of formation, and the degree of exposure. However, all duricrusts have undergone very advanced forms of chemical change and so consist essentially of the least mobile (most stable) substances. These two aspects are described below.

The deep-weathering profile

The typical deep-weathering profile is the 'duricrust' profile, consisting of, among other layers, an *indurated* (hardened), a mottled and a pallid zone, overlying weathered and/or unweathered bedrock (Faniran, 1970*a*; Faniran and Areola, 1978; Faniran and Ojo, 1980). Duricrust is the term used for the indurated zone of the typical deep-weathering profile; it may be essentially aluminous (bauxite), ferruginous (ferricrete or iron-oxide ore), siliceous (silcrete), calcareous (calcrete), or a combination or two or more of the above – for example, laterite (consisting of silica and the sesquioxides of iron and aluminium).

The typical laterite profile has been described in several places in the literature (cf. Thomas, 1974, pp. 53–54; Faniran and Aerola, 1978, pp. 165–67; Faniran and Ojo, 1980, pp. 283–84). The description in Table 3.3 is based on Faniran (1968) and Thomas (1974) (cf. also Figure 3.9). Although the terminology of laterite is not standard (cf. Goudie, 1973; Faniran, 1970*b*; Thomas, 1974; cf. also Table 3.4), the authors still prefer this term to any of its proposed substitutes, mainly on account of its neutral connotation in terms of mineralogical composition, and, perhaps more importantly, because it is the oldest and the most consistently used term in the literature (cf. Faniran, 1969*b*). The definition below, modified from Faniran (1968) confirms the appropriateness of the term:

a mass of vesicular, vermicular, concretionary, pisolitic or massive

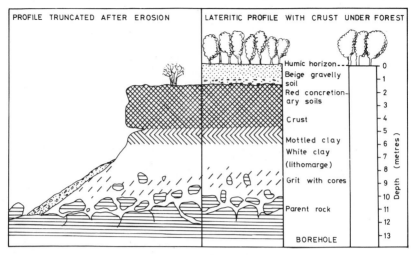

Fig. 3.9 Typical lateritic profile in Sudan

TABLE 3.3 Typical laterite profiles

(A) Laterite profile in sandstone

Depth in metres

0.0–0.6: Variable depth of rather structureless pisolitic soil, with colour varying from yellow-brown to dark-brown (brown-podzolic soil), and pH varying between about 5 and 4. The soil has a high gravel and sand content and is low in silt and clay. The soil (which may be absent) is underlain, where present, by massive and highly indurated pisolitic (upper-indurated-zone) laterite. The laterite sample contains highly spherical and smaller pisolites than do the lower portions. They generally form the hardest rock samples and also the most dehydrated. Petrographical study shows that the quartz size and content are smaller and the sesquioxide content is higher than in the lower portions of the profile.

0.6–1.4: Concretionary (pisolitic) middle-indurated-zone laterite. The constituent pisolites are bigger and softer than those present in the upper portions, but smaller and harder than those below. Hand specimens are generally less compact, and their observed colours vary from predominantly dark-red matrix and lighter-brown pisolites at the top to darker-brown in both pisolites and matrix below.

Quartz grains are generally coarser, and the contents of silica and quartz tend to increase relative to their respective proportions in the upper-indurated zone. Hand specimens need proper impregnation and careful handling for purposes of satisfactory thin and polished sectioning of them.

1.3–2.0 Pisolitic (nodular) lower indurated-zone laterite. Pisolites are much bigger and softer than in the upper portions of the profile. They are in many cases rather

	indistinguishable from the matrix, which they resemble both in physical and in petrographical characteristics. Hand specimens need several rounds of impregnation for satisfactory thin sections: they do not take on good polish and are thus of little use in polished-section work. The colour of the hand specimen is mostly dusky to dark red.
2.0–3.5	Transitional indurated/mottled zone. Nodules are rare and ill-defined or poorly developed. Colour of hand specimens are mainly red to dusky red (red earth) with some mottling effect. Hand specimens, except where they were exposed and indurated, proved impossible to cut into thin sections; they could not be impregnated by the conventional methods.
3.5–5.0	Mottled zone, comprising a thin layer of mottled rock, exposed on the floor of the quarry and a thicker simply ferruginized bed underneath. The mottled zone may be essentially white kaolinite (or white clay) encased by sesquioxide material, or vice versa. The ferruginized bed is mainly sesquioxide-cemented quartz.
5.0–5.3	Pallid zone: rather ill-defined in general. It can, however, be tolerably well defined and even very conspicuous, particularly in fine-grained sandstone.
5.3	Unweathered sandstone with or without ferruginization.

(B) Laterite profile in shale

0.6	Pisolitic (brown-podzolic) soil, presumably derived from dehydrated middle or the lower indurated-zone laterite.
0.6–2.0	Lower-indurated-zone laterite composed essentially of large and small lumps of apparently compact, but rather soft and variably weathered, laterite. The laterite lumps were perhaps formed by induration subsequent to exposure. They crumble under slight pressure, particularly when wet. A hand specimen may be cut into thin sections and polished sections if properly and patiently impregnated: the polished section does not, however, provide much information, since the constituent materials are mostly fine-grained and earthy.
2.0–6.0	Mottled zone or red clay (red earth). The material may or may not be mottled. It hardens on exposure; it is distinguished from laterite by the fact that this material, unlike what is here called laterite, contains no concretionary (colloform) textures. The material also occurs below typical laterites and is underlain by the pallid zone.
6.0–14	Pallid zone consists essentially of white clay with bleached shale fragments occurring along the exposed section.

TABLE 3.4 Terminology of laterite

Iron clay (India)	Moco de hierro (Venezuela)
Brickstone (India)	Ironstone (Nigeria)
Canga (Brazil)	Plinthite (USA)
Murrum (Uganda)	Cabook (Ceylon)
Ouklip (S. Africa)	Picarra (Brazil)
Ferricrete (southern Africa)	Eisenkruste (Germany)
Laterite (India)	Pisolite (Australia)
Cuirasse (France)*	Krusteneisenstein (Germany)
	Mantle rock (Ghana)

*'Cuirasse' is a term used to denote cemented material that is resistant and solid; 'carapace' is used for cemented material that is friable.

Source: Goudie, 1973

indurated material consisting essentially of iron oxide, aluminium oxide, and quartz. Its hardness varies with age, position in profile, parent rock, constituent minerals and degree of post-duricrust disintegration.

A similar definition (of laterite) is contained in Du Preez (1949), quoted by Thomas (1974), while Mohr and Van Baren (1959) among others have argued strongly in support of retaining in the scientific literature both the term *laterite* and the process of its formation, *lateritization*, both of which indicate 'that weathering tends to assist the accumulation of iron and aluminium [in] and the leaching of the bases' (p. 353) from the weathered mantle. At the same time, however, it is important to point out that the original 'laterite' of Buchanan is much closer structurally to the mottled zone of the ideal laterite profile, whose material can be cut into bricks when wet, dried and/or burnt, and then used for house building among other things. The real laterite is already indurated and so has to be dug up or quarried and shaped with the hammer or other metal instruments to produce the 'building' stone. Other common uses of laterite blocks include road stones, stone hedges, stone gods or goddesses and hardcore in many civil engineering constructions (cf. Faniran, 1973).

The 'classic' profile (Table 3.3, Figure 3.9) is of course very rare for various reasons. Among these are:

1 post-duricrust destruction by subsequent weathering and removal processes; and

2 absence in the original profile due to rock type and/or 'accidents' in the course of formation.

Details of these are contained in Faniran (1968; 1974) and Thomas (1974); they are also tied up with the mode of formation of the laterites.

A full discussion of laterite genesis as well as the evolution of the typical laterite (deep-weathering) profile is not feasible here; readers are therefore referred to existing sources, including Allen (1948), Prescott and Pendleton (1952), Mohr and Van Baren (1959), Sivarojasingham *et al.* (1962), Maignien (1966), and Thomas (1974). The main areas of dispute and/or discussion are as follows:

1 the origin of the sesquioxide (especially iron) concentrated in the profile's indurated zone;
2 the causes and processes responsible for the zonation or stratification associated with the typical laterite (deep-weathering) profile; and
3 the causes and processes responsible for the hardening as well as the structural characteristics (including the formation of concretions, pisolites, and so on) of the indurated zone.

The following are some of the views expressed in respect of each of these problem areas, starting with the iron concentration in the indurated and mottled zones:

Origin of iron in laterites (duricrusts)

This problem arises because of the marked disparity between the percentage of iron in unweathered rocks compared to typical laterites. Of the several possible sources of this enrichment process, the following have been most seriously canvassed:
1 Addition from above, effected through precipitation by percolating water;
2 Capillary rise of sesquioxide-rich soil water, particularly effective under seasonal rainfall and fluctuating groundwater conditions;
3 Precipitation at the upper part of the water table;
4 Lateral movement of sesquioxide-bearing solutions and their precipitation in 'basins' or the lower slopes of valleys;
5 Residual (accumulated) iron and aluminium during the process of denudation – for example, during the development of Trendall's (1962) 'apparent peneplains'; and
6 Enrichment by removal of other constituents, especially in basic and ultra-basic rocks. This process of relative enrichment was thought by Harrison (1934) to be responsible for the development of true primary laterites on elevated sites in Guyana. The same explanation can be offered in the case of similar materials overlying extensive surfaces of African plateau, destruction of which is now being associated with the younger formations at lower altitudes (the so-called secondary laterites).

All these sources can be collapsed into between three and four as follows:
1 Lateral flow from neighbouring iron-rich rocks – for example, older and reweathered laterites, basic rocks or secondary sources. The case of intruded basic igneous rocks is well illustrated in the Northern Sydney district of Australia (Faniran, 1971b), among other places, while the case of older duricrusts has been mentioned above.
2 Vertical upward movement of iron by upward ionic diffusion process from the permanently saturated pallid zone.
3 *In situ* weathering by the common pedological process of leaching, leading to the downward transfer and precipitation in the indurated zone.
4 Vertical (downward) transfer of residual iron (concretions and other fragments) during the process of erosion (peneplanation).

The zones of the deep-weathering profile

Of all the reasons so far advanced for the stratification of the typical deep-weathering profile, the combined influence of groundwater and topography is the most important. According to Ollier (1975) there are at least five groundwater belts, namely:

1 Belt of soil moisture;
2 Intermediate belt;
3 Capillary fringe and belt of fluctuation;
4 Belt of discharge; and
5 Belt of stagnation.

These different belts are associated with different weathering environments. Zones 1–2 and part of zone 3 are associated mainly with oxidation, which is conducive to the formation of the indurated zone. In zones 3–4 the fluctuating groundwater produces conditions suitable for the operation of both the process of oxidation (during aeration) and reduction (during saturation), leading to the formation of the typical mottled zone. Zone 5 is the zone of stagnation or permanent saturation, where the reduction process predominates leading to the formation of the typical pallid zone.

The topographic factor is relevant to the extent that it influences the nature and course of groundwater movement. Thus, where topography produces a generally well-drained environment, the three zones – indurated, mottled and pallid – are rare. Rather, the sesquioxides are distributed throughout the profile to produce a typical red earth (Krasnozem) profile (Faniran and Areola, 1978).

Available evidence points to wide variations both in the total depth of weathering in general and in that of individual zones from place to place. Among the factors responsible for this spatial variation are rock type (Faniran, 1968) and post-duricrust destruction (Faniran, 1974). Others include climate, especially rainfall; topography and age, although a number of them are still being discussed and disputed (Thomas, 1974). With respect to rock type, Faniran (1958) reported, for the Sydney area, that the controlling factor was the quartz grain size, the area being predominantly one of sedimentary rocks. The situation was described thus:

> Depths appear to be shallowest in coarse-grained sandstones and probably deepest in shales and other fine-grained rocks. In the former, the mean depth . . . is about 30 feet (9.2 m); varying between 20 feet (6.1 m) and 40 feet (12.2 m). In the latter . . . the depth may be more than 100 feet (30.5 m). [Faniran, 1968, p. 38]

And for the Nigerian situation, the same author wrote as follows:

> The thickness of the various zones seemed to reflect rock type . . . the thickest zone occurred in the sedimentary-rock areas and the thinnest in the Basement Complex country. The mean thickness of the indurated zone in the former areas was 11.3 ± 8.21 m, but only 5.2 ± 1.9 m in the latter area . . ., in the alluvium-covered areas of the Jos

Plateau, the indurated zone was about 10 m thick. . . . The indurated zone formed just about 10 per cent of the entire profile in sedimentary-rock area and between 30 and 40 per cent in the alluvium-covered areas. [Cf. Faniran 1974, p. 25; cf. also Table 8.1.]

The profiles described so far are either laterite or ferricrete – that is, aluminous and/or ferruginous duricrusts. A summary of other observations made on these materials is contained in Table 8.2a.

The induration and structural characteristics of laterites

The common explanation given for the hardening of laterite is exposure following the removal of the topsoil. The processes involved are described in terms of crystallization and dehydration (Faniran, 1971b). The process of dehydration is particularly prominent in iron-oxides; thus the hardest laterites (duricrusts) are also those with the greatest iron contents. The following possible reasons have also been adduced:

1 Changes in the degree of hydration of the oxides of iron and aluminium present in the duricrust (cf. Harrison, 1910; Sherman et al., 1952; Faniran, 1968, 1971b);
2 Development of crystallinity in goethite and/or haematite (Sivarojasingham et al., 1962);
3 Cementation by solution and recrystallization of the carbonate in calcretes;
4 Cementation by the dehydration of iron and aluminium compounds in laterites and related materials;
5 Formation of calcium silicates from calcium carbonate and opaline silicas;
6 Reduction of plasticity by reaction between calcium carbonate and clay minerals to form calcium aluminates;
7 Increased deposition of carbonate by soil water under intense surface evaporation and heating; and
8 The presence of suitable cementing material, especially silica.

In connection with the last point, Flach et al. (1961) claimed that 'as little as 10 per cent Si can cement a soil horizon effectively . . . hence it is a factor in the great strength of even relatively poorly cemented silcretes' (Goudie, 1973). The presence of quartz, particularly the coarse-grained types, is also important. This has been demonstrated, again, in the Northern Sydney district (Australia), where the laterites have been distinguished according to those found in

1 shale, in which quartz is scarce and quartz-grain diameters are generally less than 0.1 mm, averaging less than 0.05 mm;
2 fine-grained sandstone with generally coarser and more abundant quartz (calibre $0.1 - 0.2$ mm); and
3 coarse-grained sandstone with quartz forming 50 per cent of the sample and averaging 2 mm in size.

Of these, the most common on the present landscape are the third group, which have been preserved on account of their hardness.

The structural characteristics of laterites have also been observed to

63

be related to the constituent quartz minerals. The vermicular and vesicular types have been observed to be common in fine-grained rocks, such as shales and basic igneous rocks, and the medium-grained types (for example, sandstones, granites and so on), forming most of the concretionary types. Where rocks with medium-sized grain form the massive-type duricrusts, it has been found that this is partly due to low sesquioxide–quartz ratios. The position of the indurated zone relative to the movement of groundwater and the associated precipitation of sesquioxide 'films' or rings that eventually result in concretions is also important in the explanation of the physical characteristics of duricrusts. Whereas some rocks seem to weather directly into concretionary, including pisolitic (plinthic) duricrusts, it appears that most of the concretionary types occur on low relief surfaces which have taken full advantage of the fluctuating groundwater level (Faniran, 1968, 1971b).

Petrographical considerations

The petrography of a rock, whether weathered or not, concerns the investigation of the physical (including colour, textural and structural) chemical and mineralogical characteristics of such a rock, with a view to its effective description and classification. This type of study, especially the detailed type, is rare as far as the products, of deep weathering are concerned. Exceptions are those of the laterites of the Northern Sydney district (Faniran, 1968, 1970a, b, 1971a, b, c) and elsewhere (Von Gaertner, 1963; Maignien, 1966; Pullan, 1967); those of the world calcretes (Goudie, 1971a, b, 1973); the silcretes of South Africa (Visser, 1964); and, of course, the world deposits of bauxites (Cooper, 1936; Harder, 1952; Gordon and Tracey, 1952; Gordon and Ellis, 1958), ironstones/iron ores (Hazell, 1958; Jones, 1958) and manganese (Service, 1943; Nagell, 1962). Of these the most widespread and perhaps the most typical product of deep chemical weathering is the laterite/ferricrete. The bulk of the following discussion is therefore devoted to these materials, based on the personal experience of one of the co-authors of this book, in Australia and Nigeria.

Physical characteristics

The most important physical characteristics of rocks are colour, texture and structure. These vary both within and between the different zones of the deep-weathering profile. In the indurated zone, the most important morphological forms are the 'concretions', which include forms such as the nodules, pisolites, oolites and so on, and the generally massive matrix which cement them. Of course there are 'massive' duricrusts as well, which are essentially iron-cemented sandstones. Softer types contain other structural forms such as vermicules and vesicules. Mottled zones are characteristically soft multi-coloured whereas pallid zones are more uniform colourwise, just like the indurated zone.

The colour system used here is that of the Rock Colour Charts (Deford, 1944; Goddard *et al.*, 1963). The most common colours in the laterite/ferricrete group of deep-weathering products are black to dark and brownish-red in the pisolitic section of the concretionary (indurated) layer; yellow to brown in the matrix section of the same concretionary layer; red, yellow, brown and grey of the mottled zone and white to greyish-white of the pallid zone. The ordinarily weathered rocks are mostly dullish grey (Faniran, 1968, pp. 325–31).

The morphological study of laterites by standard petrographical techniques shows that:

1 Textures such as nodules, pisolites, oolites, pellets and birds' eyes[*] appear to have developed their most perfect forms in crystalline rocks such as sandstones and granites. Examples are the fossil laterites of the Sydney district of Australia, formed particularly on the Hawkesbury sandstones (Faniran, 1968, 1970), and those in Nigeria, Uganda, and Zambia, formed on crystalline basement rocks. Even in these rocks, the frequency and perfection of these textures decrease with depth. In fact it is essentially on the basis of this textural characteristic that many indurated zones have been subdivided into upper, middle and lower indurated zones with the last zone hardly distinguishable from some mottled zones which underlie them (Table 3.3). The examples considered above are mostly the primary types. In the secondary and re-worked types, these textures, where present, appear to have been derived directly from former (primary) profiles and worked into the matrix. Other textural forms identified in laterites are:
   a) replacement patterns where sesquioxides appear to have replaced silica and kaolinite cement of crystalline rocks. This is what seems to have happened in the case of many ferruginized rocks, including ironstones and some ferricretes, with basically massive textures;
   b) voidal textures or intergranular spaces, usually but not always filled by uncemented or loosely cemented and sometimes iron-stained quartz grains; and
   c) cement textures, characteristic, again, of ferruginized rocks. The term 'cement' also applies to the matrix component of concretionary duricrusts.
2 Quartz appears to be concentrated more in the matrix than in the concretions of the fine-grained rock laterites, and vice versa in coarse-grained rocks. The portion with the least quantity of quartz is the ring or banding structure surrounding the pisolites and separating them from the surrounding matrix. The quartz grains in the indurated zones in laterites and ferricretes alike show evidence both of scratching and of fracturing, an indication of the intensity of the weathering process.
3 Structural modification of the original rock varies both in different

---

[*]These terms describe the hardest parts of laterites, bauxites, ironstones, etc., which are collectively called in some part of the literature concretions or colloform textures as distinct from the 'matrix'. In the case of the birds'-eye texture, there is no recognizable structural boundary with the matrix, except for the black colour, which gives it the name.

---

parts of the profile and with rock type. By far the most advanced forms of change occur in the indurated zone where it is almost impossible to see any semblance of the original rock. By contrast, many pallid zones and, more importantly, the underlying weathered rock, still retain many features of the original rock. Thus many pallid-zone shales show typical bedding and striation structures as do their fresh unweathered counterparts, whereas pallid-zone sandstones and granites are essentially white or grey quartz grains with intergranular or interstitial kaolinite, sometimes with slight staining by iron and/or other impurities.

The mottled zone in shales, claystones and basic igneous rocks are similar, being mostly amorphous-looking hydrated oxides of iron and alumina (limonite) derived directly from deep chemical weathering. The dehydration and crystallization of these materials may have led to the formation of some laterites and bauxites (Sivarajasingham *et al.*, 1962).

Unlike the physical characteristics, the chemistry and mineralogy of laterites and related rocks have always been incorporated in most studies of these materials, the first chemical analyses appearing as early as the late 1800s (see Maigmen, 1966; Faniran, 1968, p. 55). The advantage of this type of data is the ease with which they can be manipulated, especially in an effort to classify the materials into types (Dury, 1969; Faniran, 1970*b*; cf. also Figure 3.10). Because the data are also useful in an

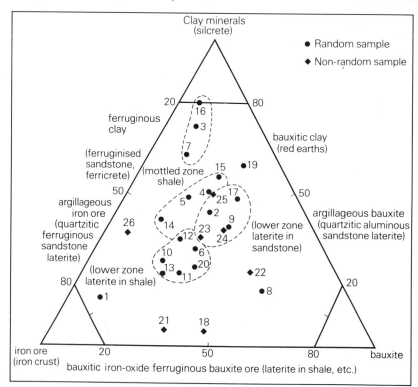

Fig. 3.10 Classification of duricrust

TABLE 3.5 Duricrust chemical analysis: textural variations

Sample Reference			Chemical Contents %						
Sample	Specimen	Texture	SiO_2	FeO_2	Al_2O_3	TiO_2	MnO_2	Others	Total
8	212	Bulk	19.24	24.90	55.30	0.20	0.06	0.24	10.000
21	212	Birds' eye	5.70	59.80	34.10	0.40	Tr.	—	100.00
22	212	Matrix	25.70	25.10	49.20	0.25	Tr.	—	100.25
9	173	Bulk	39.80	23.80	35.30	0.10	0.11	0.89	100.00
25	173	Deep-red pisolite	49.50	24.20	26.00	0.01	Tr.	0.29	100.00
23	50	Deep-red pisolite	35.80	35.00	29.00	0.07	Tr.	—	100.37
24	217	Black pisolite	38.60	26.20	34.70	0.20	Tr.	0.30	100.00
25	215	Matrix	37.20	55.80	6.90	0.52	Tr.	—	100.42

TABLE 3.6 Chemical composition of ferricretes (%)

	a	b	c	d	e	f	g	h
SiO_2	2.70	5.4	19.83	0.62	–	17.08	16.4	35.6
Al_2O_3	9.60	44.3	12.47	10.54	31.37	20.83	23.5	18.8
Fe_2O_3	74.00	23.1	47.80	74.43	19.12	40.18	45.0	33.9
FeO	–	0.2	–	0.65	–	–	0.5	–
CaO	Traces	0.2	0.13	0.02	0.10	–	0.1	–
MgO	0.72	Traces	0.10	Trace	–	–	0.1	–
Cr_2O_3	0.45	–	–	–	–	–	–	–
TiO_2	0.68	1.4	0.86	3.91	1.12	1.8	1.4	
H_2O+	12.40	24.5	10.49	9.60	9.10	11.05	13.3	10.1
MnO_2	–	0.1	–	nil	–	–	0.1	–
Na_2O	–	–	0.09	nil	–	–	0.2	–
K_2O	–	–	0.24	nil	–	–	0.1	–

a Dunite-derived crust in West Africa (from Maignien, 1966).
b Syenite-derived bauxitic crust from West Africa (*ibid.*).
c Terrace crust from Kankan, Guinea (Maignien, 1958).
d Dolerite-derived crust from Dahomey (Alexander and Cady, 1962).
e Granite-derived crust from Dahomey (Alexander and Cady, 1962).
f Trap-derived crust from southern India (Fox, 1936).
g Dolerite-derived crust from Guinea (Sivarojasingham *et al.*, 1962).
h Granite-derived crust from Samaru, Nigeria (*ibid.*).

Source: Goudie, 1973

economic appraisal of the deposits, massive data have emerged that cannot be reviewed in a work such as this. Those undertaken by one of the co-authors are given in Tables 3.5 and 3.6, while Table 3.7 is taken from Goudie (1973). Illustration of the trend shown by the data in Table 3.5 is given in Figure 3.11. The following are some of the points that emerge from the available analysis:

1 The amount of the sesquioxides (especially iron and aluminium) is high, relative to the others, in some cases as high as 80 per cent or more. The most common amount is, however, between 50 and 65 per cent, but in some sandstones and conglomerates the amount of sesquioxide may be quite low, especially if the material involved consists of coarse-grained quartz with sesquioxide cement.
2 Apart from iron and aluminium, other chemical components of laterites are kaolinite, titanium, chromium, vanadium, manganese and nickel, any of which, like iron and aluminium, can occur in quantities which make exploitation for economic purposes worthwhile (Faniran, 1971a).
3 Quartz is usually an important constituent, especially in acidic crystalline rocks such as granites and sandstones. Indeed, as observed earlier, some ferricretes are quartz sands and gravels cemented with relatively small amounts of sesquioxides or iron (goethite, haematite).
4 Alkalies and alkaline-earths are almost entirely absent, having been leached out of the profile.
5 The free alumina in these materials generally occurs in the form of

TABLE 3.7 Chemical analysis of Nigerian duricrusts

Sample no.	Insoluble material	SiO₂	Al₂O₃	FeO	Fe₂O₃	TiO₂	MnO	P₂O₅	H₂O	Total
1.	7.20	38.75	18.95	0.77	16.07	0.52	0.07	3.30	14.50	100.13
2.	5.34	51.24	15.88	0.41	16.27	0.64	1.57	0.25	8.82	100.12
3.	11.20	40.43	18.46	0.54	16.37	0.79	0.22	0.42	11.37	99.80
4.	11.92	34.85	21.16	1.43	15.51	2.02	—	0.46	12.29	99.84
5.	10.75	38.39	20.33	0.46	16.55	0.73	0.03	0.44	12.13	99.81
6.	3.08	50.39	18.20	0.32	15.51	0.89	0.03	0.15	11.57	100.14
7.	9.33	36.87	17.60	0.37	16.97	0.91	0.03	1.09	16.55	99.72
8.	9.58	38.45	18.59	0.54	16.65	0.84	0.37	1.15	13.51	99.68
9.	10.08	42.36	17.01	0.54	16.65	0.96	0.12	0.83	11.43	99.98
10.	5.10	50.12	17.14	0.41	16.78	1.01	0.37	0.62	8.17	99.99
11.	13.08	36.83	20.38	0.32	19.68	0.58	0.06	0.57	8.53	100.03
12.	6.04	52.06	17.41	0.28	14.15	0.84	0.02	0.22	8.95	99.97
13.	5.05	54.60	16.51	0.36	14.42	0.77	0.10	0.38	7.79	100.04
14.	10.08	49.71	18.87	0.45	10.22	2.10	0.06	0.35	8.23	100.07
15.	5.62	43.56	19.91	0.19	19.81	0.62	0.08	0.45	9.69	99.93
16.	8.53	45.44	19.30	0.32	10.96	1.17	0.08	0.29	13.81	99.90

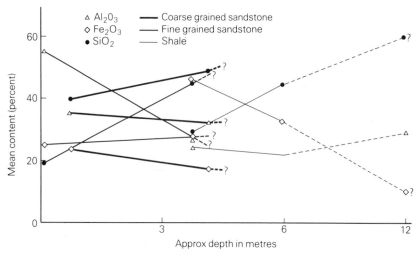

Fig. 3.11 Chemical analysis of duricrust trends of major elements in profile in the Sydney district, Australia

gibbsite ($Al_2O_3.3H_2O$), though boehmite and diaspore may be encountered in some profiles; similarly, iron occurs as goethite (FeO.Oh), ilmenite or haematite (α–Fe_2O_3) and/or maghemite (γ–Fe_2O_3) (Faniran, 1971c). Silica occurs as quartz, chalcedony, opal or phytolithic material, while kaolinites are the most common clay minerals with occasional illites, halloysites, and vermiculites. Contained water ranges between 8 and 15 per cent (Table 3.8).

6 In ferricretes, Fe_2O_3 may be as high as 80 per cent and in alcrete (bauxite), Al_2O_3, may reach 60 per cent. In laterites, both Fe_2O_3 and Al_2O_3 are high and may together make up more than 80 per cent of the components. The combined silica content is generally low but may exceed 20 per cent, made up essentially of kaolinite. Other constituents are TiO_2, M_nO, R_2O_3 and MgO (cf. Table 8.4c).

TABLE 3.8a Heavy-mineral separates: analysis of heavy magnetic grains

Sample	Magnetite (?)		Maghemite		Ilmenite		Others		Total	
101	0	0	63	92.6	2	3.0	3	4.4	68	100
162	0	0	40	78.5	2	3.9	9	17.6	51	100
215	0	0	15	100	0	0	0	0	15	100
212	18 ?	8.6	190	91.4	0	0	0	0	208	100
235	4 ?	2.0	187	98.0	0	0	0	0	191	100
217	6	2.6	227	97.4	0	0	0	0	233	100
147	0	0	2	100	0	0	0	0	2	100
226	0	0	24	100	0	0	0	0	24	100
211	1	5.2	12	63.0	6	31.6	0	0	19	100

TABLE 3.8b Laterite (X-ray diffraction) mineralogical analysis

Sample no.	Sample specimen no.	% Mineral content				
		Gibbsite $Al(OH)_3$	Goethite $Feo(oH)$	Hematite Fe_2O_3	Kaolin $Al_2Si_2O_5(Oh)_4$	Quartz SiO_2
		$\pm$ 7	$\pm$ 7	$\pm$ 10	$\pm$ 5	$\pm$ 10
1	101 Sp.	30	15	5	25	25
2	101 P.	20	10	25	25	20
3	101 M.	35	15	0	25	25
4	211 Sp.	5	30	5	25	25
5	211 M.	5	25	5	25	40
6	2112 Sp.	40	20	5	10	25
7	215 Sp.	35	5	20	30	10
8	215 H.	20	5	35	30	10
9	212 Birds'-eye	15	10	70	0	5
10	212 D.R.M.	50	30	10	5	5

Abbreviations

Sp.	= Specimen or bulk sample undifferentiated
P.	= Pisolite sample
M.	= Matrix sample
H.	= Heavy mineral grain sample
D.R.M.	= Deep Red Material (Matrix)

7 The constituent minerals of the laterites studied vary in their proportions and combinations in different rock types and in different sections (portions) of the profile. These variations appear to correlate very strongly to the mean size of the constituent quartz grains. Also, the generally high iron contents compared to alumina suggest that the process of ferruginization with age, as described among others by Grubb (1954), has taken place; alternatively, the relatively lower alumina may be due to the kaolinization and silicification of gibbsite in ageing laterites.

We have gone to this great length in order to show that the process of deep weathering in the humid tropics has yielded, and is probably still yielding, products with unique characteristics which mark them out as worthy of study. As a matter of fact, this material is so unique and important that it is being studied by many of the sciences, including pedology (Faniran, 1978), geology (Faniran, 1971a; Faniran and Omorinbola, 1980) and, of course, geomorphology.

There is, however, one weakness in this kind of study: most of the materials studied are fossil or relic. Indeed, it is yet to be established empirically that the process of deep weathering is taking place today in the humid tropics nor do we know the rate of its operation. This is due to the complexity of factors conducive to the process, as well as to the great gap in knowledge. Whereas processes such as podzolization have been reproduced in the laboratory, similar success is yet to be achieved in the case of lateritization. Consequently, most of the information used in this and other works have been inferred from the study of the products. This is not necessarily bad, and is largely in tune with the practice in a large area of geomorphology. It is only very recently that we have started to measure and simulate process. An extension of this innovation into the study of deep weathering will go a long way to elucidate the relationship between the humid tropical environment, the operative weathering processes and the resulting products or materials.

References and further reading

Allen, V. T. (1948) 'Formation of bauxite from basaltic rocks or Oregon', *Econ. Geol.* **43**, 619–29.

Bloom, A. L. (1969) *The Surface of the Earth*, Foundation of Earth Science Series, Prentice-Hall, Englewood Cliffs, New Jersey.

Butzer, K. W. (1976) *Geomorphology from the Earth*, Harper & Row, New York.

Cooper, W. G. G. (1936) 'The bauxite deposits of the Gold Coast', *Gold Coast Geol. Surv.*, **7**.

Deford, R. F. (1944) 'Rock colors (review)', *Bull Amer. Assoc. Petrol Geol.* **28**, 128–36.

De Swardt, A. M. J. (1964) 'Lateritisation and landscape development in parts of equatorial Africa', *Zeits. für. Geomorph.*, **8**, 313–33.

Douglas, I. (1980) 'Climatic geomorphology. Present day processes and landform evolution problems of interpretation', *Zeits. für Geomorph.*, N.F. Suppl. Bd. **36**, 27–47.

Du Preez, J. W. (1949) 'Origin, classification and distribution of Nigerian laterites', *Proc. 3rd Internl. West African Conf.*, Ibadan, 1949, 223–34.

Dury, G. H. (1969) 'Rational descriptive classification of duricrusts', *Earth Sc. J.*, **3** (2), 77–86.

Dury, G. H. (1971) 'Relict deep weathering and duricrusting in relation to the palaeoenvironments in middle latitudes', *Geog. J.*, **137** (4), 511–22.

Faniran, A. (1968) 'A deeply-weathered surface and its destruction', Ph.D. thesis, Univ. of Sydney, N.S.W.

Faniran, A. (1969a) 'Duricrust, relief and slope populations in the Northern Sydney District, N.S.W.', *Nig. Geol. J.*, **125**, 53–62.

Faniran, A. (1969b) 'On the use of the term *laterite*', *Prof. Geogr.*, **21** (4), 231.

Faniran, A. (1970a) 'Landform examples from Nigeria, No. 2: the deep-weathering (duricrust) profile', *ibid.*, **13** (1), 87–8.

Faniran, A. (1970b) 'The Sydney duricrusts: their terminology and nomenclature', *Earth Sci. J.* (N. 2), **4** (2), 117–28.

Faniran, A. (1971a) 'Implications of deep-weathering on the location of natural resources', *Nig. Geol. J.*, **14** (1), 59–69.

Faniran, A. (1971b) 'The parent material of Sydney laterites', *J. Geol. Soc. Aust.*, **18** (2), 159–64.

Faniran, A. (1971c) 'Maghemite in the Sydney duricrust', *Amer. Mineral.*, **55**, 925–33.

Faniran, A. (1973) 'Geography and the construction industry in Nigeria', *Nig. Geog. J.*, **16** (2), 151–65.

Faniran, A. (1974) 'The extent, profile and significance of deep-weathering in Nigeria', *J. Trop. Geog.*, **38** (1), 19–30.

Faniran, A. (1978) 'Deep weathering, duricrusts and soils in the humid tropics', *Savanna*, **7** (1), 61–5.

Faniran, A. and Areola, O. (1978) *Essentials of Soil Study, with Special Reference to Tropical Areas*, Heinemann, London.

Faniran, A. and Ojo, O. (1980) *Man's Physical Environment*, Heinemann, London.

Faniran, A. and Omorinbola, E. O. (1980) 'Evaluating the shallow groundwater reserves in basement complex areas: a case study of south-western Nigeria', *J. Mining Geol.*, (in press).

Fripiat, J. J. and Herbillon, A. J. (1971) 'Formation and transformation of clay minerals in tropical soils', in *Soils and Tropical Weathering*, Proc. Bandung Symposium. UNESCO, 1969, 15–24.

Goddard, E. N., Trask, P. D., Deford, R. K., Roue, O. N., Singewald, Jr., J. T. and Overbeck, R. M. (1963) *Rock color chart*, Geol. Soc. Amer., New York.

Gordon, M. Jr., and Ellis, E. M. (1958) *Geology of the Arkansas Bauxite Region*, U.S. Geol. Surv., Prof. Paper 299.

Gordon, M. Jr., and Tracey, J. T. Jr., (1952) 'Origin of the Arkansas bauxite deposits' in *Problems of Clap and Laterite Genesis* (Symp.), American Inst. Metall. Engin., 11–34.

Goudie, A. *et al.* (1970) 'Experimental investigation of rock weathering by salts', *Area*, **1** (4), 42–8.

Goudie, A. (1971a) 'Calcrete as a component of semi-arid landscapes', Ph.D. thesis, Univ. of Cambridge.

Goudie, A. (1971b) 'Climate, weathering, crust formation, dunes and features of the Central Namib desert near Gobabeb, S.W. Africa', *Madoqua*, **2**, 15–31.

Goudie, A. (1973) *Duricrusts in Tropical and Sub-tropical Landscapes*, Oxford Research Studies in Geography, Clarendon Press, Oxford.

Grubb, P. L. C. (1954) *Bauxite from Jarrahdale, Western Australia.*, C.S.I.R.O. Australian Min. Invest. Rept. No. 884.

Hallsworth, E. G. and Costin, A. B. (1953) 'Studies in pedogenesis in New South Wales IV: the ironstone soils', *J. Soil Sci.*, **4**, 24–46.

Harder, E. C. (1952) 'Examples of bauxite deposits, illustrating variations in origin', in *Problems of clays and laterite origin* (Symp.), Amer. Inst. Min. Metall., 35–64.

Harrison, J. B. (1934) *The Katamorphism of Igneous Rocks under Humid Tropical Conditions*, Imperial Bureau of Soil Science, Harpenden.

Harrison, J. B. (1910) 'The residual earth of British Guiana commonly termed "laterite" ', *Geol. Mag.*, **7**, 439–52, 488–95 and 553–62.

Hazell, J. R. T. (1958) *The Enugu Ironstone, Udi, Onitsha Province*, Rec. Geol. Surv. Nigeria, (1955), pp. 43–58.

Jenny, H. (1941) *Factors of Soil Formation*, McGraw-Hill, New York.

Jones, H. A. (1958) *The Oolitic ironstone of the Agbaja Plateau, Kabba Province*, Rec. Geol. Surv. Nigeria, 20–42.

Keller, W. D. (1957) *The Principles of Chemical Weathering*, rev. edn. 1961, Lucas, Columbia, Mo.

Loughnan, F. C. (1969) *Chemical Weathering of the Silicate Minerals*. Elsevier, New York.

Lukashev, K. I. (1970) 'Lithology and geochemistry of the weathering crust', Israel Program for Scientific Translations, Jerusalem.

Maignien, R. (1966) *Review of Research on Laterites*, UNESCO Natural Resources Res. IV, Paris.

Mohr, E. C. J. and Van Baren, F. A. (1959) *Tropical Soils*, NV Uitgeverij W. van Hoevo, The Hague.

Nagell, R. H. (1962) 'The geology of Sierra do Navio manganese district, Brazil', *Econ. Geol.*, **53**, 481–98.

Ollier, C. D. (1975) *Weathering*, 2nd edn., Longman, London.
Pedro, G. (1968) 'Distribution des principaux types d'alteration chimique à la surface du globe', *Rev. Geogr. Phys. Geol. Dyn.*, **10**, 457–70.
Playford, P. E. (1954) 'Observations on laterite in Western Australia', *Aust. J. Sci.*, **17**, 11–14.
Polynov, B. B. (1937) *Cycle of Weathering*, trans. A. Muir, Murby, London.
Prescott, J. A. and Pendleton, R. L. (1952) 'Laterite and lateritic soils, *Comm. Bur. Soil-Sci. Tech. Commn.*, **47**.
Prider, R. T. (1966) 'The laterised surface of Western Australia', *Aust. J. Sci.* **28**, 443–51.
Pullan, R. A. (1967) 'A morphological classification of lateritic ironstones and ferruginised rocks in Northern Nigeria', *Nig. J. Sci.*, **1**, 161–73.
Reiche, P. (1943) 'Graphic representation of chemical weathering', *J. Sed. Petr.*, **13**, 58–68.
Reiche, P. (1950) 'Survey of weathering processes and products', *Univ. of New Mexico, Publ. Geol.*, **3**.
Ruxton, B. P. and Berry, L. (1957) 'Weathering of granite and associated erosional features in Hong Kong', *Bull. Geol. Soc. Amer.*, **68**, 1263–92.
Sanches Furtado, A. F. A. (1968) 'Altération des granites dans les régions intertropicales sous différents climats', *9th Intern. Congr. Soil Sci.*, Adelaide, Trans, IV, 403–9.
Service, H. (1943) 'The geology of the Nsuta manganese ore deposit', *Gold Coast Geol. Surv.*, **5**, 5–28.
Sherman, G. D. (1952) 'The genesis and morphology of the alumina-rich laterite clays', in *Problems of Clay and Laterite Genesis*, Amer. Inst. Min. Metall., pp. 154–61.
Sivarojasingham, S., Alexander, L. T., Cady, J. G., and Cline, M. G. (1962) 'Laterite', *Adv. Agron.*, **14**, 1–60.
Thomas, M. F. (1974) *Tropical Geomorphology: a study of weathering and landform development in warm climates*, Macmillan, London.
Trendall, A. F. (1962) 'The formation of "Apparent Peneplains" by a process of combined laterisation and surface wash', *Zeits. für Geomorph.*, N.F., 6, 183–97.
Visser, J. N. J. (1964) *Analyses of Rocks, Minerals and Ores: Republic of S. Africa*, Geol. Surv. Handbook, No. 5.
Van Gaertner, H. R. (1963) *Geochemistry of Lateritic Deposits: some general patterns of lateritic weathering*, U.N. Seminar Proc. Geochemical Prospecting Methods, No. 21, 67–75.
West, G. and Dumbleton, J. J. (1970) 'The mineralogy of tropical weathering illustrated by some West Malaysian soils', *Quart. J. Eng. Geol.*, 3(1), 25–40.
Woolbough, W. G. (1927) 'The duricrust of Australia', *J. Proc. Roy. Soc. N.S.W.*, **61**, 24–53.

4 Removal processes in the humid tropics

The literature on removal processes in the humid tropics is replete with generalizations and inexactitudes. Although the major factors are often identified – namely, deep tropical weathering, abundant soil moisture, high temperatures and vegetation – the nature and extent of their joint and/or respective influences are not often clearly stated. An example of this type of description will suffice:

> In the humid tropical regions (selvas) temperatures are consistently high and there is abundant moisture and luxuriant vegetation . . . . Chemical weathering is rapid and intense. Few fragments of a size greater than sand survive weathering. The rivers lack . . . the tools used to abrade the bed and banks. . . .
>
> Other factors also mitigate against strong stream action . . . . The vegetated slopes and banks resist erosion . . . though precipitation is high and there is commonly a dense network of permanent rivers and streams which have incised deep valleys . . . running water is not the dominating component of the morphogenetic system . . . indeed degradation of the tropical humid lowlands is apparently among the slowest known. (Twidale, 1976, p. 273)

One other feature of the literature on this topic is its brevity, where it is mentioned at all in geomorphological texts. Thus Twidale (1976) devoted half a page to a discussion of this aspect, out of the eight or so devoted to the topic 'Rivers and Running Water in the Tropics'. The rest was devoted to 'The Arid Tropics', particularly to 'pediments'. Twidale's book *Analysis of Landforms* is 572 pages long.

For his part, Butzer (1976) observed as follows:

> A basic trait of these environments [savanna and rainforests] is intensive chemical weathering and a deep, continuous soil mantle on gentle to intermediate slopes . . . .
>
> Savanna slopes retreat by a special form of tropical pedimentation, through chemical undermining, throughflow eluviation, and sheetwash. . . . Rainforest environments favour valley incision in deep regolith, aided by subterranean mass movements and leading to multiple, small planation surfaces along the margins of high-order valleys through repeated tributary convergence. (pp. 404–5)

Here again, only indirect reference is made to removal processes in this environment, and that in connection with tropical savannas on about eight pages of the 450-page book. Garner (1974) subsumed this topic under 'humid geomorphic systems and landforms', devoted, except in the mention of tropical soils, almost entirely to the humid temperate environment and systems.

What appears to permeate these woolly and generalized statements is a dearth of empirical studies, which have only been appearing since the late 1950s. Butzer (1976) has thus described the tropical landscape assemblages as remaining poorly understood because many problems remain oversimplified and tropical landscapes themselves are elusive.

Recent empirical studies in the humid and sub-humid tropics have, however, provided both framework and material for a systematic description of the removal processes in these areas, just as has been done for the sister processes of weathering (Chapter 3) and deposition (Chapter 5). This framework is simply that of establishing the nature of removal processes in the region – that is, identifying the various forms in which weathered and other loose and transportable surface and near-surface materials are transported over the earth's surface. These processes are often discussed under the term *denudation* and are here illustrated by means of the available information from both the literature and field work.

Denudation (removal) processes in tropical environments

If there is one phase in the gemorphological process which is difficult to circumscribe it is the removal or transportation phase. This is because it is involved in aspects of the preceding phase of rock weathering, while it operates sometimes *pari passu* with the process of deposition. At the same time, however, this is also the easiest process to observe, especially where motion is fast enough to be monitored or measured. It is also the major factor of landform to the extent that weathered rock debris has to be removed (detached and transported) and redeposited in order to make landforms. It is this process, sometimes referred to as 'denudation', which is responsible for the degradation and aggradation of land surfaces, the two key events in land formation. Accordingly, denudation occupies a dominant position in the geomorphological literature, especially those aspects described as physical geology or dynamic geomorphology.

The processes which are often listed and described in this connection – that is, processes which shape the terrain in warm, moist and, more often than not, vegetated environments – include differential weathering, including deep chemical weathering, groundwater solutioning, chemical denudation, fluvial activity along surficial water courses, sheetflows and mass wasting. They are here grouped into two categories:

1 Chemical processes by solutioning and chemical denudation generally; and

2 physical processes under the influence of gravity and running water.

In addition, wind and ice feature among the physical processes, but these are restricted in scope. Wind is important along some coastlines while ice caps occur on a few mountain peaks in the region under reference. Rock weathering has already been discussed (see Chapter 3), and coastal processes are treated in Chapter 12. Ice is 'foreign' or erratic to humid tropical environments. We are thus left with the chemical denudational processes and the physical removal processes by gravity (mass wasting) and running water, which can be taken as characteristic processes of the region. The character of most of these processes is strongly influenced – indeed, governed – by the physical setting described in Chapter 2. This includes the climatic, vegetational and geochemical (weathering) environments, which may be summarized as follows:

1 The area is characterized by generally abundant rainfall and high temperatures, which favour, in suitable rocks and topographies, deep chemical weathering leading to the production of thick regoliths of varying depth, internal structure, texture and constitution.

2 There is, except where disturbed by man or other agents, a continuous ground cover of plants – grasses, shrubs, trees – which differ in density, structure and physiognomy.

3 More water reaches the vegetated surface than is utilized there, so that there is a net movement of surplus water both through the regolith by percolation and through flow downslope into adjacent water channels and other depressions, carrying with it the products of weathering, erosion and biologic activity.

The regimen for landform genesis being described here has perhaps most appropriately been referred to as the moist vegetated setting (Garner, 1974). Nevertheless, these situations or characteristics are not invariable. Bedrock outcrops (inselbergs and so on) occur widely in places, especially over crystalline basement rocks, while vegetation breaks, both artificial and natural, may be important in places. Artificial breaks are caused mainly by human activities of clearing for agriculture, settlement and other purposes; while natural breaks are associated with slopes too steep to retain soil or provide extensive root attachment, water courses, landslide scars (including gully and other badland topographies), erosional or structural escarpments, newly emerged surfaces (such as volcanic terrain, emerged coasts and sandbeaches), and naturally impoverished sites (such as by soil poisoning). In the last case, the example of pyritic shale reacting with rainwater to form acidic ground locally toxic to plants is a case in point. Such conditions, as most of the other natural courses of vegetation breaks, are usually temporary, as they 'heal' with time and so permit vegetation to be re-established. In other words, and except in permanently disturbed areas of human settlement, the vegetated setting is the rule and the breaks the exception in the humid tropical geomorphological system, and more so for the denudational or removal processes sub-system.

The vegetated setting is particularly relevant in the geomorphology of the humid tropics because of the well-known role of plants in:

1 Absorbing the energy of falling precipitation to the extent of essentially eliminating rainsplash erosion in certain circumstances;

2 Frictionally resisting surface water flow effectively everywhere except

in the vegetation breaks and open places such as rock outcrops and water courses;

3 Aiding infiltration of water by loosening the ground surface and sub-surface, generating root passages, improving soil structure and providing favourable habitats for burrowing animals;

4 Using water and other released rock elements in a variety of plant metabolic and other processes associated with plant functioning, including osmosis, photosynthesis, transpiration and biogeochemical processes;

5 Invigorating ground solvency as well as regolith and bedrock leaching through additions of organic acids and carbon dioxide;

6 Influencing soil fertility and capability to sustain plant growth and ground cover;

7 Shielding the soil surface from erosion by running water and wind;

8 Inhibiting sediment transportation through sediment trapping in swamps, thickets and so on; and

9 Shading or otherwise insulating the ground surface from short-term temperature extremes.

These influences are further considered below in relation to the dominant geomorphic processes identified earlier in this chapter.

Chemical denudation

The discussion in Chapter 3 indicates, among other things, that rock weathering, although often perceived as an *in-situ* process, also involves the movement and translocation of dissolved ions which are later immobilized at new sites. This is partly what is being described here as chemical denudation, suggesting, as earlier indicated, that the dividing line between the process of *in-situ* weathering and that of the removal of the weathered product is definitely tenuous. The process bears very stong links with *in-situ* weathering to the extent that the removal of the mobile elements in the weathering equations is essential for the continuation of the weathering process. It can, therefore, be regarded as a natural and important corollary of the chemical weathering process. Nevertheless, the process, in form of solutional removal and mechanical eluviation of weathered products, appears to have noticeable influence on land formation, especially the process of land-surface lowering.

The process of chemical denudation or solutional removal is expected to be very active in the humid tropics. This viewpoint is based, apart from those arguments already listed for the moist, vegetated regimen, on the preponderance in the area of a favourable combination of thick and saturated deeply-weathered rock mantle and thick vegetation cover which, among other things, leads to a large proportion of stream flow being groundwater base discharge, delayed return, or base flow (Douglas, 1975; Ogunkoya, 1979). However, the empirical evidence for this phenomenon is yet to be established beyond reasonable doubt. Reported observations range widely both for ion concentrations (Table 4.1) and for the ratio of dissolved to suspended load (Table 4.2).

TABLE 4.1 Dissolved mineral content in tropical rivers

(a) *Dissolved silica in PPM*

Source	Location	Value	
		Range	Mean
Davis, 1964	Africa	–	24
Douglas, 1969	Queensland, Australia and Malaysia	10–25	–
Grove, 1972	Rivers Niger, Benue, Logore Senegal (West Africa)	5–17	10
Trendall, 1962	Borehole, Uganda	–	50

(b) *Dissolved load, as percentage of suspended load*

Source	Location	Value	
		Range	Mean
Gorbel, 1957, 1959	World	34–70	–
Rougerie, 1960	Ivory Coast	60–95	–
Gibbs, 1967	Amazon r., S. America	–	32
Depetris and Griffin, 1968	Rio de la Plata S. America	–	35
Ogunkoya, 1979	Upper Owena, south-western Nigeria	6.8–0.65	–

TABLE 4.2 Suspended and dissolved load: Upper Owena, Ondo State, Nigeria

Station	Dissolved load (tonnes/km²)	Suspended load (tonnes/km²)	Ratio (Dissolved/Suspended)
Okun	34.9	17.1	2.1:1
Opapa	77.9	25.1	3.1:1
Alura	18.7	11.9	1.6:1
Olotun	10.9	1.6	6.8:1
Etiokun	6.9	1.2	5.8:1
Ohoo	29.5	7.2	4.1:1
Erinta	19.2	29.5	0.6:1
Mogbado	15.3	5.5	2.8:1
Erin	33.6	13.8	2.4:1
Arosa	1.5	0.4	3.8:1
Anini	8.6	2.6	3.2:1
Okorkoro	10.6	7.1	1.5:1
Ofi	13.6	5.6	2.4:1
Apon	11.0	2.9	3.8:1
Okunro	31.9	6.6	4.8:1
		Total	3.3:1

Source: based on Ogunkoya, 1979

Recent findings suggest, however, that the suspicion regarding over-estimation of this process may not be well founded and that the theoretical postulates may be largely valid.

This point is best illustrated by a recent study in the Upper Owena basin of south-western Nigeria, where the following observations were made:

1 Sediment yield varies from 0.4 tonnes/km² to 25.1 tonnes/km² with an average of 1.2 tonnes/km² for suspended sediments and from 1.5 tonnes/km² to 77 tonnes/km², with an average of 21.6 tonnes/km² for dissolved solids.

2 The values for suspended sediments are low compared with known results from other humid tropical areas. Conversely, dissolved solids are generally much higher than in other humid areas (Tables 4.1, 4.2).

3 The most important factors governing the production of dissolved solids – namely, geology, relief, basin shape and percentage of area under settlement – confirm the importance of sub-surface flow in the area, which explains the generally high values of the dissolved load of the streams.

4 Baseflow is an important feature of the rivers, especially those on quartzites.

Mechanical eluviation

Eluviation is a pedogenic process involving the movement of soil material (particulate, colloidal, suspension and solution material inclusive) from one place to another within the soil profile. The process operates in humid environments essentially with surplus surface water and well-structured soil material, which situation is found over large parts of the humid tropical lands. As such, the process encompasses both the solutional activities described above, in which colloids and solutions are involved, and the mechanical process, in which particulate material is involved. Both are effected by groundwater movement and so will be expected to be facilitated by free drainage conditions. In the former case of the solutional process, the material is often lost into drainage water (leached), mobilized in parts of the profile, or gets involved in the biogeochemical cycling process. In the latter case of mechanical eluviation, the material involved is almost invariably retained in the profile, except where it is transported by surface flow into the base-level rivers, where it forms a part of the suspended sediment load.

Nevertheless, and as in the case of the solution process, the selective removal of the fine (silt and clay) fractions of the soil profile by groundwater has been shown to constitute an important geomorphological process especially in humid tropical environments. An example often cited is the catenary relationship commonly found along many hillslopes in this area, indicative of this kind of movement – that is, clear evidence that fine material has been removed from the upper and accumulated in the lower members of the catena. In some cases (as observed earlier), some of the fine material gets transported out of the catena profile into the adjoining stream, where it forms part of the stream-suspended load.

The process we are discussing here is a variant of the leaching process, this time involving particulate material. It is suspected to be the major cause of the unexpectedly low silt and clay fractions of some tropical regoliths (Ruxton, 1958; Ruddock, 1967). According to Ruxton (1958) this process is rapid on steep hill-slopes, especially where rainfall is both adequate and frequent. The example recorded by Ruxton in parts of the Republic of Sudan consists of a sequence of material starting from a coarse, bimodal regolith and changing progressively, first to an upper pediment characterized by moist incoherent debris, then to a dry, clay-rich, compacted surface deposit on the pediment.

Another example of this process is provided by Nossin (1964) from Malaysia, where removal of fines was thought to be the explanation for the downslope movement and eventual settling of the granitic corestones, similar to the winnowing explanation of the desert reg or grey billy (Figure 4.1). It is, however, not easy here to establish precisely the agent responsible for the removal of the intervening fine material, which could have been affected by running water and/or wind, especially at exposed locations.

Fig. 4.1 Inselberg and footslope corestones in the savanna area of Nigeria

Finally, mechanical eluviation has been associated with natural tunnels, cracks and burrows. Sub-surface through flow is thought capable of exploiting such passage ways and so remove fine particles, in addition to colloidal and solutional material, thus widening such spaces and sometimes causing cave-ins. This is similar to what happens in limestone areas, as shown in Chapter 11. Unlike the solution process, measurements of the mechanical eluviation process are not easy to come by.

Physical removal processes

Unlike the chemical processes, which almost invariably operate within the regolith and so affect the sub-surface, the physical processes

Plate 1 Tor and corestones overlying the basal surface of weathering in granite on a new road cutting south of Jos, Nigeria (M. F. Thomas)

considered here operate essentially at the earth's surface. They are mainly of two categories: those involving the massive movement of the regolith downslope, almost invariably under the influence of gravity; and those involving external agents, particularly running water.

Mass movement

Mass movement, also called mass wasting or mass wastage, comprises all gravity-induced movements of slope material, particularly that resulting from weathering, but excludes those in which such material is carried directly by transporting media such as water, wind and ice. The latter process is referred to as mass transport. In nature, the two processes of mass wasting and mass transport merge into each other, so that (as in several other cases of the classification of natural phenomena) the distinction becomes arbitrary.

Mass movements exhibit great variety, which reflect the diversity of the factors responsible for their origin. The most important of these factors – namely, the geology (tectonic geology), slope gradient, climate, the nature of the waste mantle that is being moved, the type and density of the groundcover, land use, and the degree of human interference – have already been described in the literature (cf. Faniran, 1981), while different systems of classification have been proposed. Among the first attempts at classification was that by Sharpe (1938), based on three criteria:

1 the nature of water (or ice) present in the moving mass;
2 the nature of the movement itself; and
3 the speed of the movement.

TABLE 4.3   Mass movement on slopes

Creep	1 Shallow, predominantly seasonal creep
	(a) Soil creep
	(b) Talus creep
	2 Deep-seated continuous creep; mass creep
	3 Progressive creep
Frozen ground phenomena	4 Freeze-thaw movements
	(a) Solifluction
	(b) Cambering and valley bulging
	(c) Stone streams
	(d) Rock glaciers
Landslides	5 Translational slides
	(a) Rock slides; block glides
	(b) Slab, or flake slides
	(c) Detritus, or debris slides
	(d) Mudflows
	(i) Climatic mudflows
	(ii) Volcanic mudflows
	(e) Bog flows; bog bursts
	(f) Flow failures
	(i) Loess flows
	(ii) Flow slides
	6 Rotational slips
	(a) Single rotational slips
	(b) Multiple rotational slips
	(i) In stiff, fissured clays
	(ii) In soft, extra-sensitive clays;
	clay flows
	(c) Successive, or stepped rotational slips
	7 Falls
	(a) Stone and boulder falls
	(b) Rock and soil falls
	8 Sub-aqueous slides
	(a) Flow slides
	(b) Under-consolidated clay slides
Mass movement involving sinking	9 Subsidence caused by
	(a) Mining
	(b) Marine erosion
	(c) Sub-surface erosion, or piping
	(d) Sub-surface solution
	(e) Melting of ground ice
	(f) Chemical changes
	(g) Outflow of lava
	(h) Old land movements
	10 Settlement caused by
	(a) Consolidation
	(i) Surface loading
	(ii) Lowering of ground-water level
	(iii) Underdrainage
	(b) Collapsing grain structure

By contrast, Hutchinson (1968) used the mechanism of movement, morphology of the mass and the rate of movement. He also separated the mass movement occurring on slopes from those involving the sinking of the ground surface (cf. Table 4.3).

The humid tropical environment has been identified as one of those most prone to mass movement, mainly on account of the usually thick zone of incoherent slope material with properties of low shear strength, high voids ratio, and high clay content, all resulting from the operation, over a long geological period, of the deep-weathering process. The process (mass wasting) is further enhanced by the heavy and protracted rainfall, particularly during the wet season, when the regolith usually becomes saturated and overloaded with water, which raises the pore-water pressure to a point where failure occurs. Another important factor is the seasonality of the climate, which facilitates the wetting and drying process, known to be capable of inducing downslope movement of the regolith. The presence of clays (especially the montmorillonitic type) in the regolith, of a sharp basal surface beneath the regolith, and of lateritic crusts on top, also contribute to the mass movement process in humid tropical environments. The swelling clay accentuates the heaving process and so further lowers the shear strength of the slope material; the sharp basal surface is thought to be capable of inducing sliding; while the usually well-jointed and permeable lateritic crusts (duricrusts), apart from admitting water, by way of the cracks and joints (which, among other things, aids the processes of eluviation and sapping of underlying fine material) also exert pressure, by virtue of their weight, over the underlying clay horizons. Finally, and very importantly, though not restricted to the humid tropical environment, tectonic activity – earthquakes and other forms of earth movement – induce mass wasting on slopes. These various factors are discussed in relation to specific examples of mass movement in what follows. The Hutchinson (1968) system is followed as appropriate.

Creep

The term *creep* is often used rather loosely to describe all very slow and imperceptible deformation or movement of slope material, regardless of the causative mechanism and the material involved. Sub-types of this process include the shallow, predominantly seasonal or mantle creep involving either the soil (soil creep), or other slope material (talus creep); the deep-seated, continuous (mass) creep; and what has been described as progressive creep. Of these, only the first category – mantle creep – has been widely reported. It is also the one that is commonest under the humid tropical environment.

Creep is essentially a near-surface phenomenon, being felt to depths of only a few metres. In fact, most of the available field measurements restrict it to depths of less than one metre. Although effective with all grades of materials and affecting the overlying vegetation and man-made structures, Lewis (1974, 1976) has shown that the tree-root system in particular, especially in rainforest environments, affects the movement pattern. Measurements in El Yunque National Forest, among other places, in Puerto Rico, show two zones of maximum

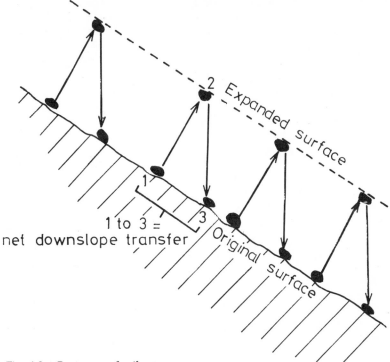

Fig. 4.2a Processes of soil creep

movement: one in the near-surface root zone and the other below it (Figure 4.2a). This is a major departure from the theoretical model in which the rate of movement tends to decrease from surface to the flow base where it reaches zero. However, it confirms the traditional viewpoint to the extent that creep processes do not shear at the base and so do not abrade the buried surface.

The main thrust of Lewis's study is to modify Kirkby's (1967) original soil-creep model, which tends to approximate the drier/normal weathering environment of the humid temperate/savanna regions (cf. Figure 4.2b) by incorporating the slow flow or throughflow component. The original soil creep model states that

$$C_{(z)} \propto C_o M_{sinB \ cosB} \frac{1 + b^2}{b^2} ebz \ \dots \ (4.1), \ where$$

$C_{(z)}$ = net soil creep at depth z
B = slope gradient
M = moisture change at depth z, assumed to decrease exponentially,
b = constant, and
$\propto$ = sign of proportionality.

This situation seems to be drastically changed in the rainforest high rainfall area where through flow resulting from a high infiltration/percolation rate produces a saturated soil below the root zone, which results in

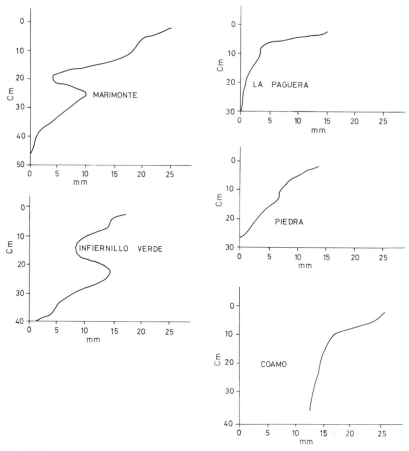

Fig. 4.2b Models of soil creep in Puerto Rico (after Larry Lewis, 1976)

reduced soil strength. This gives rise to the second maximum zone of creep. The double maximum condition has been modelled by Lewis (1976) as follows:

$$C_{(z)} \propto M_{\sin B \cos B} \frac{1 + b^2 - g_z^2 + k_z 3}{b^2} \, C^{-bz}, \ldots (4.2)$$

where the coefficients g and k reflect the soil strength and moisture conditions at depth z, and the other signs are as in equation (4.1) above.

Whatever their pattern or shape, however, the geomorphological role of creep and other slow flow processes is hardly visible: it can only be inferred indirectly from a variety of surface phenomena, including curved trees; tilted fence telegraph poles and monuments; broken and/or displaced walls and foundations; crooked 'stonelines' (quartz veins) displaced along the valley-side slope; roads, railways, and so on moved out of alignment; marked slope convexity; smooth, rounded concavo-convex landscapes; stone/gravel accumulations at slope bases, in soil profiles and at stream beds, and so on (Figure 4.3).

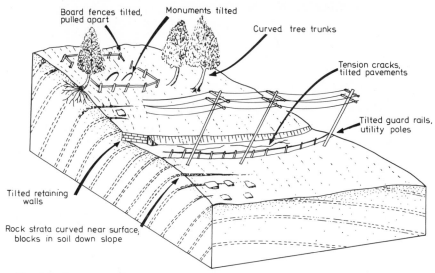

Fig. 4.3 Common effects of soil creep

Moreover, much difficulty attends the field measurement of this slow, imperceptible process, which takes years and much care and patience to yield any meaningful result. The attendant problems, shown below, will probably explain the paucity of empirical data on creep processes:

1 the non-representativeness of the conditions under which measurements are made, especially those in which an identifiable regolith column is inserted in the natural profile;

2 the difficulty of reaching bedrock, which is the ideal reference point for measurement – this is particularly difficult in the humid tropics where regoliths measure tens, sometimes hundreds, of metres, and even when this base can be reached, the disturbance caused in the process is certain to affect the result;

3 interference by man and animals at experiment sites;

4 the dearth of qualified personnel in several countries, especially in the humid tropics; and

5 the inclement physical (climatic, vegetational, pest) environment, again of the humid tropics.

This is particularly true of the humid tropics where the Puerto Rican data (Lewis, 1974, 1976) are among the first to be recorded. Carson and Kirkby (1972) did not include any measurement from the humid tropical region in their summary (Table 10.1, p. 288), while Thomas (1974) included two examples, from northern Australia and Malaysia (Table 4.4). There is therefore no gainsaying that more work needs be done in this environment in order to validate the results that are being provided in the more recent literature (Lewis, 1976).

Landslides

Another major category of mass movement, which is very prominent in

TABLE 4.4   Measurements of creep

Location		Speed (cm³/cm/yr⁻¹)	Source	Remarks
Marimonte, Puerto Rico	(1)	7.50	Lewis (1974)	
Marimonte, Puerto Rico	(2)	8.00	Lewis (1976)	Rainforest
Infiernillo, Puerto Rico	(1)	8.50	Lewis (1976)	Location
Infiernillo, Puerto Rico	(2)	9.10	Lewis (1974)	
Coamo, Puerto Rico		10.40	Lewis (1976)	
La Paguera, Puerto Rico		2.13	Lewis (1976)	Savanna
Piedra, Puerto Rico		3.43	Lewis (1976)	Location
Northern Territory, Australia		4.40	Williams, quoted by Thomas (1974)	Rainforest
Kuala Lumpur, Malaysia		12.40	Eyles and Ho (1970)	Location

the humid tropics, is that described as landslides. These are relatively rapid (perceptible) downslope movements, sliding or falling of relatively dry, weathered debris as a result of failure. In contrast to creep, in which there is generally a continuous gradation between the stationary and moving mass, and in which movement seems to be tolerably constant over relatively large (extensive slope) surfaces, the movement in landslides takes place characteristically on discrete surfaces which sharply define the moving mass.

Landslides may be indistinguishable from the rapid-flow processes; in fact, mudflows, bog flows and the like are recognized as sub-systems of landslides (Table 4.3). In nature, what appears to happen is for mass movement to grade from creep through slump, slip and slide to flows, depending on the controlling factors. The result is that, except in restricted locations, it is variations of this continuum, or succession, that occur most commonly in the landscape, and very many good examples have been described. The best examples are those associated with retreating duricrust-capped hill-slopes such as those of parts of Uganda (Pallister, 1956; Doornkamp, 1970); Sudan (Ruxton and Berry, 1961); Nigeria (Moss, 1965; Ologe and Shepherd, 1972); and Sierra Leone (Thomas, 1974). In Southern Nigeria, for example, Moss (1965) observed two main types of landslides. Slumping was associated with sapped duricrust-capped slopes, the sapping process being associated with the removal of the underlying weathered (mottled and pallid zone) sandstone by the combined processes of mechanical eluviation (see above) and simple flushing, including basal sapping, by both ground and surface flow. Rotational slips were common on longer and gentler slopes, especially where the movement also involved the regolith itself (Figure 4.4). The latter example has also been observed in the Sula mountains of Sierra Leone, among other places, where earthflows were also observed. Thomas (1974) aptly described the situation as follows:

> The highly permeable laterite admits much ... rainfall, leading to the development of underground channels beneath the duricrust and to an increase in pore-water pressure within the regolith. Cambering and collapse of the laterite occurs widely and the regolith periodically suffers shear failures resulting in slumping and earthflows (p. 108).

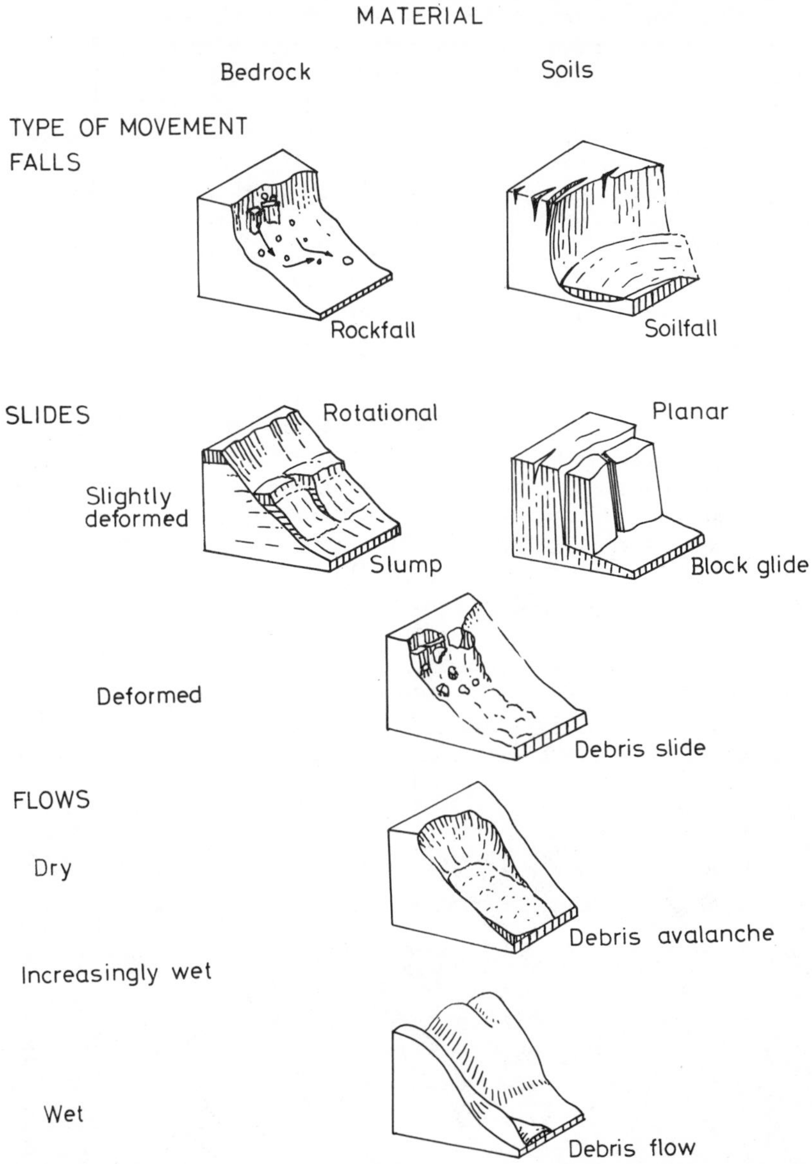

MATERIAL

Bedrock   Soils

TYPE OF MOVEMENT
FALLS

Rockfall   Soilfall

SLIDES   Rotational   Planar

Slightly
deformed

Slump   Block glide

Deformed

Debris slide

FLOWS

Dry

Debris avalanche

Increasingly wet

Wet

Debris flow

Fig. 4.4 Complex landslides

However, it appears that slides and flows, as opposed to slumps, are more common in less deeply-weathered material, but again in high-rainfall, high-drainage density and deeply-dissected, steep-sloped terrains. Among the earliest described examples are those from Hawaii, where Wentworth (1943) reported shallow slides and debris avalanches (cf. Sharpe, 1938) over wide areas, which he thought were capable of explaining the knife-edged features associated with the major range divides.

The examples describe by So (1971) from Hong Kong, like those by Temple and Rapp from Tanzania, have not been linked with specific landform features as have those of Hawaii, but are no less spectacular, particularly in the number involved. The Hong Kong example numbered 700 while the Tanzanian one were more than 1 000. Both sets resulted from heavy and concentrated rainfall – June 1966 in Hong Kong, and February 1970 in Tanzania. However, whereas the Hong Kong examples appear to cast some doubt on the much acclaimed protective role of forest vegetation in the humid tropics (Thomas, 1974, p. 110), the Tanzanian case appears to confirm this role (Temple and Rapp, 1972; Faniran, 1981). The two viewpoints seem to apply. Where slopes are steep and tree roots shallow, the effect may be minimal or, in fact, negative, because fallen trees can actually trigger off a landslide by allowing water into the regolith, which can collect and lead to an earthflow. Jeje (1979) recently described a situation similar to this in the forest belt of south-western Nigeria but it is still too early to assign cause and effect in a case such as this, because landslides, including slumps and flows, can also cause trees to fall and render their protective roles ineffective.

Perhaps the best and most instructive illustration of the protective role of vegetation cover can be obtained from gullied terrains. Whereas landslides, slumping and earthflows are very common features of the

Plate 2 Earth pillars in Kalahari sands, Shaba, Zaire (M. F. Thomas)

gullies of both south-eastern Nigeria and the Zaria area with depleted vegetation, the bluffs of the Eleiyele, Ibadan, gully were found to be relatively stable, under the protection of tree roots and foliage during the greater part of the year when the teak foliage is on (Faniran and Areola, 1974).

Rockslides or boulder slides are also quite common in the humid tropics, and Burke and Durotoye (1970) recently described a classic example from south-western Nigeria (Figure 4.5).

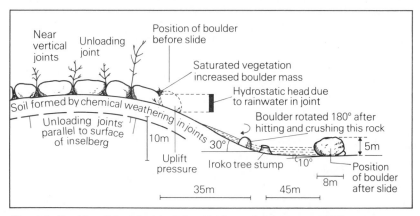

Fig. 4.5 Boulder slide, Nigeria (after Burke and Durotoye, 1970)

Subsidence

Subsidence, in contrast to horizontal mass movement of regolith on slopes, involves predominantly vertical downward movement. It is not as widespread as the other forms, but it can be quite spectacular, being generally more deep seated. The rate of sinking is particularly important in this respect, the most spectacular ones being the most rapid.

The most important causes of subsidence include mining, especially the adit (seam) type, sub-surface erosion or piping, sub-surface solution, surface overloading, underdrainage melting of ground ice and other forms of physical and chemical changes in the sub-surface (Table 4.3). Of all these causes only snow or ice melt is not particularly significant in the humid tropics. Solution pits of different origins are by far the most important, particularly in limestone (Karst) and some duricrusted (deeply-weathered) areas. In Gombe, Northern Nigeria, for example, bridges, houses and other structures continue to sink gradually over subterranean limestone; just as cambering and other forms of surface collapse mark many duricrusted surfaces of Sierra Leone (Sula mountains), Nigeria (Jos plateau) and Uganda, among other places.

The role of man as a geomorphic agent is universal and has been noted in connection with subsidence. This is particularly true of mining areas of intense activity and urban development. The case of the sinking structures at Gombe, Northern Nigeria, has already been mentioned above, while the disasters in the South African coal-mining region have been widely reported.

Surface wash

While the chemical denudational processes and the physical removal processes in this chapter tend to witness their most intensive (spectacular) forms in the very wet rainforest areas of the humid tropics, surface wash is much more widespread in its effect, and in fact appears to be most effective in the drier more open parts. And although important in virtually all forms of geomorphological processes, it is as an agent of mass transportation of rock materials at the earth's surface that the geomorphological significance of water is most noteworthy. This role is played through the agency of two interrelated processes of raindrop impact of splash erosion, and runoff, or slope and channel erosion, by way of overland and channel flows.

The major role of raindrop or splash erosion is to dislodge (detach) rock and/or soil particles from their original resting positions. This process is best understood by considering the momentum of a single raindrop falling on a sloping surface. The downslope component of this momentum is transferred in full to the soil surface. This has two major effects: first, it provides a consolidating force on the particles on which it lands and so compacts the soil; secondly, it imparts a velocity to some of the soil particles, which are thus launched into the air. On landing, these particles transfer their own downslope momentum to other particles, leading to a repetition of the jumping or saltating process. The compaction process leads to the formation of surface crusts, usually only a few millimetres thick, but effective enough to reduce infiltration and increase surface flow. The particle-lifting process provides the surface flow with soil/rock particles. One other way in which the dispersion of soil particles occurs, in addition to the direct impact of falling raindrops, is by

Plate 3 The path of debris flow on a ridge side near Ile-Ife, Oyo State, Nigeria (L. K. Jeje)

slaking, a process of soil breakdown by the compression of air ahead of a wetting front as raindrop starts to infiltrate a dry soil surface (cf. Bryan, 1969). All three processes of compaction, splash and slaking participate in the removal of slope materials.

Overland flow or run-off occurs during a rainstorm when either soil moisture storage (in the case of prolonged rain) or the soil infiltration capacity (in the case of intense rain) is exceeded. The hydraulic characteristics of runoff or overland flow have been quantified in terms of its Reynolds number (Re) and its Froude number (F). The Reynolds number is defined as:

$$\text{Re} = \frac{vr}{v} \ \ldots \ \ (4.3) \text{ and}$$

The Froude number as

$$F = \frac{v}{\sqrt{gr}} \ \ldots \ \ (4.4), \text{ where}$$

r   = hydraulic radius or flow depth
v   = kinetic viscosity of water
g   = force of gravity.

The Reynolds number is an index of the turbulence of flow: the greater the turbulence, the greater is the erosive power generated by the flow. Similarly, when the Froude number is less than 1.0, flow is described as tranquil, without erosive power; whereas values greater than 1.0 denote supercritical or rapid flow, which is more erosive.

A very important factor in these and other hydraulic relationships is the flow velocity, which must attain a threshold value before erosion (removal) commences. As shown in Figure 4.6, there are different

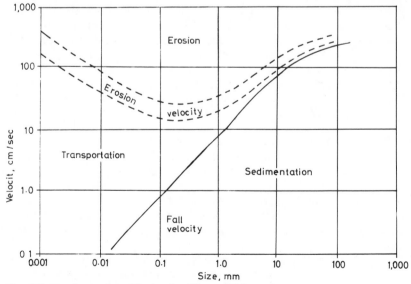

Fig. 4.6 Erosion – deposition and calibre model

thresholds for grains of varying sizes. The trend of this variation conforms to some sort of a parabola, the threshold increasing with grain size, but also high for very fine (clay) materials. The velocity of flow itself is dependent upon several factors of the flow depth or hydraulic radius, the roughness of the surface, and the slope. This relationship is commonly expressed by the Manning equation:

$$v = \frac{r^{\frac{2}{3}}S^{\frac{1}{2}}}{n} \ldots (4.5), \text{ where}$$

v = velocity of flow
s = slope, or gradient angle (tan θ, sin θ),
n = Manning coefficient of roughness, and
r = hydraulic radius.

Another important factor is the spatial extent of flow. According to Horton (1945), overland flow covers two-thirds or more of the hillsides in a drainage basin during the peak period of a storm, the flow itself, resulting from the intensity of the rainfall, being greater than the infiltration capacity of the soil. However, recent studies have shown that overland flow may not be so widespread within a drainage basin. For one thing, in well-vegetated areas, such as the humid tropics, flow may occur very infrequently and cover very limited areas of the basin, closest to the stream courses (Kirkby, 1967). It has also been shown that this area, often referred to as contributory area, is not static: it expands and contracts in response to rainfall (or snowmelt) and other drainage physiographic characteristics (see Chapter 7).

This takes us to the constraints offered by the environment to the natural removal processes. These constraints, in the humid tropics, relate essentially to both the weathering and the vegetation characteristics. The weathering environment here refers to deep chemical weathering, the products of which consist essentially of fine-grained material with limited erosive ability. It relates, therefore, mainly to the runoff (overland-flow) component of the process of surface removal. By contrast, the vegetation environment affects both the raindrop impact and the runoff components of the removal processes. A third environment of relevance is the topographic or slope environment. Thus, Morgan (1979) recently made the following observation, in connection with the factors controlling the process of soil erosion:

> The factors controlling the working of the soil erosion system are the erosivity of the eroding agent, the erodibility of the soil, the slope of the land and the nature of the plant cover (p. 15).

The erosive agent in humid regions is water in the form of rainfall. This, as observed earlier, occurs partly through the detaching power of raindrops striking the soil surface and partly through the contribution of rain to runoff.

The ultimate goal in studying rainfall characteristics is the derivation of some sort of index with which to predict erosivity. The best available expression of rainfall erosivity is that based on the kinetic energy of rain, in which case the erosivity of a rainstorm is a function of its intensity (amount per unit of time), duration and mass (or total volume), as well as of the diameter and velocity of the raindrops. Studies of tropical rainfall (cf. Hudson, 1963) indicate that the established relationship on raindrop size characteristics (varying with the intensity) holds only for rainfall intensities of up to 100 mm h^{-1}, after which the median drop size decreases with increasing intensity. Also, it is well known that the drop size characteristics of convectional (tropical) and frontal (mid-latitude) rain are different, as are those of rain formed at the warm and cold fronts of a temperate depression. It is, therefore, understandable that the models derived vary with area (cf. Morgan, 1979, pp. 19–20; Oyegun, 1980). It is only those models derived specifically for humid tropical conditions (cf. Hudson, 1965; Oyegun, 1980, p. 54) that should be applicable. Thus Hudson's erosivity index proved more reliable in Zimbabwe than the Weischmeier and Smith (1958) model, while the Oyegun (1980) model:

$$R = \frac{E1_{15}}{100} \ldots \quad (4.6), \text{ where}$$

R = rainfall erosion factor
E = storm energy
1_{15} = maximum or 15-minute intensity of rainfall

appears further to improve the prospects of predicting soil loss from rainfall characteristics in humid tropical areas. This will be most essential in these regions where qualified personnel are lacking and the nature environment repressive.

In the meantime, most geomorphological studies in the humid tropics tend to hold the rainfall factor constant and to lay emphasis on the other factors – the soil or material being removed, slope and vegetation cover. The argument seems to be that tropical rainfall, if there is anything so called, is not necessarily restricted to the geographical limits of the humid tropics. Its influence may, therefore, be subsumed under the other environmental factors. The succeeding discussion is in two parts: raindrop impact or splash erosion, and runoff or flow erosion.

Raindrop impact or splash erosion

Much of our understanding of the temporal, spatial and intensity (degree) aspects of soil erosion emanates from empirical studies in which an array of data on soil loss and the presumed controlling variables are collected and subjected to statistical (correlation and regression) tests. One expected result of this kind of approach is a multiplicity of variables considered important and worthy of study. With respect to rainfall, Douglas (1976), for example, lists seven variables which have been used to explain spatial variations in erosion, while Oyegun (1980) listed four. Among the indices already used are rainfall intensity, measured in terms of maximum 5-minute, 10-minute, 15-minute, 30-minute intensities, and so on; number of storms; length of storms; total rainfall; raindrop size (maximum, mean, median, etc.); and rainfall concentration. Most of these studies have come from extra-tropical regions (cf. Morgan, 1977, 1979). It is only recently that such experiments have been tried in humid tropical environments. The following account is based on the result of an on-going research based at the Department of Geography, University of Ibadan, Ibadan, and the International Institute of Tropical Agriculture (IITA), also at Ibadan.

The initial interest in field observation of soil erosion in vegetated hill-slopes resulted from a study conducted in 1974, of a gully developed in a teak plantation established purposely to check erosion and sediment yield within the catchment of the Eleiyele water reservoir (Faniran and Areola, 1974). A few years earlier a study of the infiltration parameters of various land-use surface, including forested surfaces in the Ibadan urban and suburban areas, had started in the same department (Akintola, 1974). The IITA also started a runoff plot experiment in the early 1970s. The results so far published show that:

1 runoff may be low to nil on vegetated surfaces;
2 rainfall reaches the surface, including vegetated ones, both directly, in openings and during the early rainy season when the canopy is nil, and indirectly, through leaf drops and flows on tree trunks and branches;
3 infiltration rates are high on vegetated surfaces; and
4 splash erosion tends to act uniformly on slopes and its geomorphological effects limited in places where stones, tree roots or other hard objects selectively protect the underlying soil to produce features variously referred to as splash pedestals, soil pillars and earth or rock pillars. Earth pillars are widespread under many forests, especially old sites of settlements where broken pots and other hard substances act as protective cover to the soil.

Perhaps the most informative and also most thorough of the Ibadan studies is that recently completed by R. Oyegun for a doctor of philosophy degree of the University of Ibadan. The major contributions of this study to our knowledge of the mechanism and factors of removal in humid tropical environments include the following:

1 Different surfaces within a given area have different rates of soil erosion even when the area is monolithologic, of similar slope and subjected to the same rainfall.
2 The sediment yield rates for the three surfaces studied were estimated at
   (a) 2 809 $\times$ 10^{2}kg/ha per annum for the shaded grass surface;
   (b) 1 742 $\times$ 10^{2}kg/ha per annum for the plantation surface; and
   (c)   492 $\times$ 10^{2}kg/ha per annum for the unshaded grass surface.
   This means that tree-covered surfaces are not necessarily protected from erosion, and in fact may be the most susceptible to it.
3 Following from (2) above, the surfaces that require urgent attention are those under trees, where cover of grass or litter is below 90 per cent.
4 Trees taller than 4 metres have little impact on rainfall energy impact because drops falling from their crowns reach the ground at their terminal velocities.
5 Soil aggregates may be very unstable under tree-shaded surfaces, hence erosion may be quite effective on such surfaces.
6 The vegetation on the ground in form of grass or dead litter may offer adequate protection to the soil, especially if they reach 90 per cent

cover.

7 Grass at about 90 per cent cover is very effective in controlling soil erosion under tropical rainfall conditions.

8 The aspect of tropical rainfall that best predicts sediment yield is the total monthly rainfall caused by storm rainfall, especially those with mean maximum intensity of 15 minutes.

Finally, the Ibadan studies, like the others on the same topic, emphasize the unity of the removal processes on the earth's surface. Thus, it is almost impossible to separate splash from flow erosion, since both form phases of one continuous process, starting with the detachment of rock particles and their deposition somewhere else. In fact, the detachment stage (splash) forms a very limited aspect in terms of the time and the total load involved in the mass transportation process. It is therefore not surprising that the bulk of the discussion of water erosion is based on the runoff/surface flow component, with splash erosion mentioned only in passing.

Runoff or flow erosion

When the rates of infiltration and through flow are incapable of coping with that of precipitation, water begins to collect in surface depressions, and later (where there is sufficient gradient) flow as rills, sheet, stream or channelized runoff. Rill wash and sheet wash are often included in what is here called overland flow, while all concentrated flows, whether temporary or permanent are termed stream or channelized flow. Whichever form this excess surface water takes to drain itself from the land surface, its erosive power, which is closely related to its ability to remove and transport rock and soil particles, depends on a number of environmental factors. These factors include:

1 the type of flow – whether rill, gully, sheet or channel;
2 the hydraulic attributes of the flow–depth, capacity, competence, etc.;
3 the bed and bank material;
4 the nature of the vegetation cover, including the extent of man's interference; and
5 the effective gradient and/or slope.

The Food and Agriculture Organization (FAO) (1965) listed these factors as relative to soil erosion slightly differently, as follows:

1 physiographic factors – slope steepness, length and curvature;
2 climate – rainfall (quantity, intensity, energy and distribution), and temperature changes;
3 soil – detachability and transportability, aggregation and surface sealing, depth, water-holding capacity;
4 cover or vegetation – cultivated, fallow, forests.

The factors are then used to obtain a model, often referred to as the Universal Soil Loss equation:

$$E = R.K.L.S.C.P., \ldots (4.7), \text{ where}$$

E = average annual (soil, sediment) loss
R = rainfall factor
K = soil erodibility factor
L = Length-of-slope factor

S = Steepness-of-slope factor
C = cropping-and-management factor
P = conservation-practice factor.

The FAO studies were concerned mainly with soil erosion which takes place on slopes: the above equation cannot therefore form a sufficient basis for the study of erosion by running water which also – and principally, too – involves channel flow. Nevertheless, it forms a basis for discussing the slope-wash component of the runoff process.

Slope wash

The term used for this type of process is *overland flow*, which may take several forms, including:

1 *Laminar flow*: This is characteristically a low-velocity movement of water, measurable in fractions of a millimetre per second, where water moves in parallel layers, each shearing past the other. By definition the shearing stress, y in a laminar flow situation at a point y distance above the flow bed and at a flow velocity v, is:

$$y = U\frac{dv}{dy}, \ldots (4.8), \text{ where}$$

U = viscosity, and

$\dfrac{dv}{dy}$ = change in velocity from one layer to the next, also called the velocity gradient (cf. Figure 4.7).

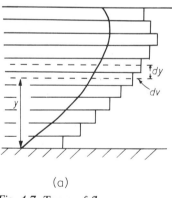
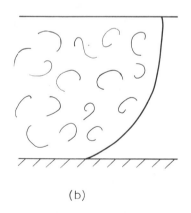

(a)   (b)

Fig. 4.7 Types of flow

As such, laminar flow cannot support solid particles in suspension. It is also typically non-erosive. Laminar flow is very rare in nature, except in the case of what Butzer (1976) calls laminar surface flow, which is equivalent to a film of water, seldom more than a couple of millimetres deep, and whose erosional activity is limited to the removal of ions, colloids and some clays in solution or suspension. Laminar flow is also suspected to occur at the vegetated fringes of quasi-sheetfloods.

2 *Turbulent overland flow*: This characterizes localized concentrations of water developed on an uneven surface, particularly as flow (runoff)

gains momentum on steeper and mid-slope sections. Turbulent flow effects mechanical erosion and participates in the mass transportation of surface materials.

3 *Surging overland flow*: This is associated with irregular surges of water in hillside rills. It is also capable of mechanical erosion.

The type of overland flow which can also be defined in terms of the intensity or characteristics of flow reflects several factors, prominent among which are:

a) the intensity and duration of precipitation;
b) the rate of interception by vegetation and evaporation;
c) the rate of infiltration and through flow into the soil;
d) the nature of the vegetation cover;
e) the nature of the regolith or soil mantle;
f) the nature of the surface in terms of roughness;
g) the steepness, length and aspect of slope.

These same factors affect the effectiveness of runoff generally, but slope wash or overland flow in particular, as an agent of mass transportation.

The present state of our knowledge of these factors is, however, still unsatisfactory. It is as yet too early to make many definitive pronouncements as to the direction and extent of each factor. Even such obvious factors as forest cover have been disputed, not only in connection with the incidence of landslides, but also in connection with runoff erosion. Although theoretical postulations tend to point to the protective role of vegetation, similar views have also been supported by empirical studies and rigorous statistical analysis (cf. Ebisemiju, 1976), data have been and are still being provided from other empirical/experimental studies pointing to the need to exercize some restraint (Fournier, 1960; Young, 1972; Faniran and Areola, 1974; Oyegun, 1980).

Among the reasons usually advanced for the effectiveness of slope wash under forest, particularly tropical forest, are:

1 the occurrence of canopy openings, through any one or combinations of clearing, tree fall, leaf fall (in deciduous trees) and soil poisoning leading to forest depletion;
2 bare ground surface under thick tropical forest, whether in terms of undergrowth or in terms of tropical leaf litter;
3 occurrence of intense rainfall both as high incidence short storms and as prolonged heavy rainfall.

These conditons are important not only in high-relief, steep-slope areas (Ruxton, 1967) but also on pediment surfaces (Oyegun, 1980).

The relatively high rates of soil loss from humid tropical areas have also been related to the effects of human interference (Selby, 1974; Douglas, 1975), assigning the figures to the process of accelerated erosion. However, this viewpoint seems to be punctured by studies such as those by Eyles and Ho (1970) and by Young (1972), both in Malaysia. Both studies reported high rates of surface removal from slopes under mature rainforest, even on slow-angle slopes. Similar conclusions can also be inferred, among others, from both the hydrograph of many tropical streams which rise quickly to peak flows at the onset of rains (Figure 4.8) as well as by the high silt/clay load carried by these streams during storms. In short, it is difficult, in the light of available evidence, to disagree with

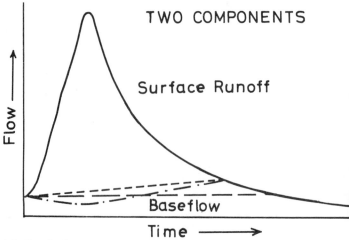

Fig. 4.8 Unit hydrograph — tropical rivers

Thomas (1974, p. 117), who concluded that slope wash should not be discounted as an important process under rainforest. This is further borne out by Table 4.5.

The effectiveness of slope wash has not been disputed to the same extent in the drier areas of the humid tropics – namely, the savannas – as in the rainforest belt. Both raindrop impact and slope wash are believed to be more effective than in the former areas, in conformity with both theoretical and empirical renderings. Theoretically, savannas are believed to be more open and therefore less protective of the ground surface. Accordingly, both the Langbein and Schumm (1958) model and the Fournier (1962) model recognize the savannas as areas of high erosion (Figure 4.9).

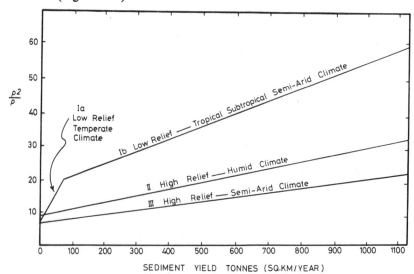

Fig. 4.9 Global erosion (after Fournier, 1962)

One significant support for this viewpoint has come from calculations of runoff from catchments in different climatic locations (Ledger, 1964; Okechukwu, 1974). These figures suggest, among other

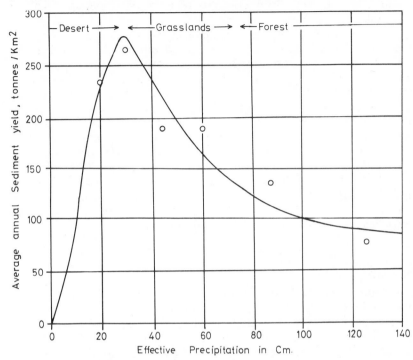

Fig. 4.10a Variation of sediment yield with effective precipitation (after Langbein and Schumm, 1955)

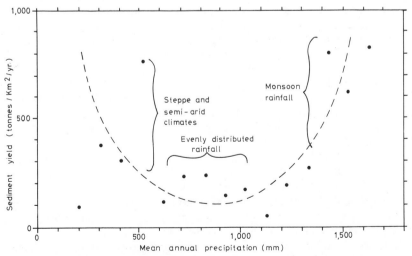

Fig. 4.10b Parabolic relationship between sediment yield and mean annual precipitation (after Fournier, 1962)

things, that peak flows in small streams are determined by vegetation cover, being highest in dry regions and lowest in high forest areas (Figure 4.10). Thus relatively high runoff coefficients are recorded in savannas, which are associated with effective removal of slope and channel material.

It is necessary, however, to sound a note of warning at this stage, to the effect that we do not yet have enough data from which to make conclusions. Indeed, the study by Oyegun (1980) suggests that a grass mat can provide surer protection than tree cover. Whereas sediment yield from a forested surface was estimated at $1\,742 \times 10^2$kg/ha, that from a grass-covered surface was only 492×10^2kg/ha. Moreover, it does not always happen that high discharges are accompanied by high rates of removal. The conclusion here, therefore, seems to be that the suggestion that the maximum denudation occurs in the intermediate savanna zones (Fournier, 1960) cannot yet be decided on the basis of available evidence. The available information on the rates of surface wash in both rainforest and savannas is summarized in Table 4.5. It is quite interesting that, except for the isolated case of Darwin's (1967) study, the theoretical construct has yet to be borne out by observation.

Stream erosion

This component of the removal process is obviously the most important geomorphologically, being the major conduit pipe or conveyor belt for all the materials removed from dry land and transported to the ocean basins and other receptacles. Thus, it has been recognized for a long time by geomorphologists that a stream is effective as a geomorphic agent only when it has the ability to remove sediment. This the stream does in various ways, as follows:

1 As solution load, consisting of materials – namely carbonates, sulfates, chlorides, oxides chemically dissolved in the moving water and subsequently precipitated at lake, sea and other depository beds.
2 Suspension load, or fine-grained particles with limited mass carried in mechanical suspension by agitated (turbulent) water. The major components of the suspension load of streams are clay or silt-sized material and colloids.
3 Saltation load, consisting of intermediate-sized particles too heavy to be held in suspension but which bounce along the stream bed. They consist of sand-size particles and are invariably restricted to the bottom part of the stream.
4 Traction or bed load, consisting of coarse sand and pebbles too heavy for saltation, but which get dragged along the stream bed, especially during high (flood) flows.

Although the measurement of stream sediment load is not an easy task, this has been done for several streams of the developed world. The following observations can be made from the available data:

1 The dissolved load comprises from less than 1 to more than 99 per cent of total load, excluding traction or bed load. The tendency is for this component to increase under the influence of increased pollution (by

TABLE 4.5 Observed rates of surface wash

Source	Location	Rates Denudation in mm 1 000 yrs	Volumetric movt. in cm²/cm/yr
(a) Rainforest			
Rougerie (1956)	Ivory Coast	500–1 500	–
Rougerie (1960)	Ivory Coast	1 500–3 500	–
Roose (1967)	Senegal	80–500	–
Young (1972)	Malaysia	–	1–4
Rapp et al. (1972)	Tanzania	260	
Kessel (1977)	Guyana	275–323	–
(b) Savanna			
Roose (1967)	Senegal	1.6	–
Darwin (1967)	Australia (N.T.)	500–11 000	–
Fournier (1967)	Ivory Coast	20	–
Williams (1968)	Australia (N.T.)	40	–
Williams (1969)	Australia (N.T.)	10–200	–
Williams (1973)	Australia (N.T.)	54–56	–
Townshend (1970)	Brazil (Mato Grosso)	–	0.6–9.6
Rapp et al. (1972)	Tanzania	400	–

103

agricultural, industrial and other wastes) associated with modern industrialized societies.

2 Suspended sediments account for between 50 and 100 per cent of the material transported mechanically by streams.

3 The measurement of both saltation and traction (bed) loads has not been met with the type of success associated with that of solution and suspended load.

4 Finally, and perhaps most importantly, the obtained ratios are subject to very considerable fluctuations both temporally and spatially. Suffice it to say that vast quantities of earth material are constantly being transported from land to sea annually as to warrant special interest by geomorphologists.

The ability of a stream to perform the mass transportation assignment is usually defined by two concepts – capacity and competence. While capacity expresses the total potential weight of sediment load of various calibre in a stream, competence refers to the weight or size of the largest particles that the stream can possibly move along its bed. Both components depend on factors relating to the flow, load and channel physical characteristics of the transporting medium – the stream.

Again, because of the dearth of information regarding this process in humid tropical regions, much controversy surrounds the efficacy or otherwise of the process of stream erosion generally and that of stream transportation in particular. The prevailing rock weathering mechanism(s) determine the type of the dominant sediment load. Another constraint on stream erosion, especially bank erosion, is the nature of the vegetation cover, including tree-root density. Given, therefore, the popularly held views about the humid tropical environment – intensely (deeply weathered) rocks, fine-grained (clay and silt) soils, and thick forest cover over wide areas – the possible ineffectiveness of stream erosion in humid tropical regions can be easily surmised. The idea has been summarized by Thomas (1974) thus:

> Studies of weathering and inspection, if not measurement, of stream bedloads show that crystalline rocks typically do not yield many fragments of pebble or coarse gravel size. Core boulders . . . appear to disintegrate into sand and clay as a result of a breakdown of the rock fabric due to weathering. Much of the weathered material reaching the streams by means of creep or wash processes is already broken down into fine products, and the rivers may carry only fine gravel, sand and suspended clay particles, with an almost total absence of coarse bedload. Under such conditions erosion by tropical streams is relatively feeble and may become further inhibited by the nature of the stream regimes which are highly seasonal.

Thomas (1974) also summarized the views of several (mostly European) geomorphologists who appear to have subscribed to the above-stated position. They include Büdel (1957), Birot (1958, 1968), Tricart (1972) and Louis (1964). Other evidences provided by Büdel to support this position include the claim (1) that stream valleys in tropical areas are characteristically indistinct features and (2) that the stream sediments display the same grading as the regolith over which they flow. It is also pertinent to note that Speight's (1965) observations in Papua tend to corroborate this idea, which is still strongly held by many present-day geomorphologists.

Nevertheless, this position seems to run counter to experiences elsewhere. In the first instance, many tropical rivers, especially those of fourth and higher orders, flow, in basement complex areas, directly on bedrock, while the depth of the weathered rock increases with distance from such channels to the water divide (Omorinbola, 1979; Faniran, 1974a; Faniran and Omorinbola, 1980). One implication of this phenomenon which, in the middle Oshun valley of southern Nigeria, has been successfully fitted to a simple regression equation of the form:

$Y = 3.4420 + 0.2217$ D (4.9), where
Y = depth of weathering and
D = distance from river channel: the regression coefficient of 0.86, significant at the 1 per cent level of probability

– suggests, among other things, that removal processes in the hillslopes and interfluves are not as rapid and active as those in the adjoining channels.

Another possible pointer to active stream erosion in the humid tropics is the evidence of river sediment load, especially dissolved and suspended load. These figures, as well as the denudation rates based on them, compare favourably with those of other areas where river erosion is believed to be active, even when allowance has been given to the influence of human interference (Douglas, 1967, 1975).

However, perhaps the most convincing evidence of effective river action in humid tropical regions is that provided by the stream courses themselves. The features of these stream courses, which are described in detail in Chapter 7, include little or no sediment, except perhaps in braided portions, development of groves, riffles and pools and other evidence of direct river erosion on fresh riverbed rock in major stream channels and the common occurrence of incized, ingrown and river-capture phenomena within several drainage basins. For example, one of the two spectacular features of plains, which form about 80 per cent of the total surface area of virtually the whole of lowland Africa, is the incision by virtually all rivers, including small-order types like the first-order Yemoja stream in the University of Ibadan Botanical Gardens, into its flood plain. Faniran (1974b) once observed as follows about this feature in the Ibadan region:

In almost every case, even in small streams, such as the Yemoja, which flows through the University Botanical Garden, the river has cut into its bed, the depth varying from 1 m to 3 m . . .. The down cutting appears to be still in progress and can be attributed to a recent drop in the sea level, following a major change of climate . . . streams have utilized such weak areas, resulting in features which appear to illustrate classical river capture phenomena . . ... Gullying is a common phenomenon espeically in the settled parts. The most spectacular example of gullying yet observed ... is around Eleiyele water reservoir .... Riffles and pools have been observed along the Ona [river], north of the University of Ibadan campus [pp. 6–7] ....

Similar views about this landform in several other places have been expressed – e.g., in parts of northern Nigeria by Thorp (1970), and in Upper Ogun basin, Oyo State by Osunade (1977) – while examples of waterfalls, cataracts, rapids, knickpoints, gorges, canyons, ravines, and gullies, all spectacular erosional features, have been described from different locations in tropical Africa (Faniran, 1981). In short, there is evidence for active fluvial erosion and efficacious stream removal in humid tropical regions. This is particularly so in the dry-forest/savanna regions where most of the cited examples are from, but also in the forest regions. What is lacking is empirical field data on the process compatible to those obtained for creep processes, for example. Nevertheless, and as observed for a number of other geomorphological processes in the humid tropics, the stream erosion process should not be discounted but should be given the attention it deserves.

Water erosion operates in several other forms in the humid tropics. The most important ones – fluvial waves erosion and karst processes, are discussed separately in succeeding chapters.

Conclusion

The humid tropics is now witnessing a noticeable increase in the number of geomorphological research projects, the results of which are turning overboard a number of popularly held views about the operating geomorphological processes. Among these are studies of deep chemical weathering and its contribution to the denudational processes in the humid tropical environment; the effect of tree roots and through flow on the process of soil creep and the study of the sediment load and channel features of tropical streams. The effect of human interference, including urbanization, is also receiving attention, so also are the possible influences of ground cover of different types. The results are still coming in, so that it is still too early to make many definitive statements. What is at present apparent is that there is a wide scope for geomorphological research in the humid tropics. The number of research workers is too small to cope with this situation, which promises to provide invaluable information for man in his continued closer relationship with the earth's surface. One can only hope that the situation will continue to improve, especially in connection with the removal processes in the humid tropics.

References and further reading

Akintola, F. O. (1974) 'The parameters of infiltration equations or urban landuse surfaces', Ph.D. thesis, University of Ibadan.

Birot, P. (1958) 'Les domes crystallines', *Centre National de la Recherche Scientific (C.N.R.S.) Mem. Documents*, **6**, 34.

Birot, P. (1968) *The Cycle of Erosion in Different Climates*, Batsford, London.

Bryan, R. B. (1969) 'The relative erodibility of soils developed in the Peak District of Derbyshire', *Geog. Ann.*, **51**, 145–59.

Büdel, J. (1957) 'Die Doppelten Einebnungsflächen in den feuchten Tropen', *Zeits. für Geomorph.* N.F., **1**, 201–88.

Burke, K. and Durotoye, B. (1970) 'Boulder slide on an inselberg in the Nigerian rainforest', *J. Geol.* **78**, 6, 733–5.

Butzer, K. W. (1976) *Geomorphology from the Earth*, Harper and Row, New York.

Carson, M. A. and Kirkby, M. J. (1972) *Hillslope Form and Process*, Cambridge Univ. Press.

Corbel, J. (1957) 'L'érosion climatique des granites et silicates sous climats chauds', *Rev. Géomorph. Dyn.*, **8**, 4–8.

Corbel, J. (1959) 'Vitesse de l'érosion', *Zeits. für Geomorph.*, **3**, 1–28.

Darwin, W. R. B. (1967) *Northern Territory (Australia) Administration, Water Resources Branch Annual Report 1966–67*.

Depetris, P. J. and Griffin, J. J. (1968) 'Suspended load in the Rio de la plata drainage basin', *Sedimentology*, **11**, 53–60.

Doornkamp, J. C. (1970) 'The geomorphology of the Mbarara area' sheet SA-36-1, *Geomorph. Rep.*, No. 1, Dept. of Geography, Univ. of Nottingham; also Geological Survey Mines Dept. Uganda.

Douglas, I. (1967) 'Natural and man-made erosion in the humid tropics of Australia, Malaysia and Singapore', in *Symposium on River Morphology*, Bern.

Douglas, I. (1975) 'The impact of urbanization on fluvial geomorphology in humid tropics', in *Dynamical Geomorphology in Tropical Regions*, J. Alexandre (ed.), Univ. of Zaire Press.

Douglas, I. (1976) 'Erosion and climate: geomorphological implications', in *Geomorphology and Climate*, E. Derbyshire, (ed.), J. Wiley, London.

Ebisemiju, S. P. (1976) 'The structure of the interrelationship of drainage basin characteristics', Ph.D. thesis, Univ. of Ibadan.

Eyles, R. J. and Ho, R. (1970) 'Soil creep on a humid tropical slopes', *J. Trop. Geog.*, **31**, 40–2.

Faniran, A. (1974a) *The landforms of the Ibadan region*, Dept. of Geography, Univ. of Ibadan (mimeo).

Faniran, A. (1974b) 'The extent, profile and significance of deep weathering in Nigeria', *J. Trop. Geog.*, **38**, 19–30.

Faniran, A. (1975) 'Rural water supply in Nigeria's basement complex; a study in alternatives', *Proc. 2nd World Congress on Water Resources*, New Delhi, **III**, 89–100.

Faniran, A. (1980) *African Landforms*, Heinemann, Ibadan (in press).

Faniran, A. and Areola, O. (1974) 'Landforms examples from Nigeria: No. 7, a gully', *Nig. Geog. J.*, **17** (1), 57–60.

Faniran, A. and Omorinbola, E. O. (1980) 'Trend surface analysis and practical implications of weathering depths in basement complex rocks of Nigeria', *J. Mining Geol.*, **23**, 113–26.

Food and Agricultural Organization of the United Nations (1965) *Soil Erosion by Water*, FAO Agricultural Development Paper No. 81, Rome.

Fournier, F. (1960) *Climat et Erosion: la relation entre l'érosion du sol par l'eau et les précipitations atmosphériques*, Presses Univ., Paris.

Garner, H. F. (1974) *Origin of Landscapes*, Oxford Univ. Press, London, New York.

Gibbs, R. J. (1967) 'The geochemistry of the Amazon river system, part I', *Bull. Geol. Soc. Amer.*, **78**, 1203–32.

Horton, R. E. (1945) 'Erosional development of streams and their drainage basins: a hydrophysical approach to quantitative morphology', *Bull. Geol. Soc. Amer.*, **56**, 275–370.

Hudson, N. W. (1963) 'Raindrop size distribution in high intensity storms', *Rhod. J. Agric. Res.*, **1**, 6–11.

Hudson, N. W. (1965) 'The influence of rainfall on the mechanics of soil erosion', M.Sc. thesis, Univ. of Cape Town.

Hutchinson, J. N. (1968) 'Mass movement', in *The Encyclopedia of Gemorphology*, R. W. Fairbridge (ed.), Reinhold, New York.

Jeje, L. K. (1979) 'Debris flow on a hillside along Ife-Ondo road, Western Nigeria', *J. Min. Geol.* **16** (1), 9–16.

Kessel, R. H. (1977) 'Slope runoff and denudation in the Rupununi Savanna, Guyana', *J. Trop. Geog.*, **44**, 33–42.

Kirkby, M. J. (1967) 'Movement and theory of soil creep', *J. Geol.* **75**, 359–78.

Langbein, W. B. and Schumm, S. A. (1958) 'Yield of sediment in relation to mean annual precipitation', *Trans. Amer. Geophys. Un.*, **39**, 1076–84.

Ledger, D. L. (1964) 'Some hydrological characteristics of West African rivers', *Trans. Inst. Br. Geog.* **35**, 73–90.

Lewis, A. L. (1974) 'Slow movement of earth under tropical rainforest conditions', *Geology*, **2**, 9–10.

Lewis, A. L. (1976) 'Slow movement in the tropics: a general model', *Zeits. für Geomorph.* N. F. Suppl. Band 24, 132–44.

Louis, H. (1964) 'Uber rumpfflächen – und talbildung in den wechselfeuchten tropen besonders nach studien in Tanganyika', *Zeits. für Geomorph.*, **8** (Sonderheft), 43–70.

Morgan, R. P. K. (1977) 'Soil erosion in the United Kingdom: field studies in the Silsoe area, 1973–75', *Nat. Coll. Agr. Engng, Silsoe, Occasional Papers 5*.

Morgan, R. P. K. (1979) *Soil Erosion*, Longman, London.

Moss, R. P. (1965) 'Soil development and soil morphology in a part of south-western Nigeria', *J. Soil Sci.*, **16**, 192–209.

Nossin, J. J. (1964) 'Geomorphology of the surrounding Kuanton (Eastern Malaya)', *Geol. Minjbouw*, **45**, 157–82.

Ogunkoya, O. O. (1979) 'Hydrology of third-order basins in the Upper Owena drainage basin of south-western Nigeria', Ph.D. thesis, Univ. of Ife.

Okechukwu, C. C. (1974) 'Fluvial geomorphic interrelationships in some river catchments in the Nigerian pre-Cambrian basement complex', Ph.D. thesis, Univ. of Ibadan.

Ologe, K. O. and Shpherd, I. (1972) 'Landform examples from Nigeria, No. 5', *Nig. Geog. J.*, **15** (1), 83–6.

Omorinbola, E. O. (1979) 'A quantitative evaluation of groundwater storage in basement complex regoliths in south-western Nigeria', Ph.D. thesis, Univ. of Ibadan.

Osunade, M. A. (1977) 'An analysis of the physiograhic and the parametric methods of land evaluation', Ph.D. thesis, Univ. of Ibadan.

Oyegun, R. O. (1980) 'The effects of tropical rainfall on sediment yield from different landuse surfaces in sub-urban Ibadan', Ph.D. thesis, Univ. of Ibadan.

Pallister, J. W. (1956) 'Slope development in Buganda', *Geog. J.*, **122**, 80–7.

Roose, R. (1967) 'Dix années de mesure de l'érosion et du ruisselle en Sénégal', *Agronomia Tropical*, **22**, 123–52.

Rougerie, G. (1956) 'Etudes des modes d'érosion et du façonnement des versants en Côte d'Ivoire equatoriale', *Premier Rapport de la Commission pour l'étude des versants, Union Geogr. Intern. Rio de Janiro*, 136–41.

Rougerie, G. (1960) *'Le façonnement actuel des modèles en Côte d'Ivoire forestière'*, Mem. Inst. Francaise Afrique Noire 58, Dakar.

Ruddock, E. C. (1967) 'Residual soils of the Kumasi district in Ghana', *Géotechnique*, **17**, 359–77.

Ruxton, B. P. (1958) 'Weathering and subsurface erosion in granite at the Piedmont Angle, Balos, Sudan', *Geol. Mag.*, **95**, 353–77.

Ruxton, B. P. (1967) 'Slope wash under mature primary rain forest in Northern Paupa', in J. N. Jennings and J. A. Mabbutt (eds), *Landform Studies from Australia and New Guinea*, Cambridge Univ. Press.

Ruxton, B. P. (1968) 'Rates of weathering of Quaternary volcanic ash in north-east Papua', *9th Intern. Cong. Soil. Sci. Trans.*, **4**, 367–76.

Ruxton, B. P. and Berry, L. (1961) 'Weathering profiles and geomorphic position on granite in two tropical regions', *Rev. Géomorph. Dyn.*, **12**, 16–31.

Selby, M. J. (1974) 'Rates of denudation', *N.Z.J. Geog.*, **56**, 1–13.

Sharpe, C. F. S. (1938) *Landslides and Related Phenomena*, Columbia Univ. Press, New York.

So, C. L. (1971) 'Mass movements associated with the rainstorm of June 1966 in Hong Kong', *Trans. Inst. Br. Geogr.*, **53**, 55–66.

Speight, J. G. (1965) 'Flow and channel characteristics of the Angabunga River, Papua', *J. Hydrol.*, **3**, 16–36.

Temple, P. H. and Rapp, A. (1972) 'Landslides in the Mgeta area, western Uluguru Mountains, Tanzania', *Geograf. Ann.*, **54A**, 154–93.

Thomas, M. F. (1974) *Tropical Geomorphology: a study of weathering and landform development in warm climates*, Macmillan, London.

Thorp, M. B. (1970) 'Landforms' in *Zaria and its Region*, M. J. Mortimore (ed.), Ahmadu Bello University, Zaria, Dept. of Geography, Occasional Paper **4**, 13–32.

Townshend, J. R. G. (1970) 'Geology, slope form and process and their relation to the occurrence of laterite in the Mato Grosso', *Geog. J.*, **136**, 392–9.

Tricart, J. (1972) *The Landforms of the Humid Tropics, Forests and Savannas*, Longman, London.

Twidale, C. R. (1976) *Analysis of Landforms*, J. Wiley, New York.

Williams, M. A. J. (1968) 'Termites and soil development near Brocks Creek, Northern Territory', *Aust. J. Sci.*, **31**, 133–4.

Williams, M. A. J. (1969) 'Rates of slope wash and soil creep in northern and southeastern Australia, a comparative study', Ph.D. thesis, Australian National Univ., Camberra.

Williams, M. A. J. (1973) 'The efficiency of creep and slopewash in tropical and temperate Australia', *Austr. Geog. Studies*, **11**, 62–78.

Weischmeier, W. H. and Smith, D. D. (1958) 'Rainfall energy and its relationship to soil loss', *Trans. Amer. Geophys. Un.*, **39**, 285–91.

Wentworth, C. K. (1943) 'Soil avalanches on Oahu, Hawaii', *Bull. Geol. Soc. Amer.* 54, 53–64.

Young, A. (1972) *Slopes*, Oliver & Boyd, Edinburgh.

5 Deposition

A disproportionate part of research efforts in geomorphology at the present time is concentrated on erosional processes and forms (see Chapter 4). Yet the study of deposition is equally of crucial importance, as it concerns, when extended to include topics such as the process of sedimentation, source of sediments, methods of transportation, the environments in which sediments are laid down, the changes taking place in such sediments from the time of their production to their ultimate consolidation, the structures developed in them, and the resultant landforms. Deposition is also important as it constitutes a common area of interest to geomorphologists, geologists, hydrologists, engineers and pedologists. For example, the work of Smyth and Montgomery (1962) in south-western Nigeria shows that the study of deposition is a *sine qua non* to a clearer understanding of the genesis of soils in the humid tropics.

In discussing deposition in the humid tropics, an important consideration is the scale at which to examine the problem, as it can be discussed either on a regional or/and on a local scale, as attempted by Thomas (1974). In this chapter emphasis is placed on both scales, but such regional depositional forms as the various aggradational surfaces recognized in tropical Africa and other parts of the humid tropics by King (1962) will not be discussed.

Depositional agents

All erosional agents – namely, running water, mass movement, waves, winds and moving ice – also constitute depositional agents. However, except on high mountains such as Mt Kilimanjaro, and Mt Kenya in East Africa, Cotopaxi, Ancohuma, and Illampu in the Andes, and others in Colombia and Venezuela, glaciation does not constitute an important geomorphological agent in the tropics; it will therefore not be discussed here. Similarly, wind can only be regarded as of minor geomorphological importance in the humid tropics as it is largely covered by vegetation. In this chapter, therefore, attention will be focused mainly on the depositional features resulting from running water and mass movement, with a brief comment in connection with wind deposition.

Types of deposition

The most conspicuous form of fluvial deposition in the humid tropics, and indeed under other climatic regimes, include deposits associated with floodplains and deltas. Floodplain deposits range from channel to channel margin and overbank deposits; the latter are usually associated with distinct landforms like natural levees and backswamp. Delta deposits are graded from minor reservoir and lacustrine sedimentation to the enormous volume of sediments associated with features like the Niger delta. Deposits like alluvial fans can form at any point along a river provided the conditions for their formation are satisfied; and where several such fans coalesce, they readily form piedmont alluvial plains.

In most cases in the tropics, deposits by the processes of mass movement merely involve the downslope transfer of material to form fans or terracettes. However, where mass movement processes like slides and landslips constitute important geomorphological processes, such fans may cover relatively large areas. Apart from fans, mass movement processes have also given rise to scree accumulation around highlands and erosional residuals all over the tropics. Stonelines and the overlying sediment found all over the tropics have also been ascribed to the action of mass movement processes, especially soil creep.

Wind action is not particularly prominent on coastal areas in the humid tropics, so that coastal dunes are relatively absent. Wherever found, such dunes are inconspicuous and confined to narrow strips on the coastal zone. Large coastal dune fields, such as those at the mouth of the Senegal river, are possibly relict. However, such relict dunes are now in the process of reactivation into barchans, traverse dunes and nebkas, as is the case in southern Madagascar.

Fluvial deposits and landforms

Water flows in channels and as unconfined runoff, and in both instances it transports and deposits sediments. In most cases, deposition occurs where the runoff discharges, when especially its competence and/or capacity, decrease abruptly. This decrease is controlled by several interacting factors relating to stream discharge itself, stream gradient, channel characteristics and flow pattern.

If the discharge of a stream decreases, or the stream is overloaded by sediments at any particular stretch of the channel, the stream will lose the competence and capacity to transport the sediments and so will deposit part of the sediments. Gradient imparts the force of gravity on moving water, thus increasing the ability of the water to erode and transport. If, for any reason, there is a change to a gentle gradient without an increase in discharge, a loss of energy occurs and subsequently deposition of sediments ensues. Wide, shallow channels also encourage deposition, as these increase frictional drag and resistance to flow at the channel floor leading to a decrease in the available energy to transport sediments. Rough channel beds and banks equally encourage deposition for the same reason. Also, where channel bank materials are weak and easily eroded, there is a tendency towards channel widening, overloading of discharge with sediments and subsequent deposition of part of this load. The decrease in flow velocity which occurs when a river enters a large body of water – for example, a lake – also leads to deposition.

Alluvial deposits

Most rivers in the tropics cover their bedrock floors with alluvium, which Bloom (1978) defines as unconsolidated sediment of relatively recent geologic age. The same author also makes a distinction between active and old alluvium.

Active alluvium

This is defined as the sediment capable of being transported by the stream flowing on it so that its thickness depends on the depth of flood-scour of the modern channel, and the maximum grain size which the stream can competently transport at a flood-stage. Any deeper sediment can be regarded as relict (or old) alluvium (Figure 5.1).

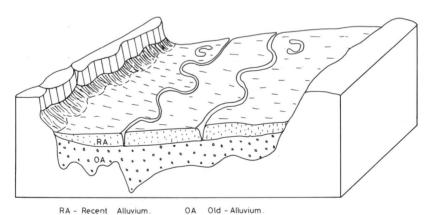

RA - Recent Alluvium.   OA   Old – Alluvium.

Fig. 5.1 Types of alluvium

Leopold *et al.* (1964) have shown that there is a decrease in alluvium grain size with distance from the upper parts of a stream, with such sorting resulting mainly from attrition and abrasion which reduce the grain size of bedload alluvium in a downstream direction.

Alluvial deposition can occur at any section of the longitudinal profile of a stream; however, it is more often associated with the floodplain, which is an erosional form resulting from lateral erosion and deposition by a meandering river. Erosion takes place at the concave curves of the meanders, whereas deposition occurs simultaneously on the convex side downstream. At concave bends, the valley slope is undercut, leading to slumping and other mass movement. As this occurs at all the meander bends, and as meanders naturally migrate downstream, the valley floor is continuously widened to form a floodplain.

Most rivers in the humid tropics have widened their valleys in this manner, at times right to the heads of the first order streams. These wide valley floors are invariably veneered with alluvium varying both in thickness and in composition. The alluvium along the smaller streams in south-western Nigeria, averaging about 1 m in thickness, has been classified into three types by Smyth and Montgomery (1962). These include the ill-drained, very clayey sand varying in colour from pale yellow to bluish grey underlying the backswamp; the coarse 'washed builders' sand' type; and those composed of fine, clean, washed sand found on the banks and beds of these rivers. All three types often contain gravel and occasional lenses of silt.

Floodplains along larger rivers, apart from being underlain by thicker alluvium, often contain the types of deposits outlined in Table 5.1.

Floodplain morphology

While the floodplain itself is formed by both erosional and depositional processes, most of the features on floodplains are formed principally through the depositional processes of lateral and vertical accretion. Examples of these features are point bars, levees, meander scroll and so on.

The most important features of lateral accretion are *point bars*, which are formed when currents in meandering channels undercut the concave stream bank, erode the material and deposit it in the form of a submerged bar on the convex side downstream. Point bars are thus in the form of low ridges separated by shallow swales, both of which together act as indicators of channel migration. Wolman and Leopold (1957, p. 96) are of the opinion that such lateral accretion may account for between 80 and 90 per cent of the floodplain deposition. Vertical accretion occurs mainly through flooding. When flooding occurs, flood water, highly charged with a very heavy concentration of suspended sediment load, overflows on to the floodplain, where the velocity declines as a result of both increase in flow width and increase in friction. As decline in velocity leads to a corresponding decline in the competence and capacity of the flood water, deposition occurs by the simple process of sediments settling out of the nearly stagnant water.

Levees are formed as a result of the sudden change in velocity at the edge of the flooded channel. Coarse sediments are deposited nearer the channel, while finer particles are deposited further away. Rapid vertical accretion of the levee can lead to a situation where a depression exists between it and the bluff of the floodplain. Such a depression, subject to regular innundation, is known as the *backswamp*. As the levees grow, there is a tendency for the river bed to rise, and it may rise several metres above the adjacent backswamp, a condition favouring severe flooding of the backswamps. Where the levee is breached during flooding, a *crevasse* is formed through which a *splay* (see Table 5.1) spreads from the channel to the backswamp.

TABLE 5.1   Genetic classification of floodplain sediments

Place of deposition	Name	Characteristics
Channel bed	Transistory channel deposits	Primarily bedload temporarily at rest; part may be preserved in more durable channel fills or lateral accretions
	Lag deposits	Segregations of larger or heavier particles more persistent than transistory channel deposits and including heavy mineral placers
	Channel fills	Accumulations in abandoned or aggrading channel segments ranging from relatively coarse bedload to fine-grained ox-bow lake deposits
Channel margin	Lateral accretion	Point and marginal bars that may be preserved by channel shifting and added to overbank floodplain by vertical accretion deposits at top.
Overbank floodplain	Vertical accretion deposits	Fine-graded sediment deposited from suspended load of overbank flood water; including natural levee and backland (backswamp) deposits
	Splays	Local accumulations of bedload materials spread from channels on to adjacent floodplains
Colluvium margin	Colluvium	Deposits derived from unconsolidated slope wash and soil creep on adjacent valley slides
	Mass movement deposits	Earthflow, debris avalanche and landslide deposits commonly intermixed with marginal colluvium; mudflows usually follow channels but also spill over bank

Source: Vanoni, 1971

In certain cases, the land between meander loops shrinks through active erosion of the concave sections of the channels, and the corresponding deposition of point bars on the convex channel bends downstream till very narrow necks of land separate the meander loops. Eventually a narrow neck may be breached when a lower meander curve intersects an upstream section of the channel to form a *cut off* (Figures 5.2a and b). The shortening of channel length locally increases the gradient and velocity at the cut-off point. The truncated ends of the

A – Alluvial fan B – Backland Bs – Backswamp C – Colluvium
F – Channel fill L – Lag deposit LA – Lateral accretion N – Natural levee
O – Oxbow lake P – Point bar S – Splay T – Transitory bar
VA – Vertical accretion

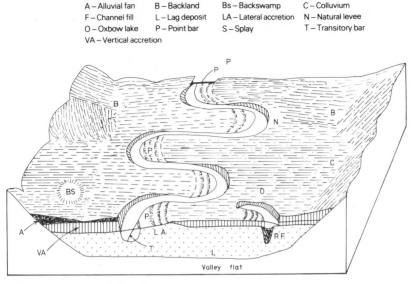

Fig. 5.2a The flood plain

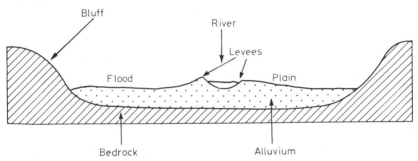

Fig. 5.2b Cross-section of a flood plain

abandoned channels are eventually sealed with bank deposits to give rise to an *ox-bow lake*. The lake may subsequently be filled up as a result of vertical accretion to form a meander trace of a *meander scroll*.

In streams heavily laden with coarse sediments and flowing on weak, easily eroded channel material, *braiding* may result. This starts with the selective deposition of the coarse material, leading to the formation of a central island which diverts flow towards the banks with the resultant increase in erosion on them. As the banks are eroded, channel depth decreases, and the concomitant increase in friction between the sediment-laden discharge and channel-bed leads to further deposition. The islands in the braided section, comprising largely of gravel with a diameter of 60–100 mm, and veneered by sand, grow mainly by the addition of sediment at their downstream end and on their lateral boundaries.

Finally, on a wide floodplain, tributaries cannot readily join the main river because of the depression formed between the levee and the valley bluff. Instead, such tributaries flow parallel to the main river for a considerable distance before they effect a link with it. This is referred to as a *deferred junction*.

Tropical rivers and floodplain morphology

Except for rivers developed entirely on sedimentary formations, most large rivers in tropical Africa, especially those descending from the ancient shield to the sea through series of rapids, have rather narrow floodplains with the typical floodplain morphology poorly developed. A typical example of a river developed entirely on sedimentary formations is the Imo river developed mainly on coastal plains sands in Eastern Nigeria. The lower course of this river in the neighbourhood of Port Harcourt occupies a floodplain about 3 km wide and displays such floodplain features as point bars, meanders, ox-bow lakes, channel cut offs, channel fillings and backswamps. The mean of the meander wavelength is 0.8 km, with a sinuosity index of 1.8 (Figure 5.3).

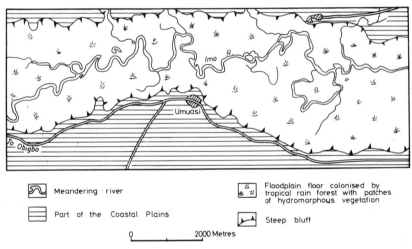

Fig. 5.3 Imo river flood plain

The lower Niger provides a better example with a distinct floodplain between Aboh and Samabri. The floodplain here is characterized by morphological features like meanders, abandoned meanders, back-swamps, meander traces, and levees among other features. The levees provide settlement and cultivation sites, as the floodplain is covered by raphia palms and are rather swampy.

Similarly, the Amazon and its tributaries like the Japura, Jurua, Ica, Negro, Branco, Purus, Madeira and Atacaxis, occupy very wide floodplains on which the rivers are braided and at the same time meandering. The Jurua is characterized by tight meander loops which are fringed by series of abandoned channels, old meander scars, swales, lakes and backswamps. As measured from Apollo–Soyuz imageries, the

Amazon in the Vicinity of Fonte Boa has a sinuosity index of 1.5 (Pinteado-Orellana *et al.*, 1978).

Old alluvium

According to NEDECO (1959), the channels of most of the large rivers in the tropics like the Niger, Benue, Nile, Olifants, Volta, Ankobra and so on are underlain by a very great thickness of alluvium, indicating the existence of buried channels and valleys along these rivers. The thickness of the alluvium underlying the Benue channel varies from 33 m near Dasin Hausa to 90 m near Makurdi, while along the Niger it is 45 m at Onitsha.

The alluvium was probably deposited as valley fills in dry periods during the Pleistocene–Holocene period, well-documented in tropical Africa (Grove and Warren, 1968; Shaw, 1976; and Selby, 1977). During the Würm glacial maximum, sea level all over the world was about 90 m below the present level (Fairbridge, 1961), during which period the larger tropical rivers excavated very deep channels and valleys which were filled with sediments during the succeeding Gamblian wet period, when the sea was 2–4 m above its present level (NEDECO, 1959).

The great thickness of alluvium associated with rivers such as the Amazon, Orinoco, Indus, Irrawaddy and the Brahmaputra, which take their sources from high mountain chains, relate to glaciation and tectonic instability in these mountain chains. Taking the Amazon as an example, during the Pleistocene glacial advances, the Andes mountains were covered by ice and valley glaciers which generated vast volumes of fluvio-glacial sediments deposited not only in the mountain valleys, where they now form terraces at different elevations, but also along the entire Amazon to a depth of thousands of metres. The period of glacial advances and regressions also witnessed several major climatic changes in the Amazon basin, from tropical dry to a forested humid climate. The dry periods coinciding with glacial advances were characterized by in-filling of the valleys to very great depths, with coarse gravel and sand, while the humid periods witnessed channel incision and partial removal of the alluvial in-filling (Garner, 1968; Pinteado *et al.*, 1978). Type areas of such alluviation include the Río Matantan valley, Sierra de Merida and Estado Barinas in Venezuela, where valley-fill thickness is estimated between 8 and 10 km, and along the Uramba river, Cordillera Oriental in south-eastern Peru, with alluvial fill of about 60 m.

Old valley fills have also been reported along small rivers in West Africa, especially along most of the rivers in the Ivory Coast, in Upper Gambia and Senegal, and along the Ogun river in Nigeria (Davis, 1964). The composition and stratigraphy of these deposits appear similar. Along the Ogun between Oyo and Iseyin, Burke and Durotoye (1971) show the deposit to consist of a bottom layer of coarse, rounded pebble and gravel about 1 m thick, overlaid by another layer composed of finer sub-angular gravel partially cemented by ferric oxide, which is in turn overlaid by wash layer. In the Ivory Coast, along the Faleme, according to Vogt (1959), the bottom layer consists of coarse, rounded gravel alternating with lenses and beds of fine gravel and sand; this layer is overlaid by another composed of rounded gravel forming lenticular beds 2–3 m thick, which is in turn overlaid by cuirassed clay sand 0.5–3.0 m thick. According to Tricart (1972), similar gravels are found in Malaya and Indonesia. Similar stratigraphic sequences were also observed under the swamps and floodplains in the Yengema–Koidu, and the Tongo Lease Areas in Sierra Leone by Thomas and Thorp (1980). These authors identified three basic strata in the exposed sections in the floodplains of Gbendewa, Tongo–Woa, Gbobora and Moinde rivers. These comprise the upper, black, organic sandy clay loam largely of colluvial and biogenic origin and subject to prolonged waterlogging and flooding in the wet season; beneath them is a sandy layer with a varying proportion of clay, which is in turn underlaid by subangular to well-rounded basal gravels of quartz, rock blocks, hard ironstone pisoliths, and heavy minerals such as corundum and diamond. The two lower strata, believed to have been deposited during periods of high floods in the past (wet periods), are respectively dated at 36 000–20 000 B.P. and 12 300–7 800 B.P.

Terraces

Terraces, *stricto sensu*, result from the erosion of floodplain materials by streams under a continually falling base level. However, they can be examined along with depositional features.

Topographically, a terrace consists of two parts: a *tread*, which is the flat surface representing the level of the former floodplain, and the *scarp*, or the steep slope connecting the tread to any surface standing lower in the valley (Figure 5.4). The tread surface is normally underlain by alluvium of variable thickness, belonging to an earlier period of valley fill events, which has been cut into during a subsequent period of fluvial erosion or down-cutting.

Howard *et al.* (1968) recognized two major types of terraces – erosional and depositional. In erosional terraces the tread is formed by lateral erosion; in depositional terraces, the treads represent the

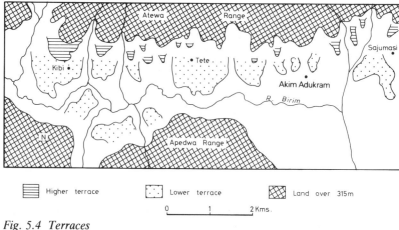

Fig. 5.4 Terraces

uneroded surfaces of a valley fill. However, we are concerned here mainly with fluvial deposition and so with depositional terraces, which may be paired or unpaired. In the latter type, treads on opposite sides of the valley are on the same level, whereas in the unpaired types, treads on opposite sides of the valley are on different levels. Both may occur along a single valley, where they could be partially obscured by colluvial material.

The development of a depositional terrace requires, among other things, a period of valley filling followed by stream entrenchment into or adjacent to the fill. Valley filling occurs when, over an extended period, the amount of sediment produced in a basin exceeds what the river system can transport. This may be caused by a glacial outwash, climatic change or changes in base level, slope or load due to rising sea level, a rising local or regional base level. Similar factors – especially, tectonic and climatic changes – trigger off valley incision.

Two major areas are ideal for this cycle of valley-filling and valley-cutting events along river channels. These are the regions affected by glaciation, where outwash is the common fill material, the amount and size of the load transported and released by active glaciers by far outstripping the ability of the melt-water rivers. In the coastal regions, depositional terraces often result when alternating cutting and filling accompany sea-level fluctuations. This was particularly prominent during the Pleistocene, when glacial and deglacial episodes caused large-scale eustatic deviations of the ocean level. Outside of these two regions, terraces, where they occur, present problems. This is particularly the case in humid tropical regions where glacial effect is minimal and the nature and extent of quaternary climatic changes are not generally agreed. In fact several workers (cf. Burke, 1967; personal communication) deny the existence of terraces in south-western Nigeria in particular. According to Burke and his disciples, the 'pediment' surface extends from the mountain (ridge) base to the incized channel bank (Burke and Durotoye, 1971).

Although Burke's position appears to have been strongly influenced by his conviction about the universality of the pedimentation and pediplanation process, the lack of convincing examples of terraces along major rivers in the humid tropical regions seems to lend support to his position. If anything, paired terraces, whether erosional or depositional, are a rarity in the areas known to the authors, although terrace patches or stretches occur along some river valleys, especially in the vicinity of large settlements, where they may represent evidence of accelerated erosion and deposition. The gullies around Zaria are cut in alluvium (Ologe, 1972) and so may represent the initial stages of terrace development in that region. Patches of terraces have also been reported along some rivers in West Africa, especially in the Ivory Coast, Upper Gambia, Ghana, and have been observed in Southern Nigeria. Buckle (1978) reports on two major terraces along the Upper Birim river in Ghana. The lower terrace between Kibi and Sajumasi is at 270–300 m above sea level, while the higher terrace is at 300–315 m. The lower terrace is covered by coarse, well-rounded gravel deposits. The river is now incized into the valley floor to a depth of between 2 m and 20 m within a distance of 10 km near Kibi. Similar terraces can be identified

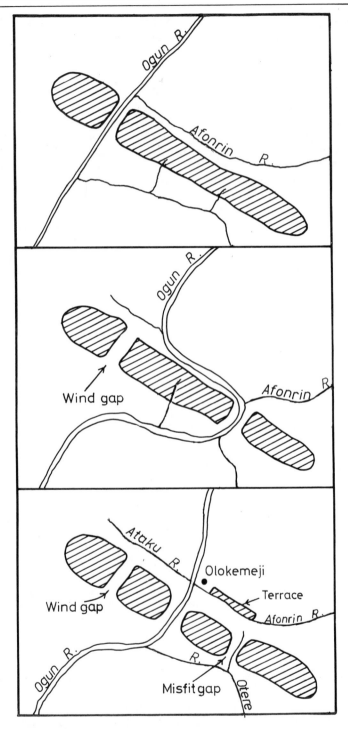

Fig. 5.5 Terrace along Ogun river (after Pugh)

downstream near Oda and Kade. Other examples of terraces in Ghana are found along Oti river near Demon and Saboba north of Ghana at 14 m, 20 m, 27 m and 33 m above the river bed.

The terraces along the Niger are found at 30 m, 45 m and 90 m above high water level near Onitsha, and at 95–105 m between Jebba and Awuru. NEDECO (1959) ascribed the formation of the three terraces to gradual down-cutting during a period of gradual tectonic uplift in the Pleistocene following the periods of valley in-filling already described.

The terrace along the Faleme in the Ivory Coast, described by Vogt (1959), is underlain by coarse, rounded gravel alternating with lenses of fine gravel and sand, in turn overlaid successively by a 2–3 m bed of rounded gravel and clayey sand 0.5–3.0 m thick. The terrace along the Ogun was studied between Oyo and Iseyin by Burke and Durotoye (1971) who found it underlaid by a layer of coarse rounded pebble and gravel up to 1 m thick. The terrace along a tributary of the same river at Olokemeji observed by one of the authors is exposed along a railway cutting for a distance of about 0.5 km, where it could be seen to be composed of rounded quartz gravel about 90 cm thick overlaid by a layer of fine sub-angular gravel. The terrace was formed in association with the capture of the Ogun river at Olokemeji. Before its formation, the Ogun river flowed in a loop around the now dissected Olokemeji ridge, depositing thick, gravelly alluvial layers on the northern side of the ridge. With the successive capture of the river by its south-flowing tributaries, which cut across the ridge, the old course of the Ogun was occupied by a smaller tributary now incized to a depth of about 10 m to expose the terrace material (Figure 5.5).

Lacustrine sedimentation

Gradual sedimentation of small bodies of water occurs wherever the surface is exposed to the depositional action of running water, and this is a universal phenomenon. However, because of certain physical and cultural factors, it is more emphasized in the tropics. Accelerated soil erosion in most places in the tropics is probably a recent phenomenon, arising largely from increasing population and the attendant increasing pressure on available land. Associated sheet and gully erosion are often readily discerned, but the insidious effects of deposition on arable lands and in adjacent water courses go unnoticed until serious damage has been done – for example, silting of dams and the resultant problems of water supply.

Accelerated sedimentation is, however, to be expected wherever soil erosion is taking place, especially in tropical climates characterized by long, dry seasons and sparse vegetal cover, which make the land vulnerable to severe erosion by rain splash, slope wash, rills and gullies. It is also important where heavy rains frequently give rise to landslides which remove large volumes of sediments from hill-slopes into adjacent river channels. Tropical cities and settlements with their characteristically unpaved and/or ill-maintained roads and drainage channels also constitute unfailing sources of large volumes of sediments. In all these places, accelerated sedimentation of reservoirs and natural lakes constitutes one of the most serious problems of water supply management, as the expected life of reservoirs is relatively very short.

Except for the pioneering efforts of Rapp (1975), Rapp *et al.* (1972) and Rust (1972), studies of sedimentation of water bodies in the tropics are rare. Most studies of this nature focus largely on eutrophication and analysis of the physical and chemical properties of the deposited sediments rather than on the rate of sedimentation.

In Nigeria, the Kainji lake on the Niger recedes regularly between February and August, and so offers a unique opportunity for the study of sedimentation rates. However, apart from Halstead's (1975) observations on the pattern of sedimentation, and the analyses of a few samples by Adegoke and Kogbe (1975), very little else has been done.

Most of our knowledge on accelerated sedimentation derives from the DUSER* research project in Tanzania in an area characterized by low rainfall totalling less than 700 m per annum, with the dry season lasting between 7 and 8 months, and where the natural vegetation comprises bush species, especially acacias, and short-lived ground flora, grasses and legume weeds.

The researchers found that in all the reservoirs studied deposition was either in the form of a delta located where the stream entered the water body, or in the form of uniform bottom sediment or both, with the latter occurring after the delta. In small reservoirs, the bottom sediments settling from suspended load transported during storm flow were found to be in form of varves grading from fine gravel to sand and silt and dark clay at the top, and composed of 2–3 per cent fine sand, 4 per cent medium sand, 3 per cent coarse sand, and 90–98 per cent clay. Adegoke and Kogbe (1975), analyzing such sediments derived from Gafara flat in Kainji Reservoir, show them to comprise 90 per cent silt and clay with little sand.

TABLE 5.2   Rates of erosion in some river basins and sedimentation in the associated reservoirs in Tanzania

River basin	Area of basin above reservoir (km²)	Relief ratio of the basin	Annual sediment yield in the basin (m³/km³)	Mean annual rate of reservoir sedimenta-tion (m³/yr)	Study period	Expected total life of reservoir
Ikowa	640.0	0.015	195	124,500	1957–69	30 years
Matumbulu	18.1	0.060	729	13,200	1922–71	30 years
Msalatu	8.7	0.045	406	3,260	1944–71	110 years
Imagi	1.5	0.076	601	927	1930–71	190 years
Kisongo	9.3	0.040	481	4,482	1960–71	25 years

Source: Modified from Rapp, 1975, and Rapp *et al.*, 1972

*Dar-es-Salaam/Uppsala Universities Soil Erosion Research Project, carried out in Tanzania between 1968 and 1972.

Rapp *et al.* (1972) also reported a large volume of sediments deposited upstream of the reservoirs in the form of sand fans in the channels, and silty-clayey layers on the floodplain. They related the rate of sedimentation to several parameters, the most important of which are basin relief ratio, population concentration, and intensity of agricultural activities in the basins, intensity of erosion and the size of the basins. Table 5.2 shows the rate of sedimentation in some of the reservoirs studied.

In the reservoirs studied, the sedimentation rate was inconstant, varying more or less with the rate of soil denudation. For instance in the Ikowa Reservoir, between 1957 and 1960 the rate was 6.1 per cent; between 1960 and 1963, 3.2 per cent; and between 1963 and 1969, 1.9 per cent of reservoir capacity.

Deltas

With or without lakes on the courses of rivers, the sediments transported by these rivers will eventually be deposited in the sea. Lacustrine deposition can thus be regarded as a temporary storage of sediments between the sources of derivation and the deposition in the sea, because with time, the lakes will be completely aggraded, and the rivers will cut channels through the aggraded stretches. With a negative base-level change, degradation along the length of the rivers will lead to the removal of most of the temporarily stored sediments. Thus most rivers enter the sea through estuaries or through alluvial deltas.

Deltas are often triangular in appearance, resembling the Greek letter Δ. The alluvial deposits usually cause the river to form series of meandering, anastomozing or even braided distributaries, with the main branches flanked by levees. The distributaries may be many, as in the case of the Niger, or few, as in the case of the Mississippi. The entire feature, which is often swampy and in the tropics covered by *Rhizophora racemosa*, is subject to tidal inundation. Most have arc-like outer edges modified by fringing sand spits, ridges and lagoons.

Based on morphology, deltas could be classified into four types (Figures 5.6 a and b):
1 the arcuate types, which have the classical arc-like appearance and are characterized by several distributaries transporting coarse and fine sediments to the sea, with longshore currents building bars at the mouths of the distributaries (for example, the Niger delta in Nigeria and the Muneru delta in Eastern India);
2 the birds'-foot type, which are rather elongated, having two or three main distributaries through which most of the sediments reach the sea, so that the rate of delta progradation is rather fast (for example, the Mississippi);
3 the cuspate types, in which alluvial deposits are spread along a straight coastline on either side of the river mouth by powerful waves, as on the Tiber in Italy;
4 the estuarine type built by rivers depositing materials in a submerged

river mouth forming sandbanks and islands, in between which are several distributaries (for example, the Mekong estuary in Vietnam, the Congo estuary, and the Amazon estuary).

Deltas vary in size, ranging from a few square kilometres to the 40 000 km² of the Niger Delta. Many of them do appear smaller than they really are; this is true of those affected by post-glacial drowning and isostatic sinking. Deltas are formed where the rivers transport a very large volume of suspended sediments, and where sediment removal by longshore drift and currents is low; deltas also form where the flow velocity of rivers is low. This happens as the river enters the sea, aided by the deflocculation of fine sediments on mixing with salt water.

The pattern of delta deposition is rather complex. Most deltas comprise three sets of deposits: the bottom set, the fore set and the top set

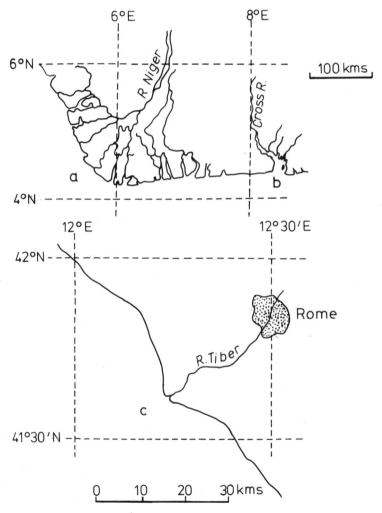

Fig. 5.6a Some common delta forms a: arcuate type, b: esturine type, c: cuspate type

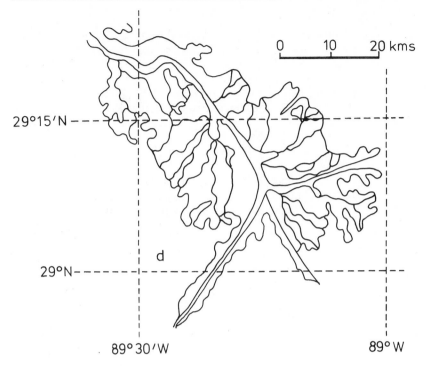

Fig. 5.6b Some common delta forms, d: digitate or bird's foot type

beds (Figure 5.7). The bottom set consists of fine sediments which mix with sea water, coagulate and settle over a wide arc surrounding the mouth of the river. Over these settle the coarser sediments subsequently brought by the river to form the fore-set deposits. These fore-set beds are inclined as they are deposited, one bed after the other. The top-set deposits are composed of fine material deposited horizontally on top of the fore-set beds near the surface, and constitute the seaward continuation of the floodplain. The Niger Delta, the largest in Africa and possibly in the humid tropics, is described in what follows to illustrate delta development in the tropics.

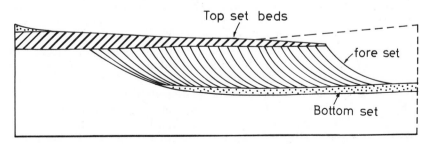

Fig. 5.7 Depositional sequence in the formation of a delta

The Niger Delta

The Niger Delta extends along the West African coast from the Benin river in the west to the Bonny river in the east (Figure 5.8). Inland, it ends at Aboh, above the difflucence of the Niger into the Nun and Forcados rivers. The region is flat and riddled with a very intricate system of anastomozing and reticulated channels, which eventually reach the sea through several distributaries, including the Benin, Escrovos, Fish town, Sengana, Nun, Bras, St Nicholas, St Barbara, St Bartholomeo, Sombreiro, Ke, New Calabar and Bonny.

The delta is divided into three morphological units by NEDECO (1959) as follows:

1 a sandy area seldom flooded along the branches of the Niger;
2 a mangrove-swamp area subject to regular tidal invasion;
3 a series of sandy beaches and ridges along the coast (Figure 5.8).

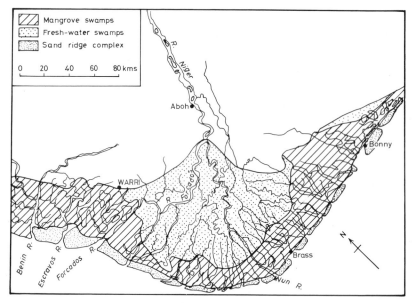

Fig. 5.8 Morphology of the Niger Delta

The sandy area extends for 120 km downstream from Aboh, and constitutes a southward extension of the Niger floodplain. The slope of the channel is very low so that the river meanders freely until a few kilometres south of Samabri, where it bifurcates into the Forcados and the Nun, both of which bifurcate again before entering the tidal flat zone. The whole area is subject to annual flooding, and so is marked by extensive levees developed along the distributaries. In fact, these levees provide both settlement and cultivation sites. A large area is under flood water for a part of the year, which areas support raphia palm.

The tidal flat is colonized entirely by *Rhizophora racemosa* and covers about 10 240 km². It varies in width from 15 km between the Middleton and Fishtown rivers to about 50 km between the Forcados and

Warri rivers. This muddy, swampy terrain is criss-crossed by innumerable, interconnected creeks, through which the flat is daily invaded by tidal water which was estimated at $50 \times 10^6 m^3$ per cycle (Adegoke, 1977). The area is uninhabitable except for the new 'islands' of terra firma on which settlements have been established.

The sandy beach ridges lying between the swamp and the open sea, trending mainly NW–SE, and E–W, vary in width from a few metres, between the Middleton and Fishtown rivers, to 15 km between the Forcados and Dodo rivers. The beach ridges were built by littoral drift generated by waves moving either eastwards or northwestwards with sediments derived from the Niger distributaries.

The build-up of the Niger Delta probably dates to the Cretaceous period; however, since the Pleistocene, the delta has developed in response to constant falling and rising sea levels. Not much is known of the pre-Pleistocene development of the delta, but its pattern of development within the last 75 000 years has been studied in some detail (NEDECO, 1959). Most of what follows is based on this study.

The total volume of the delta area between Onitsha and the present 50-fathom contour line has been built in the last 75 000 years. The underlying sediment, estimated at about 3×10^{12} m^3, is believed to have been deposited at different rates depending on the prevalent climate in the Niger basin especially during the Pleistocene–Holocene world-wide climatic fluctuations. Table 5.3 shows the rate of sedimentation at different periods.

TABLE 5.3   Rate of sedimentation in the Niger Delta since 75 000 B.C.

Period B.C.	Total sediment load in m^3	Average annual sediment load in m^3
75 000–55 000	1.2×10^{12}	6.0×10^7
55 000–30 000	1.0×10^{12}	4.0×10^7
30 000–15 000	0.3×10^{12}	2.0×10^7
15 000–10 000	0.2×10^{12}	4.0×10^7
10 000– 5 000	0.2×10^{12}	4.0×10^7
5 000–present	0.1×10^{12}	2.0×10^7

Source: Modified from NEDECO, 1959, p. 267

As is apparent from this table, the major part of the present delta was deposited in association with the rising sea level which accompanied the pluvial period of between 75 000 and 55 000 B.C., and in fact the rapid rate of deposition was continued during the succeeding drier phase. The rate of sediment deposition has been very low from about 30 000 B.C., and since about 7 000 B.C., in response to the falling sea level, the river has been more erosive than depositional and, having been incized into its channel, the incision has now reached a little upstream of Siama.

Sediment distribution on the coastal end of the delta is controlled by the pattern of longshore drift, the orientation of which in turn depends on the direction of the wind and swell in the Gulf of Guinea. The prevailing direction of wind and swell in this area is between SSW and SW, hitting the coast at right angles near the Nun and Sengana entrances, at each side of which littoral drifts are generated, one moving eastward and the other north-westwards. To maintain an equilibrium in the shape of the delta shoreline, it means that sediments must be supplied at the point at which littoral drifts are generated – that is, the bulk of the sediments must be supplied by the Nun and the Sengana. However, the shape of the shoreline has varied over geologically short periods owing to the preference given by the Niger to certain distributaries. At present, most of the sediments reach the sea by way of the outlets west of the Nun, and this may explain the planed-off stretch of the coastline east of the Nun river.

Although the bulk of the silt transported by the river is trapped in the mangrove swamp, most of the sand is deposited in an area roughly within a semi-circle with a radius of 75 km centred on Patani. However, the Ramos, Middleton, Fishtown, Sengana and Nun still transport a large volume of sand to the sea. This sand is formed into beach ridges by the littoral drifts.

Alluvial fans

These are fan-shaped masses of sediment with their apexes pointing upstream. Particularly conspicuous in arid and semi-arid environments, they are also found in the humid tropics, where they are often obscured by vegetal cover.

Alluvial fans vary in height from a few metres to more than 1 000 m, and in width from a few metres to more than 160 km. Surface slopes are normally convex owing to the partial sorting of the underlying sediments. Average slopes vary from about 3° to 15°, depending on the size of the constituent sediments; however, slope angles rarely exceed 34°, which is the angle of settling of cobbles and gravels.

Alluvial fans result from an abrupt loss of competence in a stream, especially where there is a sudden change of slope gradient along the stream course; the commonest situation is when a sediment-laden mountain stream debouches on to the lower angled foot-slope, where its velocity and energy are reduced, or where a tributary stream enters directly on to the floodplain of a main river. Alluvial fans may be found individually in inter-montane basins; however, where several of them are contiguous they may coalesce to form a piedmont slope. Whichever shape it takes, the typical fan consists of layers of stratified sediments which grade from the coarsest gravels. In ideal situations, alluvial fans grow to form cones. Examples of alluvial fans resulting from sudden change of gradient along rivers are those along the Kilombero valley in southern Tanzania, formed where tributaries cross the steep sided valley slopes into the broad valley floor especially between Utengule and Ifakira at the confluences of the Lumemo, Luri, Rondo and Rufiri with the Kilombero. Another example is the Lume fan formed by the Lume river on descending to the Semliki plain to the west of Ruwenzori Mountain.

Fans are largest and best developed where erosion takes place in a mountain region, and the river builds the fan into an adjacent basin with marked changes or differences in hydraulic properties. The bigger the catchment area of the mountain stream, the larger the fan. Consequently, among the largest alluvial fans are those formed by the Himalayan rivers as they enter the Indo-Gangetic plain. A study by Geddes (1960) identified the two major components of alluvial fans in the Himalayan examples, as follows: the fans proper and the intervening slopes of inter-fan areas. The following extract from Douglas (1977, p. 172) is based essentially on Geddes's (1960) work:

> The fans, triangular or rather segmental in plan and convex in form, have their apices at the gorge mouth from which their axes run to their bases towards the Deccan Plateau edge, or the sea. The inter-fan areas taper from the Himalayas and are slightly concave at the edge of their even slopes. Fans with gradients over 1 in 5,000, and the inter-fans, with somewhat less gradient, merge, where space allows, in a fan-forest plain with a gradient of 1 in 10,000 or less, which continues to the delta and sea.
>
> The plain of the north Bihar between the Himalayas and the Ganga consists of the three fans of the Gandah, the Kosi and, in part, the composite Mahananda–Tista fan, with the two inter-fan areas in between. These three fans exhibit a radial pattern of distributaries. Those of the Kosi have moved westward for the last 150 years, their floods and deposition . . . causing colossal damage, thus earning the river the epithet of 'Bihar's Sorrow'.

The situation described for the Himalaya region is of course true of several other regions with similar topographic and hydrologic features. For instance, in many mountain regions of Papua, large, coalescent alluvial fans, called alluvial aprons, are known to occur along the foot ridges of several rivers, including the Mambare river north-west of Kokoda (Blake and Paijmans, 1973). The fans, which are similar in both origin and morphology to those of the Himalayan region, and described as follows by Douglas (1977, p. 173), based on the Blake and Paijman (1973) study:

> . . . the debris shed by the high-relief areas is stored in the piedmont zone until it is reduced by weathering and erosion to finer sizes. Successive layers of gravel, sand, silt and clay deposited on the fans by the streams disgorging from the mountain ridge gradually extend the apex of the fan further downstream, replacing the original break of slope with a less marked break of slope further upstream and a gentler slope down to the major river which cuts the base of the fan.

Apart from these major examples, smaller examples abound in the humid tropical areas. Several of these minor examples are seasonal, as in the case of the several examples along the Ogun river in south-western Nigeria, which are built up during the period of low flow during the dry season, only to be washed and/or submerged during the ensuing wet season. Alluvial fans in humid tropical environments are also often

Plate 4 Gully head in Kalahari sands, Shaba, Zaire (M. F. Thomas)

Plate 5 Gully erosion emanating from engineering works on Jos plateau, Nigeria (M. F. Thomas)

Plate 6 Rilling and incipient gully erosion on a hill-side carrying a degraded pasture near Jos, Nigeria (M. F. Thomas)

Plate 7 Back wearing of gully heads into the hill near Jos, Nigeria (M. F. Thomas)

obscured by vegetation which quickly colonizes them and so renders them difficult to identify. Consequently, the best examples have been studied from dry environments – namely, in the California–Nevada region of the USA.

Several fans in the humid tropics are also associated with gullies or other badland topography. Both in the rainforest and alternatively wet and dry tropical areas, recent anthropogenetic disturbance of the natural vegetal cover has led to severe gully development. Examples are the gullies of the Udi plateau, and the dip slope of Enugu–Nsukka cuesta in Eastern Nigeria. However, the most spectacular types are the ravine-line gullies described as *lavaka* or *vocoroka* by Tricart (1972), examples of which are the ravines cut into deeply weathered kaolinitic clay on hillsides of Malagasy, Brazil and Hong Kong. Wherever they occur, they terminate abruptly on the foot-slopes with a narrow opening, through which huge fans comprised of micaceous and argillaceous sands and angular rubble of incoherent quartzitic rocks are deposited. Tricart calculated that in an area of 60 000 km² in the Betsiboka basin in Malagasy the mean annual deposit by lavakas was about 15 106 m³. In some places, these deposits from barriers on the courses of streams, which get diverted. In other cases, the alluvial fans are eroded and deposited in the channels of the streams to cause braiding.

Piedmont alluvial plains

These are large, regional landforms that have developed by the lateral merging of adjacent low-gradient alluvail fans at the base of dissected mountains. Such mountains are subject to epeirogenic uplift and severe fluvial erosion. The most important examples of such plains in the humid tropics include the Indo-Gangetic plain on the southern piedmont of the Himalaya mountains. Other examples occur on the eastern slopes of the Andean Cordillera, south-east of Guayaquil in Ecuador (Garner, 1968); on the southern coast of New Guinea, derived from the southern flanks of the mountain chains traversing the island (Thomas, 1974, p. 127); and on the western side of the Jos plateau, where the head-streams of the Kaduna river descend from the plateau surface. Climatic changes during the Quaternary period are known to have accelerated the formation of piedmont alluvail plains through severe glaciation of the mountains and the generation of fluvio-glacial sediments deposited on the foot-slopes.

Deposition by mass movements

The most obvious form of mass movements in the humid tropics are those related to soil creep, debris creep, slides, slumps, debris avalanche and mud or debris flow. Slides are common in the Brazilian Serro do Mer district, especially in Val Paraiba basin, as well as in Hong Kong, New Guinea, Java, and parts of tropical Africa – for example, on the steep fault scarps in Tanzania (Temple and Rapp, 1972). In most cases deposits by slides involve only minor transfer of material downslope, where such material may form fans or terracettes which are short-lived, or may be removed by wash. The fan may be relatively small and flat, as is the case on a foot-slope near Ife where a fan from debris flow covered only 300 m^2 (Jeje, 1979). Generally, individual fans are lobate in form (Shrodder, 1976). The one measured by Temple and Rapp (1972) in the Mgeta area in Tanzania has a relatively convex, long profile, with slope angle at midsection inclined at 16°, increasing to 22.5° at the distal end. The fan was composed of silty, clayey regolith with occasional boulders up to 2 m across. Fans associated with individual slides occur singly, but where avalanching is triggered by earthquakes or heavy rainfall, slide fans may coalesce, as on the Bewani and Torriceli mountains (Simonett, 1967), or on the Adelbert range (Pain and Bowler, 1973), both in New Guinea. Where several landslides and deep-seated slips occur, as in Hong Kong, giant mud and debris fans may block the entire valley system (So, 1967). In some watersheds, as on the Adelbert range in New Guinea, up to 60 per cent of the surface may be removed. The coalescence of alluvial fans resulting from more than 1 000 landslide scars has been observed on the Uluguru mountains in the Mgeta area of Tanzania (Temple and Rapp, 1972). Here the deposited mass spread fan-like over flat foot-slopes and also form an even-crested depositional surface of fine-grained material mixed with rock blocks. Within an area of 20 km^2 these slides produced 13 500 m^3 of sediment per square kilometre; while in the Nyika plateau in Malawi, Shrodder (1976) calculated that a single earthflow at Demso resulted in the deposition of 190 000 m^3 of sediments, while another at North Rumpi deposited about 525 000 m^3.

Apart from fans, mass movement, especially slides, rock creep and rock fall have given rise to scree accumulation around rock outcrops. At the foot of highlands and hills almost anywhere in the tropics are accumulations of scree composed of angular to sub-angular broken fragments of the local rocks and detrital duricrust. Where hills and ridges are separated by narrow valleys, such debris may entirely fill the valleys. However, where the hills are isolated, screes form concave slopes, with the size and frequency of scree material decreasing rapidly away from the hillside slopes. Scree slope angle varies with the angle of repose, or, as Young (1972, p. 130) prefers, the angle of settling of the constituent material; in general, the coarser the rock debris, the steeper the scree slope angle.

In humid tropical environments, scree slopes are known to range from 12° to 80° (Jeje, 1970; Young, 1972, p. 165). Unusually steep slopes of about 70° to 80° are, however, short-lived, as they are subject to reduction to flatter declivities by subsequent slides. Coarse duricrust rubble can also subtend relatively steep debris slopes, and can even completely obscure the free face on the massive laterite, as observed by Pallister (1956) in Uganda. Values of maximum slope angles on such slopes on a scarp developed on Cretaceous sandstone in the Abeokuta area of south-western Nigeria range from 8° 30' to 26° (Jeje, 1972). Except in a few cases, these debris accumulations are rather thin, possibly not exceeding 15 m at the thickest (Berry and Ruxton, 1959).

Hill-slope deposits

The literature on slope deposits in the tropics show that most slopes have experienced a succession of erosion and deposition resulting from periods of geomorphological stability and instability, related to climatic changes since the late Tertiary, but most pronounced since the Pleistocene (Fölster, 1969; Rhodenburg, 1969; Burke and Durotoye, 1971).

Although not much is known of the palaeo-environments of the present rainforest area, a better, if still poorly resolved picture of the palaeo-environments of areas at present covered by the savannas is now fairly well established. Various categories of morpho-geological evidence have been used, among which are the dating of a stratigraphic succession of alluvial deposits and deep-sea sediment, analysis of lake level fluctuations, and pollen analysis, to establish the broad pattern of climatic change in tropical Africa since the late Pleistocene, as shown in the following table.

As observed by several authors, the stable periods (that is, wet or humid phases) were marked by deep chemical weathering and the development of soil profiles, whereas the unstable (that is, dry phases)

TABLE 5.4   The broad features of climatic fluctuations since the late Pleistocene in tropical Africa

Climatic event	Period	Author
Wet phase	36 000–20 000 B.P.	Thomas and Thorp (1980)
	30 000–20 000 B.P.	Street and Grove (1979)
	c.31 000 B.P.	Burke and Durotoye (1971)
	c.20 000 B.P.	Shaw (1976)
Arid phase	20 000–13 000 B.P.	Kolla et al. (1979)
	22 000–13 000 B.P.	Pastouret et al. (1978)
	20 000–18 000 B.P.	Burke et al. (1971)
Wet phase	12 500– 7 800 B.P.	Thomas and Thorp (1980)
	13 000– 8 000 B.P.	Talbot and Delibrias (1980)
	13 500–10 000 B.P.	Pastouret et al. (1978)
	12 000– 8 000 B.P.	Fredoux (1978)
Dry phase	7 000–3 300 B.P.	Thomas and Thorp (1980)
	5 000–3 450 B.P.	Shaw (1976)
	7 000–5 000 B.P.	Burke and Durotoye (1971)
	4 500–4 000 B.P.	Talbot and Delibrias (1980)
Wet phase	3 750–1 000 B.P.	Talbot and Delibrias (1980)
	c. 3 300–c. 1 750 B.P.	Thomas and Thorp (1980)
	c. 3 250 B.P.	Shaw (1976)
	3 500–780 B.P.	Brooks (1948)
		Ollier (1972)
		Shaw (1976)
Dry phase	780–530 B.P.	Brooks (1948)
		Shaw (1976)
	430–130 B.P.	Gribben and Lamb (1978)
	1 000 B.P.–present	Talbot and Delibrias (1980)

were marked by devegetation, erosion and accumulation of sediment on slopes. However, Thomas and Thorp (1980) cautioned that gravel deposition on slopes or on valley floors could have occurred during the wet phases, possibly in conjunction with an incomplete vegetal cover.

The repetition of slope degradation and aggradation has led to the accumulation of several deposited strata on hill-slopes and on lower, flat surfaces. Butler (1959) recognized six such strata in south-eastern Australia, while the stratigraphy of an exposed soil at the head of a bog-burst at Chelinda in Malawi reveals about the same number (Shrodder, 1976); de Villiers (1965) recognized 5 distinct strata in Natal. About the same number has been recognized in south-western Nigeria by Fölster (1969). Bruckner (1955), however, recognized only three distinct strata in the lateritic deposits of south-eastern Ghana, more or less the same number recognized in parts of south-western Nigeria by Burke and Durotoye (1971). Garner (1968) also shows that floodplains in the Cordillera Oriental in Peru are often underlaid by several sequences of alluvial deposits, much the same as those in Yengema and Tongo districts in Sierra Leone, where Thomas and Thorp (1980) observed three distinct alluvial strate in the floodplains of the Gbendewa, Tongo-Woa and Moiade rivers and the Gbobora in the Yengema area.

Thus, as observed by Thomas (1974), although most slopes in the humid tropics over a short time span may be in equilibrium with prevailing environmental conditions, over a geological time scale, these supposedly stable surfaces are more or less in a state of active and constant adjustment to changing environmental conditions.

The most important sediments on hill-slopes in the humid tropics include stone and gravel layers, and hillwash sediment which most authors attribute to the combined effects of repeated past degradation and aggradation of slopes, especially since the late Tertiary period.

Stone lines

Stone lines are soil horizons with a greater proportion of stones than the material above and below them. They occur at various depths, from the surface in parts of East Africa (Ollier, 1959), to between 25 cm and 120 cm in central Western Nigeria (Smyth and Montgomery, 1962), and about 6 m in Zaire (Ruhe, 1956). They vary in thickness from about 10 cm to more than 50 cm, and may be continuous or discontinuous. Stone lines are usually composed of varying proportions of angular to sub-rounded quartz fragments, duricrust blocks and concretions, and the whole mass may or may not be cemented together by ferric oxides. The composition varies with the nature of the local rocks. On acidic granitic rocks, detrital duricrusts and concretions comprise less than 10 per cent of the stone line, with quartz fragments making up the rest; whereas on sedimentary formations like shale and ferruginous sandstone, and on basic rocks lacking in quartz veins, detrital duricrust and concretions may comprise 100 per cent of the stone line.

Stone lines are found in both temperate and tropical areas, but

Plate 8 Stone line in southern Guinea savanna of Nigeria (M. F. Thomas)

according to Young (1976), they are more ubiquitous particularly in the tropics. They have been recognized all over tropical Africa (Fölster, 1969; Bruckner, 1955; Nye, 1955; Rhodenburg, 1969; Burke and Durotoye, 1971; Moss, 1965; Ollier, 1959; Ruhe, 1956; Radwanski and Ollier, 1959; Thomas, 1974; de Villiers, 1965), in tropical Australia (Williams, 1969; Mulcahy, 1960) in Hong Kong (Berry and Ruxton, 1959), and in Venezuela (Garner, 1968).

Whenever observed, they mark lithological discontinuity within the solum, forming a lag concentration at the base of the migratory soil layer or under a colluvial mantle. In Venezuela, and on the plateaux of Guyana and Brazil, especially in Minas Gerais, stone lines occur mainly as a single horizon, while in West Africa two or three layers have been recognized by Fölster (1969), Bruckner (1955), and Burke and Durotoye (1971). So far, there is no consensus of opinion about their origin. Whereas some authors ascribe the deposition to current geomorphological and pedological processes, especially mass movement, others believe them to be relict. Several theories have been advanced to explain their origin:
1 There is the steady-state concept which seeks to show that stone lines are being formed by contemporary geomorphological and pedological processes (Young, 1976). The most important of such processes is soil creep. Nye (1955) and Smyth and Montgomery (1962, pp. 41–78)

indicate that stone lines mark the lower limit of migratory soil layer. This is especially where quartz veins are abundant in the local rocks (Figure 5.9). In such places, as chemical weathering, soil creep and slope development progress, rock fragments from quartz veins are incorporated into the creep layer (Berry and Ruxton, 1959) and redistributed as a horizon at the top of the sedentary saprolite. Fölster

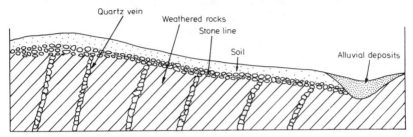

Fig. 5.9 A stone line

(1959) rejects this hypothesis, contending that creep cannot operate on the low-angled slopes in south-western Nigeria and southern Sudan where he made his observations. According to him, slope angles in such places are too low to permit the operation of creep except if there is a complete saturation of the soil by water. However, Nye (1955) has convincingly demonstrated the efficacy of soil creep in transporting the top soil (creep layer) in the Ibadan area, which is part of the area studied by Fölster in south-western Nigeria. Young (1976) contends that if this hypothesis is to be believed, then stones and gravels should eventually be evenly distributed throughout the upper part of the soil; and this seems to be the case in about 60 per cent of the soil pits dug on the basement complex rocks of south-western Nigeria by Smyth and Montgomery (1962). Soils of the Iwo, Ondo and Egbeda associations, which are the commonest in the area, have exceedingly conspicuous stony layers extending from the surface to a depth of about 1 m or more.

2 The residual weathering theory seeks to explain stone lines and gravel layers as residual products. It has been postulated that as chemical weathering progresses downwards into the rocks, rock material could be lost in solution either through vertical or horizontal groundwater drainage, whilst minerals like quartz, zircon, rutile and so on, resistant to weathering, accumulate to become embedded in mobilized iron (Young, 1976). With the exposure of the ground surface, the mobilized iron becomes oxidized and indurated. Evidence of ground loss and measurable vertical displacement has shown that this is a possible contributory mechanism of explaining stone lines on level as well as on sloping terrains under the rainforest, where erosion is mainly through chemical and mechanical eluviation.

3 Faunal pedoturbation: this hypothesis derives from the selective removal of fine soil fractions by termites, worms and ants from the sub-surface to the surface, leaving coarse quartz fragments to form a distinct horizon. The concretions found in the stone line are also formed *in situ* by the deposition of iron oxide in patches of soil where the

concentration of iron was originally unusually high (Vine, 1949) or, as suggested by Nye (1955, pp. 57–58), in the larger pore spaces of the sub-soil where a higher partial pressure of oxygen can be expected to exist. In such places, increasing concentration of iron and hardening by oxidation produces the harder mottles of incipient concretions at higher levels in the soil. The final hardening and deposition of the outer lustrous 'skins' of oxide takes place when, due to erosion, the incipient concretions reach the more porous upper horizons of the soil. Thus the concretions are being formed at the present time. Apart from Nye, Ollier (1959) and Radwanski and Ollier (1959) among others, support this theory. In support of the theory, Nye shows that the maximum particle size in soil overlying stone lines are the same as those in termite mounds. Williams (1968), apart from confirming this, even calculated that the soil above the stone line in parts of the Northern Territory in Australia could have been wholly deposited by termites in 12 000–18 000 years, much the same rate as calculated by Nye for the Ibadan area. J. Alexandre (oral communication, 1975) believes that stone lines in the Shaba province of Zaire, even in areas covered by the Kalahari sands, derived largely from termite activities. De Villiers (1965) has also shown that some stone lines in Natal are disposed horizontally along a level coinciding with the lower sub-surface limit of termitaria of *Trinervitermes spp.*

Although these theories may be useful in explaining the origin of some of the observed stone lines, they are certainly inapplicable in explaining cases where stone lines are composed of sub-rounded quartz fragments and lateritic gravels, evidence of attrition in a fluid transportation medium. Nor can they explain stone lines found at depths exceeding 2 m.

The relict hypothesis, which states that stone lines and gravel layers are of allochtonous origin, is predicated on climatic changes in the present humid tropics between the Pleistocene and early Holocene. According to Fölster (1969), the main arguments in favour of this theory are as follows:
(a) the multiple stratification complex;
(b) the rather sharp boundaries between the different layers unusual for pedogenetic horizons;
(c) inclusion of distinctly allochtonous elements in the stone line;
(d) abrupt ending of petrographic features, especially quartz veins at the stone line;
(e) occurrence of prehistoric implements of successive industries in the different layers of the complex.

The relict hypothesis has two variants:
(a) It is believed that during unstable phases of landscape development (arid periods), under the influence of rilling, gullying and mass movements, low scarps retreated in deeply weathered rocks on valley and hill-slopes so that fresh pediments were cut across older ones. Coarse debris, too heavy for water transportation, were deposited at the foot of each retreating mini-scarp, while the finer sediments were washed further downslope. As each mini-scarp retreated, more and more gravels were deposited, and these coalesced to form a sheet on the newly truncated saprolite. During the waning phases of the arid periods, finer sediments were deposited to cover the gravel layer, and

138

during the subsequent stable periods (wet periods), pedogenesis, especially cementation of the stone and gravel layer by ferric oxide, predominated. This is the theory to which Fölster (1969), Fölster *et al.*, (1971), and Rhodenburg (1969), among others, subscribe.

(b) The stone layer is supposed to have been deposited on bare rock pediments during the dry period. Deposition occurred through differential sorting by sheet flow under a desertic hydrological regime (Bruckner, 1955; Ruhe, 1956; Burke and Durotoye, 1971). The stones and gravels are supposedly derived from the physical disintegration of bare rock outcrops or older lateritic crusts located upslope of the new pediments. The presence of stone lines on valley slopes is attributed to latter dissection of pediments and deposition of gravel on valley slides (Ruhe, 1956; Burke and Durotoye, 1971).

Proponents of this theory identified some artifacts in the stone lines which have been variously dated to between 2 000 and 10 000 years B.P. Although the sub-rounded to rounded nature of the constituent stones and gravels in some of the stone and gravel layers lend credence to this hypothesis, the theory, however, poses other problems, the most important of which is the explanation of the great depth of weathering under the desert lag gravel of Bruckner, Ruhe, and Burke and Durotoye. The very occurrence of artifacts like pot sherds, and Neolithic stone implements in the stone layers is probably an indication that the palaeo-environment when the stone lines were deposited was not necessarily arid. Also as observed by Thomas (1974, p. 279), if these stone lines and gravel layers were deposited on top of bare rock pediments, under desertic climatic regimes during the later phases of the Quaternary, it is difficult, if not impossible, to account for the deep weathering often in excess of 30 m that has since taken place beneath the stone lines. Even on pediment heads around bare rock inselbergs, depth of weathering in excess of 5 m has been ascertained (Jeje, 1970). As it is well known that chemical weathering proceeds very slowly in crystalline basement rocks, the observed great depths of weathering under the present rainforest in south-western Nigeria imply a fairly long period of undisturbed wet and warm climatic regime. In fact Douglas (1969) is of the opinion that most of the areas at present covered by tropical rainforest at no time experienced a cessation of chemical weathering processes since the Pleistocene, and at no time experienced climatic regimes different from what obtain now.

A more fundamental issue about the relict hypothesis concerns the operation of geomorphological processes in areas at present under semi-arid conditions. If, as observed by the proponents of this theory, that pedimentation across valley slopes and pediments occurred during earlier drier periods, it is only reasonable to expect such processes to be very active in areas at present under semi-arid climatic regimes. This is far from being the case. As observed by Rhodenburg (1969), actively strong slope pedimentation together with the formation of stone lines and gravel layers is now absent between the Gulf of Guinea and the Sahara, taking place only in the Air plateau where, even then, it is restricted to certain areas. If these processes are inactive under contemporary semi-arid to arid conditions, there is little ground to believe that they ever operated under past arid conditions.

However, Twidale (1972) has shown an incipient gravel layer actively developing on a pediment around a quartzitic inselberg in the semi-arid environment of central Australia. The inselberg is subject to physical disintegration, and the rocks broken into smaller fragments, which are transported downslope by powerful surface wash; which also removes the finer soil fractions, leading to the concentration of the coarser particles on the surface where they form a continuous layer. In the process of further downslope transportation, the rock fragments could become sub-rounded. Thus, although no active pedimentation is now going on in semi-arid environments, severe slope wash characterizes such areas, and this was probably the most important process responsible for the formation of some of the present stone lines during the past arid phases; just as it is now occurring at present in the semi-arid areas of Tanzania (Murray-Rust, 1972). Thus it appears that no single theory can be applicable in explaining the origin of all the observed stone-line-cum-gravel-layers in the humid tropics.

Hillwash sediments

Hillwash sediments, variously called pedi-sediments (Ruhe, 1956) or drifts (Falconer, 1911), are found on all slopes wherever stone lines or gravel layers occur, except where removed by erosion. The sediments have been observed by Ruhe (1956) in the Ituri area of Zaire to cover all surfaces, summits, slopes and bottom lands, varying in thickness from a few centimetres to about 6 m. These sediments have also been observed in Nigeria right from the coastal area to the northern part of the country. Moss (1965) observes that the sediments cover about 25 per cent of the sandstone areas of south-western Nigeria. They have also been observed to cover both the interfluvial and valley slope surfaces in the Yengema–Koidu areas of Sierra Leone (Thomas and Thorp, 1980).

In the Ituri area of Zaire, the sediments are bright red, with variations of reddish yellow, brownish yellow or brownish red; friable and granular with 30–40 per cent clay, less than 5 per cent silt, with a pH of 4.5–5.0 and SiO_2/R_2O_3 ratio of less than 1.5 (Ruhe, 1956). In the Abuja area of Nigeria, the sediments contain 60–80 per cent fine sand, with the rest clayey. In most places, the sediments occasionally contain larger fractions, especially detrital duricrust and quartz pebbles.

As in the case of the underlying stone lines, opinions differ as to the mode of deposition of these sediments. Whereas some authors ascribe the deposition to the combined action of soil fauna and mass movement, others believe them to result from surface wash. Nye (1955) and Williams (1968) among others favour the theory that the sediments are derived from the breakdown of worm casts and termitaria and the downslope transportation and deposition of these sediments. Nye (1955) calculates that in an area near Ibadan under dry rainforest, on the average, 20 tons/acre of worm casts are deposited annually, while in the same area mature mounds of *Microterme Nigeriensis Sjost* could be up to 1.6 m high, weighing 2.5 tons. He further calculated that in an area free from erosion, both worms and termites can deposit 30 cm of fine sediments in 12 000 years, and where termite mounds are more numerous, the rate of accumulation could be faster.

This theory will probably account for the presence of the sediments on flat, interfluvial surfaces.

Moss (1965) proposes a slightly different theory. He ascribes the formation on the sandstones in south-western Nigeria to the disruption of the breakaway retreat of lateritic scarp through spring sapping and the flushing out of fine sand, followed by physical disintegration, chemical attrition, lateral eluviation and gradual mass wasting, especially creep in the slumped lateritic block, which eventually become finely comminuted into non-mottled, fine, sandy clay covering the pediment downslope of the lateritic scarp.

Fölster (1969) and Rhodenburg (1969) among others consider the sediments as relict, ascribing the formation to deposition during the waning period of the unstable (arid) phases before conditions become more stable. Fölster, in particular, regards the hillwash material 'as a

*Plate 9 Large termitarium (*Macrotermes natalensis*) in wooded savanna, Shaba Province, Zaire (M. F. Thomas)*

Plate 10 A close-up of Microtermes termitarium, *Shaba, Zaire (M. F. Thomas)*

product of unconcentrated wash of material mobilized on the soil surface by direct impact of rain and rill action'. This probably occurred following the cessation of the transportation and deposition of the coarser gravel layer, when the surface was probably colonized by denser vegetal cover which permits rain-wash erosion on the upper slope surfaces. Ruhe (1956) makes no such distinction as Fölster, only ascribing its deposition to the sorting action of runoff on newly cut pediments. Close to the pediment scarp, coarser debris are deposited, while further downslope finer sediments are laid down, which invariably covered the coarser sediments laid down intially close to the retreating scarp.

However plausible, this theory cannot account for the presence of these sediments on interfluvial surfaces except if there has been relief inversion of the type envisaged by Fölster (1969) for parts of south-western Nigeria.

Others, like Burke and Durotoye (1971), ascribed the formation of the hillwash sediments to the joint action of termites, worm and wash.

Wind deposition

The most important aeolian deposits comprise sand fashioned into different types of minor features, like sand ripples and sand ridges, and major ones, especially different types of sand dunes, whalebacks, sand shadows and drifts, undulations and sand sheets. Although they are most evident in tropical deserts, some of these features, especially sand dunes, are found at the margin of certain coastal plains in the humid tropics where there is adequate supply of sand and unidirectional wind of sufficient velocity.

Where dunes have formed on beaches, the deposition is more or less the same as desert dunes. The studies of Bagnold (1941) were on desert sands, but his findings are equally relevant to coastal dunes. Apart from demonstrating that sand is transported through the processes of suspension, saltation and surface creep, he also shows that the threshold wind speed for sand transportation is 4.5 m/second, above which the rate of sand movement is proportional to the cube of the wind speed. Once the wind speed is below the threshold velocity of each textural class of sand, movement of sand stops. An important factor in the reduction of wind speed is surface roughness. Thus vegetated sand surfaces, being very rough, can reduce wind speed to below threshold value, leading to sand deposition and dune formation.

Sand grains are deposited in three main ways; through sedimentation, accretion and encroachment. Sedimentation involves sand grains in suspension. In this process, sand grains fall through low velocity winds and strike the surface with insufficient force either to be moved forward, or to propel other sand grains forward. In accretion, the mass of the saltating sand moving over a sandy surface is reduced if the surface texture changes to become relatively more coarse. The sand grains may strike the surface with sufficient force to propel other grains forward, and such sand grains move forward until they are deposited in hallows on the coarse textured surface. In encroachment, sand grains involved in surface creep are held up by various obstructions.

Where formed, coastal dunes in the shape of low ridges are often arranged in rows parallel to the coast, and only occasionally do they occur in form of barchans, parabolic dunes, transverse dunes, and small dome-like mounds called nebkas. According to Boorman (1977), if the sand supply is limited, especially on exposed sites, the seaward ridge will grow to a maximum height and then, by a process of local erosion and deposition, will move landwards, either as a parabolic dune or as a complete dune ridge. The growth and stabilization of dunes depend on the existence of obstructions to reduce wind speed and to increase encroachment and accretion. Vegetation, especially that which can tolerate halomorphous soils, and can grow at a relatively fast rate to match the rate of sand deposition, constitutes the best barriers.

Coastal dunes in the humid tropics

The humid tropics are characterized by wide expanses of coastal lowlands which in most places are composed of fluvio-marine sediments. Outcrops of hard, cliff-forming rocks are limited and isolated, as at Shebro Island, Sierra Leone, Monrovia in Liberia, Cape Three Points in Ghana, and Cabo Frio and Ilheos, both in Brazil.

In most places, the seaward margin of the coastal plains are marked by the deposition of a great abundance of sand and other clastic sediments, fashioned by longshore drift into series of sand bars and sandy beaches, behind which are lagoons, depressions and swamps, as along the coast of Nigeria. Generally, these sandy beaches are not associated with dunes, and where such dunes exist at all, they are poorly developed. Except around the mouth of the Senegal river, dunes are virtually absent along the coast of West and East Africa, South-east Asia, and tropical South America. Dunes are only found where winds are not only constantly on shore but are also strong enough to move a large supply of sand inland from an extensive beach, as along the south coast of Malagasy, Bermuda, the Bahamas, Cape Verde Islands, the western coast of India, Australia, Hawaii and Ecuador (Termier and Termier, 1963). In most of these places, these dunes are rather old, having been built in the dry periods between the Pleistocene and early Holocene.

Dunes are absent from the margins of the coastal plains for various reasons. An important prerequisite for dune formation is the occurrence of sand flats at a sufficiently high level for the surface layer to dry out between the tides. Unfortunately, the sand grains on the wide beaches in most of the humid tropics are within reach of high tidal waves, and as they do not dry out during the low tide, the sand particles are damp most of the time and cannot be moved by wind. In most places, beach slopes are steep and convex; such slopes impede wind deflation by forcing the wind to push sand grains upslope. Also, winds, though constant in direction in most places, are generally of low velocity and incapable of deflation. High wind speed during the occasional tropical cyclones is negated by the effects of the accompanying heavy rains. Finally, where sand accumulates behind the present beach, it is soon colonized by dense vegetation which impedes deflation and sand movement. Two examples of coastal dune fields, one old and one modern, are examined in what follows:

The ancient dunes of Senegal

This dune field extends inland for about 200 km on either side of the lower Senegal river, and is believed to have been deposited during the dry periods between 31 000 and 21 000 B.P. and 15 000 and 10 000 B.P. (Selby, 1977). Grove (1958) believes that the sand is fluvio-marine near the coast, but inland it is aeolian. The dunes in the form of low ridges are oriented NNE–SSW near the coast, but trend ENE–WSE inland.

According to Grove, the dunes, after being subjected to repeated flooding, weathering and gullying, are now colonized by grass.

The modern dunes of southern Malagasy

As in Senegal, most of the seaward margin of the coastal plains of southern Malagasy are covered by old dunes now fixed by grass. However, modern mobile parabolic and transverse dunes formed from the partial reactivation of the old dunes are found on either side of Mandrare river, where they have developed a barrier beach, behind which there is a lagoon (Buckle, 1978). These dunes built by dominant SE winds are best developed in two large fields (Figure 5.10). South of Ambasy, in an area of about 10 km², are series of parabolic dunes and nebkas, while the trapezoidal-like area about 15 km² south of Antanandara are covered mainly by transverse dunes, nebkas and barchans.

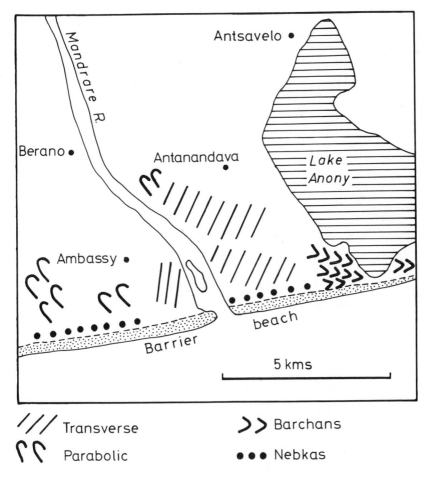

Fig. 5.10 Coastal dunes, Southern Malagasy (after Buckle)

References and further reading

Adegoke, O. S. and Kogbe, C. A. (1975) 'Littoral and suspended sediments of Lake Kainji', in *The Ecology of Lake Kainji, the Transition from River to Lake*, A. M. A. Imevbore and O. S. Adegoke (eds), University of Ife, Ile-Ife.

Adegoke, O. S. (1977) 'Prefeasibility study of a bulk ore terminal at Bonny', Department of Geology, University of Ife, Ile-Ife.

Bagnold, R. A. (1941) *The Physics of Blown Sand and Desert Dunes* (2nd edn. 1954), Chapman and Hall.

Berry, L. and Ruxton, B. P. (1959) 'Notes on weathering zones and soils on granite rocks in two tropical regions', *J. Soil Sci.*, **10**, 54–63.

Burke, K., Durotoye, B. and Whiteman, A. J. (1971) 'A dry phase south of the Sahara 20,000 years ago', *Archaeology*, **1**, 1–8.

Buckle, C. (1978) *Landforms in Africa: an introduction to geomorphology*, Longman, London.

Butler, B. E. (1959) 'Periodic phenomena in landscapes as a basis for soil studies', C.S.I.R.O. Austr. Soil Publicns., **14**, Canberra.

Cotton, C. A. (1948) *Landscape as Developed by the Processes of Normal Erosion*, Whitcombe and Tombs, Wellington, New Zealand.

Davis, O. (1964) *Quaternary in the Coast of Guinea*, Glasgow University Press.

Douglas, I. (1969) 'The efficiency of humid tropical denudation systems', *Trans. Inst. Br. Geog.*, **40**, 1–16.

Douglas, I. (1977) *Humid Landforms*, MIT Press, Cambridge, Mass.

Dury, G. H. (1963) *The Face of the Earth*, Penguin, Middlesex, England.

Fairbridge, R. W. (1961) 'Eustatic changes in sea level', *Physics and Chemistry of the Earth*, **4**, 99–185.

Falconer, J. D. (1911) *The Geology and Geography of Northern Nigeria*, Macmillan, London.

Fölster, H. (1969) 'Slope development in south-western Nigeria during late Pleistocene and Holocene', *Gottinger Bodenkundliche Berichte*, **10**, 3–56.

Fredoux, A. (1978) 'Pollens et spores d'espèces actuelles et Quaternaries des régions peri-langunaires de la Côte d'Ivoire', thesis, Montpellier.

Garner, H. F. (1968) 'Tropical weathering and relief' in *Encyclopedia of Geomorphology*, Fairbridge R. W. (ed.), Reinhold, New York.

Geddes, A. (1960) 'The alluvial morphology of the Indo-Gangetic Plain: its mapping and geographical significance', *Trans. Inst. Br. Geog.*, **28**, 253–76.

Gibbs, R. J. (1967) 'The geochemistry of the Amazon River sytem Part I: the factors that control salinity and the composition and concentration of the suspended solids', *Bull. Geol. Soc. Amer.* **78**, 1203–32.

Gribbin, J. and Lamb, H. H. (1978) 'Climatic change in historical times', in *Climatic Change*, J. Gribbin (ed.), Cambridge University Press, Cambridge, pp. 68–82.

Grove, A. T. (1958) 'The ancient erg of Hausaland and similar formations on the south side of the Sahara', *Geog. J.*, **125**, 528–33.

Grove, A. T. and Warren, A. (1968) 'Quaternary landforms and climate on the south side of the Sahara', *Geog. J.*, **134** (2), 193–208.

Halstead, L. B. (1975) 'The shoreline of Lake Kainji', in Imevbore, A. M. A. and Adegoke, O. S. (eds), *op. cit.*

Howard, A. D., Fairbridge, R. W. and Quinn, J. H. (1968) 'Terraces, fluvial – Introduction', in *Encyclopedia of Geomorphology*, R. W. Fairbridge (ed.), Rinehold, New York, pp. 1117–23.

Jeje, L. K. (1970) 'Some aspects of the geomorphology of south-western Nigeria', Ph.D. thesis, Edinburgh Univ., Edinburgh.

Jeje, L. K. (1972) 'Landform development at the boundary of sedimentary and crystalline rocks in south-western Nigeria', *J. Trop. Geog.*, **34**, 25–33.

Jeje, L. K. (1979) 'Debris flow on a hillside along Ife-Ondo road, Western Nigeria', *J. Min. and Geol.* (in press).

King, L. C. (1962) *The Morphology of the Earth*, Hafner, New York.

Leopold, L. B., Wolman, M. G. and Miller, J. P. (1964) *Fluvial Processes in Geomorphology*, W. H. Freeman, San Francisco and London.

Kolla, V., Biscaye, P. E. and Hanley, A. F. (1979) 'Distribution of quartz in Late Quaternary Atlantic sediments in relation to climate', *Quaternary Research*, **11**, 261–77.

Moss, R. P. (1965) 'Slope development and soil morphology in a part of south western Nigeria', *J. of Soil Sci.*, **16**, 192–209.

Mulcahy, M. J. (1960) 'Laterites and lateritic soils in south western Australia', *J. of Soil Sci.*, **11** (2), 206–25.

Murray-Rust, D. H. (1972) 'Soil erosion and reservoir sedimentation in a grazing area west of Arusha, northern Tanzania', *Geografiska Annaller*, 54A (3–4), 325–43.

NEDECO (1959) *River Studies and Recommendations on Improvement of Niger and Benue*, NEDECO, the Hague.

Nye, P. H. (1955) 'Some soil forming processes in the humid tropics, Part II: the development of the upper slope member of the catena', *J. of Soil Sci.* **6**, 51–62.

Ollier, C. D. (1959) 'A two-cycle theory of tropical pedology', *J. of Soil Sci.*, **6**, 51–62.

Ollier, J. E. (1972) *Climate and Man's Environment: an introduction to applied climatology*, J. Wiley, New York, pp. 315–75.

Ologe, K. O. (1972) 'Gullies in the Zaria area: a preliminary study of headscarp erosion', *Savanna.*, **1** (1), 55–66.

Pain, C. F. and Bowler, J. M. (1973) 'Denudation following the November 1970 earthquake at Medang, Papua, New Guinea', *Zeits. für Geomorph.* N.F. Supp. Bd., **18**, 92–104.

Pallister, J. W. (1956) 'Slope development in Buganda', *Geog. J.*, **22**, 80–7.

Pastouret, L., Chamley, H., Delibrias, G. and Duplessy, J. C. (1978) 'Late Quaternary climatic changes in western tropical Africa deduced from deep sea sedimentation off the Niger delta', *Oceanologica Acta*, **1** (2), 217–32.

Pinteado-Orellana, M. M., Holz, R. K. and Baker, V. R. (1978) 'Remote sensing, a tool for the hydromorphological study in drainage basin: an analysis of the Solimos River, Brazil'. *Brazilian Geographical Studies, Instituto de Geociencias Applicadas da secretaria de Estado de Ciencia e Tecnologia Minas Gerais Brazil*, International Geographical Union, 1978.

Radwanski, S. A. and Ollier, C. D. (1959) 'A study of East African Catena', *J. of Soil Sci.*, **10** (2), 149–67.

Rapp, A., Murray-Rust, D. H., Christiansson, C. and Berry, L. (1972) 'Soil erosion in four catchments near Dodoma, Tanzania', *Geografiska Annaller*, 54A (3–4), 255–318.

Rapp, A. (1975) 'Soil erosion and sedimentation in Tanzania and Lesotho', *Ambio*, **47** (4), 154–63.

Rhodenburg, H. (1969) 'Slope pedimentation and climatic change as factors of planation and scarp development in tropical Africa', *Giessner Geogr. Schr.*, **20**, 127–33.

Ruhe, R. V. (1956) 'Landscape evolution in the High Ituri, Belgian Congo', *Publications de L'Institut National pour l'Etude Agronomique du Congo Belge, (INEAC)*, Série Scientifique, **66**.

Rust, B. R. (1972) 'Structure and process in a braided river', *Sedimentation*, **5** (18), 221–45.

Selby, M. J. (1977) 'Pluvials in northern and east Africa and their relations to glacial climates in Europe', Environment Studies: Dept. of Geography and

Environment Studies, University of Witwatersrand. Johannesburg, Occasional Paper No. 19.

Shaw, D. B. (1976) 'Climate, environment and prehistory in the Sahara', *World Archaeology*. **8** (2), 133–49.

Shrodder, J. F. (1976) 'Mass movement in Nyika Plateau, Malawi', *Zeits für Geomorph*. N.F., **20** (1), 56–77.

Silva Moura, O. R., Oliveira, P. T. T. M. and Meis, M. R. M. (1978) 'Insight into the morphometry of drowned valleys', in *Brazilian Geographical Studies, op. cit.*

Simonett, D. S. (1967) 'Landslide distribution and earthquakes in Bewani and Torricelli mountains, New Guinea', in *Landform Studies from Australia and New Guinea*, J. N. Jennings and J. A. Mabutt (eds), Cambridge Univ. Press, Cambridge.

So, C. L. (1967) 'Mass movements associated with the rainstorm of June 1966 in Hong Kong', *Trans. Inst. Br. Geog.*, **53**, 55–65.

Smyth, A. J. and Montgomery, F. F. (1962) *Soil and Land Use in Central Western Nigeria*, Government Printers, Ibadan.

Street, F. A. and Grove, A. T. (1979) 'Global maps of lake level fluctuations since 30,000 B.P.', *Quaternary Research*, **12** (1), 83–118.

Talbot, M. R. and Delibrias, G. (1980) 'A new Late Pleistocene-Holocene water-level curve for Lake Bosumtwi, Ghana', *Earth and Planetary Sci. Letters*. **47**, 336–44.

Temple, P. H. and Rapp, A. (1972) 'Landslides in the Mgeta Area, Western Uluguru mountains, Tanzania, Geomorphological effects of sudden heavy rainfall', *Geograf. Ann.*, **54A** (3–4), 157–93.

Termier, H. and Termier, G. (1963) *Erosion and Sedimentation*, Van Nostrand, London.

Thomas, M. F. (1974) *Tropical Geomorphology: a study of weathering and landform development in warm climates*, Macmillan, London.

Thomas, M. F. and Thorp, M. B. (1980) 'Some aspects of the geomorphological interpretation of Quaternary alluvial sediments in Sierra Leone', *Zeits. für Geomorph*. N.F. Supl. Bd., **36**, 140–61.

Tricart, J. (1972) *The Landforms of the Humid Tropics, Forest and Savannas*, Longman, London.

Twidale, C. R. (1972) 'The neglected third dimension', *Zeit. für Geomorph*. N.F., **16**, 3, 283–300.

Vanoni, V. A. (1971) 'Sediment transportation mechanics: Q. Genetic classification of valley sediment deposits: Hydraulic Division', *J. Amer. Soc. of Civil Engineers Proc.*, Paper 7815.

Villiers, J. M. de (1965) 'Present soil forming factors and processes in tropical and sub-tropical regions', *Soil Sci.*, **99**, 50–7.

Vine, H. (1949) *Nigerian Soils in Relation to Parent Materials*, Commonwealth Bureau of Soil Science, Technical Communication, No. 46.

Vogt, J. (1959) 'Observations nouvelles sur les alluvions inactuelles de Côte d'Ivoire et de Guinée', *Acres 84 Congr. Soc. Sav.*, Dijon, 205–10.

Williams, M. A. J. (1968) 'Termites and soil development near Brocks Creek, Northern Territory', *Australian J. of Sci.*, **31**, 153–4.

Wolman, M. G. and Leopold, L. B. (1957) 'River flood plains: some observations of their formation', *U.S. Geol. Survey*, Prof. Paper 282-C, 86–109.

Young, A. (1972) *Slopes*, Oliver & Boyd, London.

Young, A. (1976) *Tropical Soils and Soil Survey*, Cambridge Univ. Press, Cambridge.

SECTION C
Landforms

The analysis in the previous section shows that, as in the temperate regions, water is the dominant geomorphological agent of landform in the humid tropics, so that all the processes operating are greatly conditioned by its occurrence, magnitude and distribution both in place and in time. Accordingly, the resulting landforms are essentially fluvial landforms.

Perhaps the most notable feature of the physical geography of the humid tropical regions is the occurrence of deep weathering, the effects of which are felt markedly in the operation of the various geomorphological processes and resulting landforms. Two aspects of this phenomenon are particularly noteworthy – the high clay content of the regolith and the duricrust. The texture and depth of the regolith have very significant implications for such processes as infiltration, overland flow, through flow and channel flow among others, while duricrusts as caprocks influence processes such as the evolution of slopes. The following chapters are an attempt to highlight these influences of the humid tropical environment.

6 The humid tropical landscape

Most of the shield areas of tropical Africa, India, Australia and South America have been tectonically stable since the Cainozoic era, and they have been subjected to continuous chemical weathering since then. As discussed in Chapter 3, and as emphasized by Thomas (1974), in certain places weathering has penetrated to great depths, to more than 90 m in quartz-diorite in Colombia, and in excess of 100 m in metamorphic rocks in Brazil. The weathered mantle has been subjected to varying intensity of erosion, especially as a result of the eustatic base-level changes during the Pleistocene and early Holocene (Fairbridge, 1961), so that most of the terrains in the humid tropics are characterized by forms resulting from differential weathering and erosion.

Terrain characteristics

With the exception of tectonically induced features such as fold mountain chains, fault scarps, fault valleys, horsts, and rift valleys found in tectonically unstable parts of the humid tropics, the terrain consists of plains on which varying proportions of the basal surface of weathering are exposed, so that the plains are diversified by different elevations. There are upland and lowland plains, the former often occurring as plateaux. In some places, the plains are separated by steep scarps or by zones of dissection.

At any particular location, the terrain can consist of two- or three-tier landscape. The highest storey may be in form of hills, ridges or even mountains that have survived denudational processes because of being underlain by resistant rocks like quartzites, phyllites, slates and so on as noted in Uganda by Doornkamp and King (1971). The middle storey may be in form of flat to gently convex interfluves alternating with broad, concave to rectilinear river valleys, while the lowest storey, usually developing along the larger rivers, comprise wide valley floors which may or may not be aggraded. The hills and mountains are often surrounded by dissected or undissected piedmont slopes, while pediments are common around the residuals. The number of storeys found in any area depends on the erosional history. For example, the landscape of Guyana and Surinam comprises five storeys separated by scarps (Eden, 1971).

This tiered terrain pattern has been observed in Malaysia. Swan (1970), studying the terrain characteristic in Johore, Malaysia, noted that 'the whole landscape consists of a bottom storey comprising extensive zones of deposition, above which rises a middle storey of dissected lowlands of gentle to moderate steepness with a relative relief within 60 m, and a top storey of steep-sided mountains, hills and ridges with occasional plateaux and high plains rising above the middle storey'. The faceted slopes of the top storey forms comprise the crest, the upper convex slope, followed by the free face inclined at more than 45°, the middle debris slope about 25°, and the lower debris slope at about 13°, followed by the piedmont slope at about 4.25°, and the alluvial toe-slope and the stream channel (cf Figure 9.6). The middle storey landscape has slopes ranging from convex through convexo-concave to almost concave especially on granites and argillaceous sediments, and predominantly convex in basic rocks. Maximum slope units vary a great deal, with a mean of 12.3° and a standard deviation of 5.2°, but these are constant on argillaceous sediments at a mean of 10.22° and a standard deviation of 1.7°.

As evident from the observations of Young (1970) and Townshend (1970), the three-tiered landscape is also ubiquitous on the shield areas of Brazil and Guyana.

In south-western Nigeria, the three-tiered landscape is most evident in the city of Ibadan where the Aremo ridge and Agodi hill, standing 65 m above the local footslopes, form the highest storey; the foot-slopes and the flat, interfluvial surfaces constitute the middle storey; while the broad valley floors along the Ogunpa and its tributaries form the lowest storey. This three-storeyed landscape can also be observed on an even grander scale in the rolling terrain around Okemessi and Effon Alaye in Ondo State. The landscapes on the sedimentary and crystalline basement rocks of south-western Nigeria further illustrate the terrain pattern in the humid tropics (see Table 6.1).

Eyles (1971) measured and analyzed five basin parameters – namely, basin area, basin relief, average basin slopes, drainage density and hypsometric integral in 410 fourth-order drainage basins in Malaysia – and classified the landscape into six major relief groups composed of (1) the high mountains, (2) the low mountains, (3) the deeply dissected low mountains, (4) isolated steep, high hills, (5) isolated steep hills, and (6) low convex hills. He showed that the first three differ significantly from the last three. For instance, group (3) differs from group (4) as follows: difference in standard deviation units in average slopes is 1.32 (8.3°), 0.75 units in basin relief (223 m), and 1.18 units in drainage density (3.4 km/km^2). He related this sharp break in the attributes of the mountains and the low-relief terrain to the absence of extensive foothill zones in west Malaysia, where granitic mountain ranges commonly rise abruptly from sedimentary lowlands. The three mountain groups cover about 32 per cent of Malaysia and occupy narrow zones to the west and east, separated by central lowlands. These latter, comprising largely of low convex hills separated by flat-floored, swampy stream courses, typify most of southern Johore, where the hills are associated with granites, arenaceous sedimentary rocks and tower karst on limestones. However, Eyles failed to give the quantitative indices of each terrain group, an omission partially rectified by the quantitative description of the erosional residuals in Johor by Swan (1970).

TABLE 6.1   Characteristics of terrain features in south-western Nigeria

Terrain	Relative relief in m	Average slope (degrees)	Maximum valley side slope	Area of occurrence (degrees)
Coastal plains	15	1.5	2.5–5.0	Mainly between Yewa and Ogun rivers but extends all over the coastal areas of Nigeria except that in Benin area relative relief is in excess of 45 m
Dissected terrain	30–45	2.5	3.5–9.5	Located on the dip slope of Ilaro formation west of Ogun river
Meko plateau (Taylor *et al.*, 1962)	45–60	2.5	12.5–30.0	Between Abeokuta and Meko
Valley and ridge topography	45–60	6.0	15.0–25.0	Found on the dip slope of Abeokuta formation between Yewa and Ogun
Flat to gently undulating terrain with low ridges and high inselbergs	15–30	–	1.0–3.0	Occurs around Oyo, covering about 20% of the basement area of south-western Nigeria
Dissected terrain with low ridges	15–45	–	3.0–8.0	Covers about 60% of the basement area of south-western Nigeria
Rolling terrain with high ridges and hills	30–60	–	8.0–10.0	Occurs mainly on quartzites and quartz-schists, and covers above 15% of the basement area of south-western Nigeria
Dissected hills	60–180	–	10.0–30.0	Covers about 5% of the basement area of south-western Nigeria, occurring mainly on older granites

As shown by Swan (1970), in Johor, areas with relative relief of less than 61 m have the following attributes: mean of slope units, 7.36° with a standard deviation of 3.11°; mean maximum slope angles, 13.05° with a standard deviation of 6.71°; mean slope length, 176.8 m with a standard deviation of 71.2 m; and mean distance between valley bottoms, 323.3 m with a standard deviation of 142.4 m. In such terrains, the commonest slope forms are convex, convexo-concave and convexo-concave-convex. Granites have higher frequency of convex slopes with argillaceous sedimentary rocks having higher frequencies of convexo-concave forms.

As the terrains described from south-western Nigeria and Malaysia contain several landforms considered characteristic of the humid tropics, these landforms are identified and examined in the remaining part of this chapter. However, before this is done, it may be relevant to define the term 'landform'.

Tropical landforms
Definition and scale perspective

According to Weaver (1965), the term 'landform' may merely imply the surface configuration of the earth's crust. This is an unsatisfactory definition, as it connotes no clear morphology or processes. Savigear (1960) defines 'landform' as a feature of the earth's surface with distinctive form characteristics which can be attributed to the dominance of particular processes or particular structures in the course of its development, and to which the feature can be clearly related. Given this definition, distinctions can be made between structural landforms related to specific geologic and geodynamic origin, and non-structural landforms resulting from exogenetic forces. Landforms have been classified along this line by Howard and Spock (1940) and by Fairbridge (1968) (see Appendices 6.1 and 6.2, pages 182 and 183–4).

As recognized by Weaver (1965), because landforms graduate from very large dimensions to smaller ones, they can be studied at various scales. The recognition of this scale perspective in landform studies led Tricart (1952) and Cailleux and Tricart (1956) to propose seven magnitudes of landforms, ranging in area from 10^7 km^2 to 10 km^2, and in extent from continental masses of geodynamic origin to smaller forms like inselbergs related entirely to lithology and exogenetic processes (Appendix 6.3). They also recognize microforms like mudflow lobes, barchans, thufurs and so on. Also it is the recognition of scale problems in landform studies that led Thomas (1969) and Young (1969) to propose a hierarchical classification of landforms, as shown in Table 6.2.

Haggett *et al.* (1965) also proposed that features, especially landforms on the earth's surface, be classified using the G-scale derived from the formula:

G = 8.7074 Log Ra ... (6.1)
where RA is the area (km^2) of the feature under study, and 8.7074 is the log value of the earth's surface (km^2).

TABLE 6.2 Landform hierarchy, as proposed by Thomas and Young

Landform order	Landform units		Scale of mapping	Examples
	Thomas	Young		
1	site	slope unit	1:10 000	Any slope unit of uniform curvatures or gradient
2	facet	–	1:25 000	Two or more contiguous slope units forming a definite feature, e.g. a valley side, or a pediment, or a scarp
3	unit landform	landform	1:50 000	Two or more contiguous facets forming a definite feature, e.g. an inselberg, or a river valley, or a ridge.
4	landform complex	–	1:50 000 – 1:100 000	Two or more contiguous unit landforms forming a definite feature, e.g. cuesta relief
5	landform system	relief units	1:100 000 – 1:60 000	A plain or range of hills or hill and ridge complex, or a dissected plateau
6	landform region	major relief units	1:500 000 and above	A fold mountain system

Unit landforms

Evident from the description of the terrain characteristics, and as shown by the detailed mapping of landforms in Johor, Malaysia, by Swan (1970), unit landforms in the humid tropics consist of erosional and depositional forms. These are forms easily recognized in the field, on topographic maps (with a scale of 1:50 000), and on aerial photographs (with a scale of about 1:40 000). Only a few of these landforms can be described here, however: some of the landforms resulting from deposition, especially on floodplains, have already been described (Chapter 5), while those related to coastal zones will be examined in Chapter 12. The erosional landforms described below include mountains, hills, ridges, valleys and pediment slopes.

Mountains

These are parts of the earth's crust rising considerably above the local surface. The local relief is usually more than 330 m, while the surrounding slopes are more than 20°. They are described as landform complexes or as major relief units wherever they occur in long chains or ranges underlain by intricately folded sedimentary rocks into which igneous rocks are ejected – as for example, the Andes system in South America.

Mountains can also occur in the form of volcanic cones in various stages of dissection, as Chimborazo and Cotopaxi in Ecuador at 6 331 m and 5 952 m respectively, Cameroon, Kenya and Kilimanjaro in tropical Africa at 4 108 m, 5 164 m and 5 861 m respectively, or they may occur as isolated fault blocks, as in the case of Ruwenzori at 5 089 m at the boundary of Zaire and Uganda.

Most individual mountains have survived as erosional residuals especially where batholiths have intruded into the shield rocks. Except where composed of very resistant rocks, these batholiths have subdued relief having been subject to severe erosion over long geological periods. Some, however, have given rise to mountains and uplands, a notable example of which is the Idanre massif, in south-western Nigeria, developed in porphyritic granite intruded into gneisses and schists. The rock, standing at 951 m a.s.l. in the Olofin peak, has a local relief above 400 m especially where it is dissected by sub-parallel system of joint-aligned streams trending in a NNE–SSW direction (Jeje, 1974). Higher than this is the Chaillu massif in Gabon, also developed in granite. Mt Ibjoundji, at 1 500 m a.s.l., constitutes the watershed between rivers draining into the Congo and the Atlantic Ocean. Other examples include the Naminli mountains (2 124 m) in northern Mozambique, with peaks rising more than 1 000 m above the surrounding plains.

Hills

These are eminences or groups of eminences rising above the level of the surrounding country and culminating in well-marked summits. Local relief is generally less than 300 m. Steep-sided types are characterized by slopes in excess of 12°, while gently sloping types have slopes less than 12°.

In the tropics, distinctions can be made among hills based on their parent rocks and on their mode of formation. *Conekarsts* (often referred to as *mogotes*, *humes*, *haystacks* or *pepino*) are conical hills separated by closed depressions about 100 m deep and 400 m wide, both of which are formed by the processes of solutional weathering and erosion in strongly jointed limestones. They are common in Jamaica where, together with the depressions, they have given rise to the cockpit topography on well-jointed white limestone (Sweeting, 1958). Conekarsts in Jamaica are characterized by steep slopes of 60–90° which occasionally overhang, and are 133–153 m high with their bases honeycombed with caves. Conekarsts have also been reported from southern Congo, where they form on dolomite on the south-western edge of the Kissenga plateau (Buckle, 1978). They are also found in western Cuba, where they have developed on old, folded limestones of the Sierra de los Organos. *Towerkarsts*, in the form of bare, limestone hills with very sharp pinnacles separated by areas of level, clay-covered plain, have also been observed in Jamaica, and in Tanzania (Cooke, 1973).

Mesas or *tabular hills*, formed by erosional dissection of a once-flat terrain or horizontally bedded rocks, are often underlain by weathered rocks or sedimentary formations like sandstone, and capped by duricrust or any other rocks resistant to erosion. They are common on basement complex rocks, as evident around Kampala in Uganda where several of them stand between 70 m and 80 m above the local surface. Tabular hills are especially common along the Jengre–Jos road and around Kanfanchan on the Jos plateau where they have developed on the fluvio-volcanic series. The hills, which are about 50 m above the surrounding plains, are composed of old alluvium overlain by duricrusts. The duricrusts are formed as a result of intense chemical decomposition of the basic lava which covered the valley floors on the plateau during the Tertiary period. The occurrence of these hills is an indication of relief inversion on the surface of the plateau as the duricrusts have protected the underlying alluvium of the old valley floors from erosion while the former higher interfluvial surfaces have been eroded to lower elevations. Tabular hills also occur on sedimentary rocks – for example, on Nupe sandstone around Lokoja where the Patti hill stands at about 250 m above the level of the Niger surface. They are also common around Itakpe. They also occur on the Gwandu formations as outliers on Sokoto and Rima formations in Sokoto State, and on false-bedded sandstones in the Nsukka area. In the latter area, Ofomata (1973) distinguishes five types – conical, domed, flat-topped, cuesta-like, and elongated – and he relates the differences in their morphology to differential lithological characteristics.

Inselbergs, which constitute one of the most typical tropical landforms, are isolated hills rising abruptly above the surrounding plains. They may occur singly as on gneisses and schists, or in clusters in granite intrusions. In plan, they may be circular, rectilinear, elliptical or complex. Castellated types are often referred to as castle *kopjes*. Among the domes varieties, distinctions can be further made between the rounded symmetrical monolithic domes referred to as bornhardts, and the asymmetrical, faceted convex types in which the lower concave units often form only an insignificant proportion of the whole characteristic of other higher residuals all over the savanna areas of the tropics. Examples of inselbergs include Zuma hill near Abuja, rising 300 m above the local plains, the Kufena hill Zaria, Kusheriki inselbergs near Birnin Gwari, Ado rock near Ado Awaiye and Olosunta hill at Ikerre, all in Nigeria. In East Africa, inselbergs are abundant in the Nwanza area of Malawi; on the central plateau of Tanzania, where the Wambengeta hill, 30 km north-west of Iringa, rises more than 450 m above the plain; near Sinda in eastern Zambia; and near Salisbury and Mtoko in Zimbabwe. They are also very common on the ancient shields of Guyana and Brazil.

Hills can also form on deeply weathered regolith. As observed by

Plate 11 Fretting and scalloping of an inselberg on the plains near Oyo, Nigeria (L. K. Jeje)

Plate 12 A small pool on Ado Awaye inselberg, Oyo State, Nigeria (A. Faniran)

157

Plate 13 Taffonis (weathering pits) on an inselberg slope (A. Faniran)

Plate 14 Basal cavity on an inselberg slope (A. Faniran)

Plate 15 Corestones in a weathering profile on a road cutting near Ondo, Nigeria (L. K. Jeje)

Tricart (1972), in the granitic and gneissic rocks of Rio, Salvador and Recife in Brazil, deep chemical weathering often exceeds 50 m. The thick regolith has been eroded at many locations into convex-sided low hills which are in the form of a hemisphere – *meias laranjas* – which stand out abruptly from the surrounding plains underlain by colluvial material. The abrupt junction between the hill-slope and the plain results from the deposition of colluvial materials derived from the hillsides, especially during the rainy season.

Like mountains, some hills in the tropics are of volcanic origin, and some of them are still active. Examples of such active volcanic hills include Nyamlagira, Nyiragongo, Teleki, Likaiya, Longonot, Meru and Oldoinyo Lengai in Central and East Africa. Several hills in the form of cumulo-domes occur in the Itasy area of Malagasy, among which are Ambasy, Antanimarina and Atsakario. These domes, typically without craters, rise about 300 m above the level of the surrounding plains. Wase hill, composed of trachyte intruded into Bima sandstone, rises abruptly to a height of about 300 m from the floor of the Benue valley and is one of the most important examples of volcanic plugs in the humid tropics.

Ridges

These are eminences with excess of length over width rising above lower-lying terrains. In the humid tropics, ridges are commonly associated with very resistant elongated rocks, especially quartzites. Two types of ridges can be observed in south-western Nigeria: the minor elongated types, often less than 1 km in length and frequently dissected into hill chains, as those around Ibadan; and the impressive systems of

TABLE 6.3   Some aspects of the morphometry of ridges around Ibadan

Ridge	Mean slope (degrees)	Standard deviation (degrees)	Height above local surface (in m)	Length (km)
Olokemeji	23.0	4.1	96	5.5
Arowosegbe	19.6	3.8	45	2.0
Ijaiye	22.9	4.3	106	10.5
Aremo	12.2	2.3	71	4.5
Ayantola (SW of Ibadan)	14.9	4.2	103	10.0

Source: Jeje, 1970, p. 59

ridges in Ilesa–Ekiti districts. Table 6.3 shows some aspects of the morphometry of the low ridges around Ibadan.

In all cases, these ridges are less than 1.5 km in width. Impressive ridges, 150–300 m above local plains, are found in the Ilesa, Effon, Ila and Osi districts in central western Nigeria. Of the three sets of ridges developed in quartzites, the best-developed can be traced for a distance of more than 200 km. To the south it comprises two branches, one originating at Ipetu-Jesa, the other at Ogotun. Both trend northwards and merge at Oke-Messi where the western arm is dissected by the headstreams of the Osun river. The merged ridges trend NNE to the Osi district in Kwara State through Ila. Another important example is the Serra da Chela ridge developed on quartzite between the coast and the Cunene river in south-west Angola.

Pediments

This term is bedevilled by imprecise definition, because most definitions connote genesis and processes, especially pedimentation or pediplanation across fresh or weathered rocks. To avoid this, some authors – for instance, Ruxton and Berry (1961), and Young (1963) – prefer to use non-genetic terms to describe the slopes at the bases of erosional residuals or scarps, substituting 'foot-slopes'. As the use of such non-genetic terms often leads to further confusion, the word has survived in most recent geomorphological literature (Jeje, 1972). Dury (1972) has proposed a non-genetic definition of the term. He defines it as 'a degradational slope, cut across rock in place, abutting on a constant slope at its upper end, and decreasing in gradient in an orderly fashion in the downslope direction'. However, it must be emphasized that pediments appear to be formed mainly on weathered rocks in the humid tropics.

However defined, pediments are gently concave erosional slopes found at the bases of most erosional residuals in the tropics; but not all slopes below such residuals are erosional, as some are piedmont slopes built of colluvial material. In the semi-arid environments, pediments are separated from their associated erosional residuals by well-defined sharp knicks, which seems not to be the case in the humid tropics, where Thomas (1969) has shown that the junction between the inselberg and the pediment can take any or all of these three forms: (1) abrupt – that is, there is a sharp break of slope between the slope of the inselberg and the surrounding foot-slope; (2) gradual transition from the slope of the inselberg to the foot-slope owing to the accumulation of talus and scree around the hill; and (3) smooth – that is, where the slope of the inselberg is rather gentle, regolith cover may persist across the foot-slopes to the surface of the inselberg.

The impression gained from the literature is that pediments are extremely wide, often several kilometres in extent from the erosional residual to the neighbouring stream lines. Although this may be true of arid and semi-arid areas where erosional dissection is limited, this is often not the case in the humid tropics where the relative extent of pediments depends on the physiographic location of the associated erosional residuals. Where the latter are located on valley sides, pediments are short or non-existent. The same is true where inselbergs occur in a clustered manner as in granites where pediments are very short, often less than 150 m. However, where the sides of the residuals mark petrological boundaries with the local rocks, pediments developed on the weaker rocks could be up to 1 km long (Jeje, 1972).

Slope angles on pediments have been measured at different locations. Swan (1970) shows that the pediments associated with the top-storey landforms in the forest-covered area of Johor, Malaysia, have a mean slope value of 4.5°. Nearly the same value of 4.5° was obtained on pediments in the forested area around Ile-Ife in south-western Nigeria. In the forest–savanna boundaries of the same area, measured pediment slopes vary between 1° and 4.5°, with a mean of 3.1° and a standard deviation of 1.17°, while in the Guinea savanna further to the north, measured slope values on pediments vary between 1° and 6° with a mean of 2.25° and a standard deviation of 1.39° (Jeje, 1970).

Valleys

These are elongated hollows found between hills, ridges or mountains or on plains. Such hollows may or may not contain any river. A striking feature of valley morphology in the humid tropics, especially on the gently undulating to rolling plains characteristic of the stable ancient cratons, is that except where rejuvenation is in progress, or where rivers are incised into hard rock formations, valleys are relatively wide, and often out of proportion to size of the associated rivers.

As observed by Young (1970), valleys in the Mato Grosso area of Brazil are characterized by wide marshy floors, more or less the same as those observed in Johor, Malaysia, by Swan (1970). Young (1969) also observed that the valleys on the 1 000–1 400 m surface in southern Malawi are 400–1 600 m wide, with maximum valley slopes varying between 2 and 8°. Also, the floors are marshy and contain hardly any well-defined river channels. Similar valleys are also found in Sierra Leone, in Tanzania and other areas of Central Africa, especially in the woodland savanna (dambos).

Probably it is in recognition of this peculiar valley cross-profile that several authors classify valleys in the humid tropics as either 'saucer' or 'bowl-shaped'. Louis (1964) recognizes the 'pure saucer-shaped' valleys with alluviated floors and gently concave slopes to be typical on planation surfaces over both crystalline and sedimentary rocks in savanna areas with less than 1 000 m rainfall in Tanzania. In areas where rainfall exceeds 1 000 mm, he recognizes the predominance of furrow or grooved-shaped, steep-sided valleys with alluviated floors, the type described as 'bowl-shaped' valleys in the forest areas of south-western Nigeria (Jeje, 1970). Wigwe (1966), analyzing more than 100 profiles at the forest–savanna boundary in south-western Nigeria on the basement complex rocks, confirms the predominance of both the 'saucer-shaped' and the 'bowl-shaped' valleys with steep concave valleys leading to duricrust-covered surfaces. Jeje (1970) equally shows the predominance of 'saucer-shaped' valleys with concave valley slopes on the basement complex rocks, and the trough or 'bowl-shaped' types on the ferruginous sandstones in south-western Nigeria. However, in both places V-form and valley-in-valley form occur locally, evidence that valley form may relate to lithology and erosional history.

Valleys in the rolling topography associated with frequent occurrence of quartzite ridges in central Western Nigeria are typcailly V-shaped, especially where they have developed along lines of weakness like joints and faults or along inclusions of weaker rocks like schists. On the sedimentary rocks to the south, valley-in-valley forms are preserved by the occurrence of duricrust on the sandstone. The faceted valley slopes contain up to 5 slope units (Jeje, 1970). Of those measured, the steep slope next to the stream is 15–40 m long and inclined at 16–45°. The succeeding valley bench covered by detrital ferruginous crust and other colluvial material is 80–170 m long, and inclined at 1–5°. This is followed by a long slope unit 120–210 m long inclined at 2–7°, which is in turn succeeded by a steep slope inclined at 10–22°, the upper part of which may be marked by the outcrop of ferruginous crust up to 15 m thick. The interfluve is typically flat (Figure 6.1).

Landform complexes

As observed by Thomas (1974), a complete and objective typology of these major landforms cannot be presented, owing to paucity of information on their important parameters like relief, slopes and drainage. In an attempt to obtain what can be described as terrain typology in the humid tropics, the land system classification concept has been applied by the CSIRO* in Australia and the DOS** in Africa. The land system, according to Christian (1957, p. 76), is 'an area or group of areas throughout which there is recurring pattern of topography, soils, vegetation and climate', so that 'a simple land-system is a group of closely related topographic units usually small in number that have arisen as the products of a common geomorphological phenomenon'. The topographic units thus constitute a geographically associated series and are directly and consequently related to one another.

* CSIRO – Commonwealth Scientific Industrial Research Organization
** SOS – Directorate of Overseas Survey, Tolworth, Surrey, England

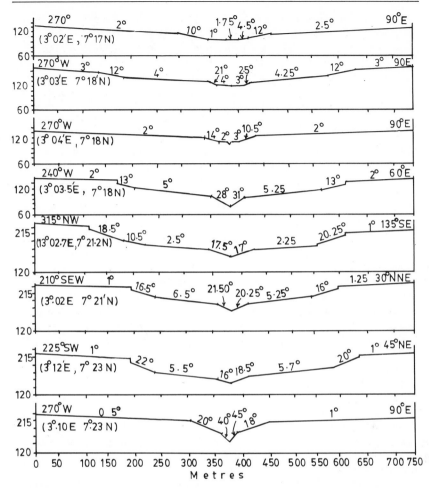

Fig. 6.1 Valley cross-sections, Ayetoro area, south-western Nigeria (Jeje, 1970)

However defined, whether in morphologic, genetic, chronologic, dynamic or parametric terms (Mabbutt and Stewart, 1965; Stewart, 1968), more than any other factors, a land system which reflects particular lithology and structures, and uniform geomorphological processes, can be regarded as the equivalence of Thomas's (1969) landform complex or landform system.

Many authors have outlined several landform complexes from the ancient tropical shields, with most of the features consisting of various types of deeply weathered erosional and depositional terrains. Thus in Tanzania, Harpum (1963) recognized hilly terrain and maturedly dissected, deeply weathered terrain. In the semi-arid area of Alice Springs in Australia, Mabbutt (1961) recognized and described three main landform complexes: the erosional weathered land surface, erosional surfaces formed below the weathered land surface, and depositional surfaces.

163

In the humid northern part of the Northern Territory in Australia, on the metamorphosed and unmetamorphosed Palaeozoic sediments, Hays (1967) recognized the coastal lowland and the inland plains, the eroded and deeply dissected plateau margin, and the deeply weathered and eroded plateau margin. The plains extend along the Daly and Victoria rivers, varying in width from 2 km to 150 km, while the dissected plateau margin is about 150 to 300 km wide descending from 500 m to 60 m at Darwin, rising again to 300 m at Arnhem Land and Carpentaria area. The plateau, covering about 60 per cent of the area, is flat to undulating; it rises from 180 m in the north to 400 m in the south, above which are scattered tabular hills rising from 15 m to 150 m and bounded by very steep scarps. The hill-slopes on the residuals exhibit weathering profiles in excess of 60 m deep comprising hard duricrusts 6–20 m thick, in turn underlain by mottled zones 15–25 m thick; the pallid zone is often absent.

Thomas (1974) outlined the major landform systems recognized by Büdel (1965) in India, most important of which are tropical plain relief, tropical mountain relief and tropical ridge relief, all in Tamilnand and Bangalore plains. The landform complexes recognized on the basement complex rocks of north-eastern Nigeria by Aitchison (Aitchison and Glover, 1972), which typify what obtain on the ancient shields in the humid tropics, are briefly described in what follows. These include erosional upland plains, mountain relief, and depositional sedimentary rocks of Nigeria; the cuesta relief is also described.

Erosional upland plains

The undifferentiated basement complex rocks west of the Adamawa mountains in Gongola State of Nigeria have ben eroded into plains, 300–600 m a.s.l., especially at the headstreams of the Ngada, Yedseram and Hawal. The most perfectly developed of these are the Askira and Tum plains, covering 1 088 km² and 3 000 km² respectively. The Askira plain at 450 m has local relief less than 30 m, so that it is flat to gently undulating. The valleys in form of broad, shallow depressions alternate with very wide and flat interfluvial surfaces underlain by concretions and ironstones. As a result of deep chemical weathering and weak fluvial erosion, rock outcrops are few. However, a few inselbergs rising 60–90 m above the plains occur locally.

The Tum plain at 400–460 m is also characterized by low relief less than 30 m, but it is studded with several massive inselbergs, rising to 150 m above the surface, while outcrops and low duricrusted hills are ubiquitous.

Mountain relief

The relief developed mainly on older granites, especially in the Adamawa and Sardauna districts, is typified by the Mandara mountains with a general elevation of 760–1 070 m, but with peaks reaching 1 220 m.

Local relief in the mountains exceeds 600 m. The dissected mountain range is characterized by very steep slopes developed on bare rocks. On the Nigerian side, the mountains are in the form of a rugged escarpment dissected into massive ridges. The crest of the ridges carry the Post-African erosion surface in Upper Hawal, with relics of ancient Gondwana, post-Gondwana and African surfaces on still higher elevations. Steep, bare rock outcrops comprise 76 per cent of the landform complex, while the flat surfaces occupy about 24 per cent.

Hill relief

The hills, composed largely of several contiguous inselbergs separated by deeply dissected terrains, are developed on older granites and volcanic plugs. The Ganda and Gongola hills, respectively covering 2 800 km² and 750 km², are found in the lower Gongola basin. The Ganda hills, which have a general elevation of 245–490 m with summits reaching above 600 m, are characterized by dissected rugged terrain with several inselbergs, bornhardts, dykes and wide areas of exposed rocks. However, in association with these hills are areas of level terrain in which wide, flat interfluves are incised by deep, narrow valleys. The Gongola hills around Kaltungo have a general elevation of 450–600 m with summits up to 900 m. The rugged hills which rise 90–300 m above the local surface cover about 76 per cent of the land surface, so that they are separated only by narrow valleys and inliers of flat surfaces, both of which together cover about 22 per cent of the land surface.

Depositional plains

These plains, developed mainly in form of piedmont slopes at the base of the Mandara mountains, especially around the village of Gwoza on the Mubi–Maiduguri road, are underlain by colluvial material. The plain covers about 260 km², and has an elevation of 395–600 m with local relief less than 30 m. The plain, underlain by gravel outwash fans which coalesce in places to form extensive sand and gravel plains, is dissected in several places by gully erosion into flat, broad, low ridges.

Cuesta relief

These are asymmetrical ridges in which the longer, and more gently inclined, slope lies parallel with the dip of resistant beds. Cuestas are both structural and erosional, as they occur where alternatively resistant and non-resistant strata are tilted so that, as a result of differential erosion, the weaker strata are eroded whereas the resistant strata form an asymmetrical ridge. A distinction can be made between a cuesta with a gentle back slope in which the dip of the rock is gentle to moderate, and a hogback in which the rock strata dip very steeply. However, the cuesta relief is usually more diversified as a result of erosion on the dip and scarp slopes to give rise to a landform complex. Examples of these include the Gambaga, Kwahu and Mampong cuestas in Ghana, and the Arochuku–Awgu cuesta in south-eastern Nigeria.

The cuesta in south-eastern Nigeria is formed on gently dipping series of sandstones, clays and shales deposited between the Cenomanian and the Senonian periods. These upper Cretaceous sediments were uplifted and folded into an anticlinorium with its axis along Okigwe–Abakaliki, flanked by a syncline on either side in the early Tertiary. The anticline has since been breached with the severest erosion on its axis to expose the underlying shales on which the Cross river plains are formed. The resulting east-facing escarpment, rising about 150 m above the local surface, coincides with a belt of relatively resistant sandstone undergoing gradual retreat as a result of the extensive erosion by the several east- and south-east-flowing rivers. Both the dip and the scarp slope of the cuesta have been subjected to intensive gullying, the former by the headstreams of Imo and Mamu rivers especially in Njikoka area, and the latter by the headstreams of the Aboine. Several low erosional residuals occur on the dip slope in the neighbourhood of Nsukka.

Erosion surfaces in the humid tropics

As mentioned earlier, the terrain on the ancient cratons in the humid tropics is comprised largely of plains at different elevations. Such plains are often referred to as erosion surfaces. An erosion surface has been defined by Rich (1938) as a surface of very low relief, often featureless, produced by sub-aerial erosion. Geyl (1961) describes it as an area of relatively flat land formed by the process of normal erosion during a period of base level stability; the surface, which stands above present base level, is in a process of destruction by retrospective erosion. Such surfaces are eroded across rocks of varying structure and lithology.

Recognition

From the literature on erosion surfaces, it appears that the recognition of these surfaces in the field depends on all or a few of the following factors:

1 the association of particular types of duricrusts with certain surfaces; older surfaces are believed to carry primary laterites, whereas younger ones carry secondary laterites (De Swardt, 1946, 1953, Chap. 9);

2 separation of surfaces by cyclical scarps or by zones of dissection; such scarps often exhibit the typical four-hill-slope elements – the waxing slope, the free face, the constant slope and the waning slope (Pugh, 1955; King, 1953; Pallister, 1956);

3 the association with superficial sediments like stone lines and gravel layers, especially on major divides (Ruhe, 1956);

4 remnants of old surfaces usually found on hills and ridges so that accordance of summits across different rock types is an evidence that such summits are remnants of a once-continuous surface (Pugh, 1966).

The above can be supplemented by map analysis (at scales of 1:25 000 or less), involving the drawing of longitudinal and cross-profiles of river valleys. Breaks of slopes unrelated to lithological and structural control are usually regarded as diagnostic of cyclic erosion (Miller, 1953). Also, analysis involving drawing of terrain profiles, superimposed or projected, and generalized contours are used in an attempt to decipher the various stages in landscape evolution and to detect erosion surfaces. The most rigorous mapwork involves morphometric analysis executed in different ways. One method involves the counting of the number of spot heights and closed contour within different height ranges to show the frequency distribution of each height range. Classes with the highest frequencies represent erosion surfaces. A variation of this method involves the imposition of regular grids on the map, the determination of the height in each square grid, and the grouping of the data for purposes of producing a frequency distribution curve. Another method involves the measuring of areas between successive contour lines, and the areas are then plotted against elevation to obtain a frequency curve. The hypsographic curve obtained by cumulatively plotting area against elevation can also be obtained. All these methods are supposed to demonstrate the existence of erosion surfaces, at least on maps.

Both the field and map methods of demonstrating the existence of erosion surfaces have obvious drawbacks. Although it is possible that erosion surfaces may be preserved by an association with duricrusts, the bounding scarp slopes are often lithologically controlled so that it may be difficult to infer any cyclic distinction between surfaces on resistant and non-resistant formations. Accordance of summits may not be useful in recognizing erosion surfaces. Hill tops, ridges, spurs and shoulders are normally subject to differential weathering and erosion even within a single erosion cycle, in which case they are bound to suffer differential loss of altitudes not easy to calculate. The amount of differential lowering will depend on the nature of the rocks concerned, especially their resistance to weathering and erosion. Stone lines can occur in association with erosion surfaces, but since they are also known to occur where unweathered quartz veins persisting in soils are broken and moved downslope under a migratory soil layer on slopes, they are not by themselves diagnostic of erosion surfaces.

The application of terrain profiles to demonstrate the existence of erosion surfaces has been severely criticized by Rich (1938), who observed that 'the projection of hill tops, ridges, and sloping surfaces in both foreground and background on to one vertical plane unduly emphasizes the appearance of horizontality, hence it is as deceptive as a view from a hill top with its accompanying psychological errors'. Craft (1933) also observed that 'the characteristic feature of the scenery as an even skyline which gives an appearance of horizontality is often deceptive as changing distance and perspective mask undulations or gradual slopes over a long distance'.

Morphometric analysis, apart from being very tedious and time-consuming, often produces rather meaningless results. Analysis involving the counting of spot heights and closed contours is misleading, as one large plateau counts equally with minor remnants of a severely dissected upland, generally giving prominence to a very small portion of the landscape. The method whose accuracy is subject to operators' variance is also flawed by the method of data presentation, which may seriously influence subsequent interpretation (Clarke, 1966). However, application of several morphometric methods – for example, altimetric frequency curves, generalized contours, projected profiles, average slopes and so on – supplemented by field measurement may demonstrate the presence of an erosion surface in any area if it is indeed present.

Formation of erosion surfaces

The concept of sub-aerial landscape evolution in a humid area has been explained by the application of three models – peneplanation, pediplanation and etchplanation.

Peneplanation

The Davisian concept of normal processes of landscape evolution which dominated geomorphic thought for a long time has been applied to explain the evolution of landscapes in all humid areas, temperate or tropical. The main points of this theory are summarised in what follows. The concept involves a rapid uplift of a region underlain by different structures, and with the irregular surface configuration representing the initial relief (see Figure 6.2). Weathering, followed by fluvial erosion,

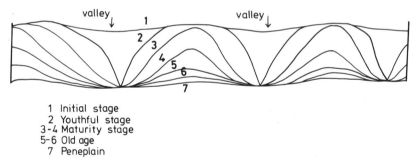

1 Initial stage
2 Youthful stage
3-4 Maturity stage
5-6 Old age
7 Peneplain

Fig. 6.2 Some stages in the process of peneplanation

emphasizes the initial relief and the emergence of consequent rivers on regional slopes. Larger rivers, because of the availability of coarse erosion tools, deepen their valleys at a faster rate than the interfluve is lowered, so that the local relief is more emphasized. When fluvial erosion reaches the base level (graded) at the river mouth, vertical incision declines at a faster rate than interfluvial lowering, leading to a decrease in the relative relief. As conditions of grade move up valleys along all rivers, valley slopes subject to weathering, slope wash and soil creep also attain grade or maturity. On the attainment of maturity, weathering and erosion continue, while the relief wastes away slowly, especially through valley widening. Whatever the structures are, a stage will be reached when all are affected by deep chemical weathering, so that eventually a surface of low relief will develop across the landscape. This surface, termed a *peneplain*, often carries erosional residuals called *monadnocks*. On such a plain, a state of dynamic equilibrium is supposed to prevail, a condition in which all the terrain elements are mutually adjusted so that no rapid transformation takes place on any part of the surface. This condition can be disrupted only if an uplift occurs or if there is a major climatic change, especially towards aridity.

Studies by other workers especially in semi-arid environments fail to confirm the Davisian concept. These researchers, especially King (1953, 1962), observed that the landscapes in both the humid or sub-humid areas are composed largely of series of surfaces separated from one another by scarps, and that the concept of peneplanation which primarily connotes slope lowering cannot account for the observed staircase terrain pattern. King (1957, 1962) crystallizes the concept of pediplanation in explaining the origin of erosion surfaces all over the tropics.

Pediplanation

Basic to this concept is King's (1953, 1957, 1962) insistence that hillsides everywhere tend to have four slope elements; convexity at the top, followed by a free face, in turn succeeded by the constant or debris slope, and then the concave slope below (Figure 9.6). These, however, may be absent in regions of low relief, zones of intense dissection where streams are closely spaced, or where the bedrocks are easily weathered. In such places, hill-slopes are mainly convexo-concave.

According to King, the pediplanation cycle can be initiated by a rapid uplift of a new surface probably of sub-continental dimensions. Following this uplift, new base levels are established and the rivers become incised and rejuvenated. The pattern of cyclical erosion depends on the size and spacing of the local rivers, the amount of uplift and the nature of the local rocks. The agencies of landscape evolution include fluvial erosion through incision, valley widening and scarp retreat, and soil creep. The dominance of each of these define the stages of youth, maturity and old age.

The youthful stage is characterized by accelerated vertical incision and the occurrence of knickpoints, falls, rapids and gorges along the rivers. Valleys are deep and steep-sided. Vertical incision continues until maximum relief is attained. With the attainment of maximum relief, and depending on the nature of the local rocks, weathering and rainwash combine to reduce the valley slopes to a stable declivity. As in the Davisian hypothesis, following the attainment of maximum relief the rivers are also graded. This condition, coupled with the attainment of stable valley-side slopes, marks the end of the youthful stage.

At the stage of maturity, rivers have ceased to deepen their valleys, but lateral erosion is very important. Hill-slopes start to retreat parallel to themselves at constant angles, leaving a low-angled pediment. Where the drainage is fine textured, the interfluves in between are rapidly destroyed, but where coarse textured, pedimentation becomes well emphasized. At the stage when pediments cover about 65 per cent of the landscape, the state of late maturity is reached. The pediments continue to widen and even regrade across one another, as those at lower levels regrade across those at higher elevations.

The residual masses are gradually reduced till the fronting pediments cover most of the landscape. Eventually the residuals are reduced to rocky eminences called *kopjes*, and the landscape is completely dominated by pediments. At this stage, the rivers meander freely over the very wide plains on which only occasional kopjes occur. This wide plain is a pediplain (Figure 6.3). Following another uplift, the cycle will be renewed, so that the terrain will be in the form of a staircase pattern.

Unlike the Davisian hypothesis emphasizing the decline of slopes as the cyclical development progresses, this model stresses the retreat of slopes and the constancy of slope declivity all over the landscape. Pugh (1955, 1966) employed this theory in explaining landscape evolution in Nigeria. However, this theory, which implies that erosion surfaces are flat, and because they are bevelled across different bedrocks are veneered by a thin soil cover, has been shown to be of doubtful validity in explaining the evolution of land surfaces on the ancient tropical shields

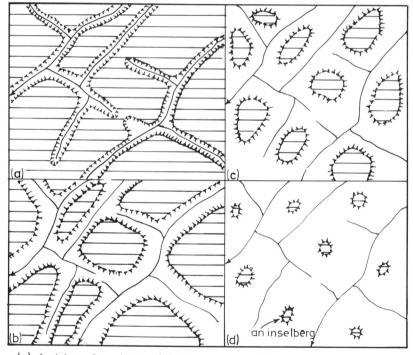

(a) Incision by rivers following an uplift

(b) Compartmentalisation of the landsurface

(c) Compartmental shrinking by scarp retreat

(d) Eventual evolution of pediplain dotted with inselbergs

Fig. 6.3 Some stages in the formation of a pediplain

characterized by deep weathering and several rock outcrops. As observed by many authors, these deeply weathered rocks are characterized by the development of younger surfaces across the weathering profiles of the older surfaces. The surfaces are not quite flat as they are usually dissected and marked by local relief of up to 45 m or more. They are also characterized by extremely irregular surface configuration arising from the variable exposure of the basal surface of weathering. They are only gently undulating where duricrusts have developed on the uneroded regolith (Thomas, 1965, 1969, 1974; Mabbutt, 1961; Wright, 1963). For instance, Mabbutt (1961) observed that antecedent weathering of the crystalline basement plateau area in the interior of Australia, followed by a change of base level, led to the development of an etchplain of considerable surficial configuration below the old plateau.

Wayland (1934), describing topographic levels in Uganda, observed that step-like series of plains in the Buganda area cannot have been produced from upper peneplains in an unstable area and particularly where deep chemical weathering has been followed by erosion. He further observed that if a lower surface develops below an upper one, it would be formed across the weathering profile of the upper surface. Thus the lower surface will have low to moderate relief. The formation of lower surfaces from the weathering profiles of older surfaces is described as etchplanation; a concept which Büdel (1957) and Thomas (1965, 1974) among others have attempted to use in explaining the evolution of the terrain on the basement complex rocks in the humid tropics. The concept is summarised in what follows.

Etchplanation

Basic to this concept is the fact that rocks in the humid tropics, especially on the shield areas, have been subjected to deep chemical weathering for long geological periods, with depth of weathering often in excess of 30 m. Because of differences in lithology and structural characteristics, the weathering front in these rocks is highly irregular and shows no relationship with ground-surface configuration. Also there are marked physical and chemical differences between the products of weathering on crystalline rocks and the unaltered rocks, especially as regards their erosivity properties.

Following uplift or a change in climate towards aridity, the rivers are incised, and the regolith is variably stripped from the unweathered rocks to reveal wide areas of the basal surface of weathering. Erosional stripping may occur through slope retreat, especially where the regolith is covered by duricrust. Slope retreat involves sheet wash, basal sapping and regolith slumping. As chemical weathering proceeds into the rocks *pari passu* with erosional stripping of the regolith, weathering can be speeded up as a result of downward migration of the water table. In such situations, a complete exposure of the basal surface of weathering does not occur. This accords with the observations of Eden (1971) on the development of etchplains in Guyana.

The nature of the etch surface depends on (1) the lithology and structure of the local rocks; (2) the intensity of erosion into the older surface; and (3) the surficial configuration of the lower surface. On the bases of these, Thomas (1965) and Jeje (1970) distinguished five different types of etchplains and surfaces in the forest-covered basement rocks of Nigeria: the laterized plain, the dissected etchplain, the partially stripped etchplain, the dominantly stripped etchsurface and the incised etchsurface.

The lateritized plain: This is gently undulating, undissected and underlain by thick regolith. Such a plain exists in south-western Nigeria on the basement complex rocks north of Shaki especially around Kishi, and on sandstone around Aiyetoro near Abeokuta. In the latter area, the plain is typified by the following features:

1 wide interfluvial surfaces, about 50 per cent of which exhibit duricrust pavements;
2 valley side slopes – (a) upper-slope profiles in the form of free faces exhibiting duricrust outcrops up to 12 m but usually about 2 m thick, (b) debris slope inclined at between 16–25° often mantled by large blocks of duricrusts, (c) mid-slope in form of terrace like benches inclined at 2–6°, and (d) followed downslope by a steep slope unit inclined at 25–45°;
3 narrow but alluviated valley floors with distinct stream channels;
4 values of drainage density (1:50 000 topographic map), average 1.1;
5 values of stream frequency (1:50 000 topographic map) average 1.0;
6 values of hypsometric Integral of 4, fourth-order basins average 48.5 per cent.

The dissected etchplain: This results from the dissection of the lateritized plain. A type area of this terrain is on the dip slope of the cretaceous sandstone between Wasimi and Aiyetoro in Ogun State, Nigeria. The depth of dissection is 45–60 m. The valleys vary in width between 0.5 and 1.5 km. The plain is typified by the following features:

1 narrow, flat interfluvial surfaces covered by duricrust and constituting about 10 per cent of the plain;
2 valley side slopes – (a) upper-slope profiles in the form of free faces exhibiting outcrops of duricrust, (b) mid-slope, rather long and inclined at 3–7° and usually covered by duricrust rubble which (c) leads to valley benches inclined at 0.5–3°;
3 wide, alluviated valley floors 0.5–1.5 km wide, frequent outcrops of basement rocks in the channel beds;
4 values of drainage density (1:50 000 topographic map), average 1.5;
5 values of stream frequency (1:50 000 topographic map), average 1.7;
6 values of hypsometric Integral of 3, fourth-order basins average 40.6 per cent.

The partially stripped etchsurface: This forest-covered surface occurs on the basement complex rocks, especially on migmatites and schists south of Ibadan in south-western Nigeria. The surface is highly irregular, as local relief varies from about 30 m to 115 m. Rock outcrops are few. In an area of 630 km² around Ilugun, only 0.66 per cent is covered by bare rock outcrops, while 1.2 per cent is covered by quartzitic ridges. Records of the Geological Survey of Nigeria show depth of weathering in excess of 30 m around Ilugun. The surface is typified by the following features:

1 quartzitic ridges, which form the most conspicuous relief feature;
2 low inselbergs generally less than 30 m high, and rock pavements

occurring on interfluvial surfaces or on the valley side slopes;

3 broad, convex interfluves occasionally carrying low rock pavements;

4 valley side slopes – mainly convexo-concave profiles, but occasionally rectilinear, slope units inclined at 5–10°;

5 values of drainage density (1:50 000 topographic map), average 1.8;

6 values of stream frequency (1:50 000 topographic map), average 2.5;

7 values of hypsometric Integral of 6, fourth-order basins average 42.3 per cent.

Dominantly stripped etchsurface: This is developed on migmatite and schists, especially in Oyo Division of Oyo State, Nigeria. Away from the inselbergs, local relief varies between 15 and 30 m around Oyo, but up to 45 m around Ijio. A significant part of the weathering front is exposed over most of the surface. In Ado Awaiye–Iseyin area (Figure 6.4), rock outcrops in the form of inselbergs cover about 7 per cent of the surface compared with about 3 per cent in Oyo. However, rock outcrops occupy a

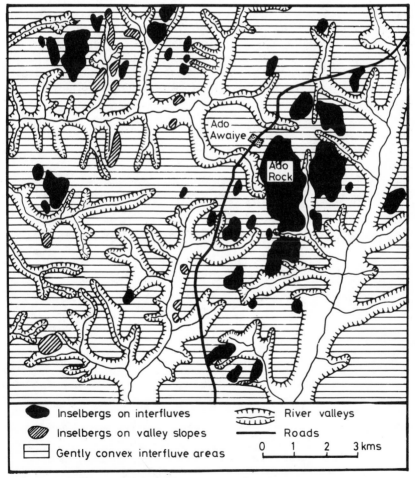

Inselbergs on interfluves   River valleys
Inselbergs on valley slopes   Roads
Gently convex interfluve areas   0 1 2 3 kms

Fig. 6.4 Inselbergs in Ado Awaiye area, Oyo State, Nigeria

larger percentage of the surface than revealed by these figures, as low, rock pavements covered by exfoliated sheets are ubiquitous all over the surface. For instance, between Aseke hill and Oyo waterworks, about 40 per cent of the surface is covered by bare rock outcrops. The surface is distinguished by the following features:

1. small to massive inselbergs, generally 30–130 m high, but up to 230 m on Ado Rock, and 320 m in Ilele hill near Ijio;
2. flat to gently convex interfluve, the former with frequent outcrops of duricrust, and the former with frequent outcrops of low rock pavements;
3. valley side slopes – gently inclined at 0.5–7.0°, but often intruded by low rock outcrops;
4. values of drainage density (1:50 000 topographic maps), average 1.9;
5. values of stream frequency (1:50 000 topographic maps), average 2.5;
6. values of hypsometric Integral of 5, fourth-order basins average 28.4 per cent.

Incised etchsurfaces: These are developed on quartzites and older granites. Examples of such surfaces on the latter rocks are found in Otan–Iree area, in Eruwa-Lalate, Idere-Tapa and Aiyete-Igangan in Oyo State, Nigeria (Figure 6.5). Away from the massive quartzite ridges and inselbergs, local relief varies from 45–150 m. The streams are deeply incised, especially along joint planes in the rocks. About 30 per cent of the surface is in the form of bare rock outcrops, especially massive

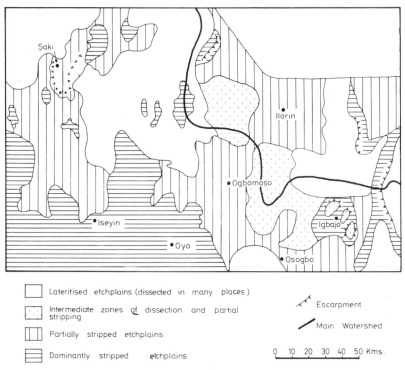

Fig. 6.5 Etchplains in south-western Nigeria (after Thomas, 1965)

inselbergs and whalebacks. Weathering depth appears to be rather shallow. In Idere-Tapa and in Otan the top soil is thin and sandy, and below it is a zone composed of hydrated biotite and felspar, and quartz

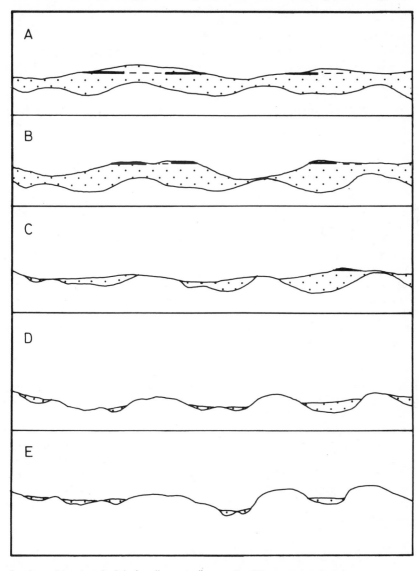

A – Lateritised etchplain (undissected)   B – Dissected etchplain

C – Partially stripped etchplain   D – Dominantly stripped etchplain

E – Incised etchplain

▬ Laterite deposit   ⌈ · · ⌉ Weathered rock

☐ Unaltered bedrock

Fig. 6.6 Different types of etchplains

fragments. The structure of the underlying granite is well preserved in the regolith above the parent rocks. The surface is typified by the following features (see also Figure 6.6):

1 large monolithic domes 60–230 m high, covered by exfoliated blocks;
2 tors and corestones located on the interfluves and on valley-side slopes;
3 broadly convex interfluves with frequent rock outcrops;
4 valley side slopes mainly in the form of converging inselberg footslopes inclined at 4–5°, but where incised, slope units vary from 10° to 25°;
5 values of drainage density (1:50 000 topographic maps), average 2.7;
6 values of stream frequency (1:50 000 topographic maps), average 2.0;
7 values of hypsometric Integral of 1, fourth-order basin is 30.2 per cent.

Major erosion surfaces in the humid tropics

Notwithstanding the mode of evolution, planation surfaces at various elevations are widespread in the humid tropics, both on the stable cratonic shields, where they have been eroded across a wide range of different lithology and structure, and on sedimentary rocks. The most comprehensive study of these surfaces is that by King (1962), who identified five main planation surfaces on most continents. These surfaces are separated from one another either by distinct scarps or by zones of dissection. Thus in most places in the humid tropics, the landscape is polycyclic.

In most cases, the oldest surfaces have virtually been destroyed as a result of the younger surfaces having been eroded into them so that only the remnants of the old surfaces are traceable either through accordance of summits or by the application of morphometric analysis.

Some of these surfaces contain structural depressions: for example, the Iullemmeden basin covering about 0.5 million km² and extending from the Sokoto area in Nigeria into Niger, Upper Volta and Mali occurs on the Gondwana surface. The Congo depression, containing sediments of Palaeozoic to Recent, and the Makarikari depression further south, occur on this surface. The Chad basin, extending from north-eastern Nigeria into Chad and Niger, occurs on the African surface. In all these places, the deposited sediments in these depressions have been eroded into series of erosion surfaces – for example, the Bateke plateau and the adjacent plain in the Kinshasa area, Zaire.

These planation surfaces have been traced in close contiguity over a small area in Nigeria between the Jos plateau and Niger river (Pugh, 1955; Buckle, 1978). The Gondwana surface occurs extensively over the Jos plateau at a general altitude of 1 200–1 350 m, especially in the Ganawuri and other hills in Naraguta area. The Post-Gondwana surface is especially well preserved on the fluvio-volcanic series consisting of decomposed basalts overlying alluvium. These series now occur as flat-topped hills at elevations between 1 091 m and 1 425 m. The plains in between these erosional residuals belong to the African erosional cycle. The plateau, which owes its existence to the occurrence of the extremely resistant younger granites and ring complexes, is abruptly terminated by steep scarps to the west, south-west and south-east. The scarp, which is about 610 m high, is structural rather than cyclical, as the African surface has been traced at lower elevations around the plateau between the towns of Gussau, Kontagora, Kusheriki, Abuja, Wase, Bauchi and Kano. The African surface occurs here at 665–755 m, especially north of a ragged scarp traceable from the Sokoto area through Birnin Gwari, to the north of the Benue valley and east of the Jos plateau (Pugh and King, 1952). South of this scarp is the Post-African surface at a general elevation of 300–330 m, especially on the Nupe sandstones along the Niger valley, rising to about 485 m near Kotangora. The Quaternary surface occurs as on the floodplain at 181–250 m along the Niger.

The African surfaces carry several massive inselbergs, which occur individually on foliated rocks like gneisses or as clusters where granitic intrusions are compartmentalized by a system of vertical and horizontal joints. Examples of these massive residuals on the African surface include the Kusheriki inselbergs near Birnin Gwari, the Dutsin Zuma inselberg near Abuja, and the Kufena hill near Zaria. On the Jos plateau, the Shere hills at 1770 m, and the Sauja hills at 1 585 m rise more than 300 m above the local surfaces. The African surface also carries several tabular hills underlain by weathered crystalline rocks and capped by duricrust. These are common between Zaria and Kano, between Birnin Gwari and Kaduna and even inside the city of Kano.

The Post-African surface on the sandstone south of the Kontagora–Zungeru–Kwali road carries several flat-topped residuals, among which are the Sunsun, Patindeji and Kochikota hills around Bida; and the Patti hill and the Agbaja plateau, standing at 250 m and 225 m respectively above the local plains.

These various surfaces have also been traced in Central and East Africa. Wayland (1934) recognized three of these surfaces in Uganda. He regarded the highest, which occurred at about 1 820 m in Ankole district, as 'a true peneplain', and the intermediate, at 1 300–1 330 m found in Mengo and Masaka districts, and the lowest, at 1 120–1 210 m extending from Lake Victoria to Lake Kyoga, as etchsurfaces. These constitute Peneplains I, II and III of Wayland. The highest, dated at Jurassic to Cretaceous, is equivalent to the Gondwana surface; the intermediate, at Miocene, is equivalent to the African surface; while the lowest, at Pliocene, is Post-African.

Ruhe (1956) not only recognized these three surfaces in the Nioka–Ituri area of Zaire, at about the same elevations as in Uganda, but he also recognized four to six lower surfaces belonging to the Quarternary period. According to him, the older surfaces cut across different lithology (granite, quartz diorite, mica and quartz schists), carry ferricretes formed *in situ*, and are reminscent of Du Preez's (1949) primary laterite, while the younger surfaces carry detrital ferricrete. The surfaces are also mantled by superficial sediments which grade into basal gravels at many localities.

McConnell (1955) recognized five surfaces in Uganda. The Ankole surfaces of Jurassic age, found on the Ankole Highlands at 1 640–1 970 m, is equivalent to the Gondwana landscape. The Koki surface, found in the Koki area of Masaka district at 1 430–1 460 m, and carrying thick duricrust cover, is equivalent to the Post-Gondwana surface, while the Buganda surface of late to Mid-Tertiary, found around Kampala, Buganda, and Busoga across flat-topped, ferricrete-covered hills at 1 300–1 330 m is equivalent to the African surface. The Tanganyika surface of End Tertiary, which occurs extensively in Tanzania, Acholi, Buganda, Bunyoro and Busoga districts at 1 060–1 150 m, is equivalent to the Post-African surface, while the Acholi surface at 910–1 000 m of early Pleistocene represented as extensive valley floors on the upper Nile, is equivalent to the Congo surface.

According to Bishop (1966), the Buganda surface is usually preserved upon quartzites of the Buganda series, while the Koki surface is developed only on the Karagwe–Ankolean metasediments, so that both surfaces may essentially belong to the same erosional cycle, separated only because of the differential resistance to weathering and erosion by the various rocks. The fact that the scarp bounding the Koki surface is lithologically determined further reinforces this observation.

Several erosion surfaces have also been recognized on the cratonic shields of South America and Australia. In Guyana, McConnell (1968, 1975) and Eden (1971) recognized five erosion surfaces, while in neighbouring Surinam three such surfaces have been distinguished by Krook (1975). Table 6.4 shows the list of planation surfaces in Guyana.

TABLE 6.4 Planation surfaces in Guyana

Surface	Average altitude	Probable age
Not named (Gondwana and Post Gondwana)	900 m	?
Kopinang (African)	600–700 m in Parakaima mountains	Late Cretaceous to early Tertiary
Kaietur (Late Cainozoic I)	400–450 m in Parakaima mountains, 200–300 m near Bartica	Late Tertiary
Rupununi (Cainozoic II)	100–140 m, rising to 180 m at some distance from the coast. Most widespread	End Tertiary
Mazaruni (Quaternary)	75 m, as valley floor plains	Quarternary and still active

These surfaces, eroded across different types of rock formations, are fairly constant in altitude and have very faint relief and gentle slopes. Each is separated from the other by pronounced scarps. The unnamed highest surface, found mainly on doleritic sills, appears structural, while the Kopinang surface, developed on dolerites, sandstones and quartzites, has been traced from Guyana to Surinam and Venezuela. It is covered by a thick layer of duricrust where underlain by dolerite. The Kaietur surface has been correlated with wide valley floors all over the Guyana shield, especially along the Potaro river on which the Kaietur Falls (214.6 m) occur, and along the Mazaruni and Upper Kurupung rivers. In most of these places, it is covered by detrital duricrust and separated from the lower surface by a scarp up to 90 m high. The Rupununi surface, which is the most widespread, occurs all over the Guyana shield, especially in the savanna zones to the north and south. It is covered by rainforest in the basins of the Kwaitaro, Essequibo and Courarntyne rivers. It extends into Brazil, Surinam and Venezuela, where it forms a vast plain at 100–150 m. However, in most of these places, the plain is studded by inselbergs and flat-crested hills, remnants of older surfaces. The lowest Mazaruni surface is found on the valley floor of the Upper Essequibo, Mazaruni, Potaro, and Cuyuni rivers, but towards the coast it forms a NE-tilting plateau which plunges below the sea following a monoclinal inflexion of the coastal region.

These planation surfaces have also been traced in Brazil (King, 1962). Relics of the Gondwana landscape are preserved as accordant summits in the Serra do Curral, Serra do Ouro Fino, Serra da Moeda on the mountains south of Belo Horizonte, and in Minas Gerais at altitudes exceeding 1 400 m. The surfaces are often mantled by thick duricrust, especially where underlain by itabirites (hematite-quartz-schists). In most of these places, the Gondwana surface is separated from the

Plate 16 'Pedisediment': transported stony soil over quartzitic bedrock, Accra Plain, Ghana (M. F. Thomas)

Post-Gondwana by steep scarps which may be up to 300 m – for example, west of the road from Belo Horizonte to Rio de Janeiro.

To the north and north-west on the southern flank of the Amazon basin, these two surfaces converge and coalesce into a smooth plain which passes under the Late Cretaceous continental sediments of the basin. The surfaces are also found on the sedimentary rocks north of the Amazon where, at elevations of 900 to 1 000 m, they pass on to the Guyana shields.

Eroded across the Gondwana surfaces is a plain of Early Tertiary age often referred to as the Sul-Americana, believed to be an equivalence of the African surface of the same age. This surface is monoclinally warped from the interior towards the eastern coast so that it declines in altitude from west to east. The surface has been traced on the cratonic shields all over South America, at elevations of 900–1 000 m on the Brazilian plateau especially in the basins of the Paraná, Tocantins, São Francisco, Uberaba and Uberlândia rivers, and in the Mato Grosso and Goias. It has also been traced all over Colombia, Venezuela, Surinam and Guyana. In north-east Brazil, it constitutes several plateau summits at 700–800 m, especially in the basins of the Piaui, Parnaiba and Rio Grande do Norte. It is, however, better preserved in eastern Minas Gerais. In all these places, it is being replaced by a younger end Tertiary surface associated with the Velhas erosion cycle.

The Velhas surface eroded into the Sul-Americana surface is the most widespread of any surface on the Brazilian and the Guyana shields. On the latter, it occurs at elevations between 200 m and 500 m, especially on the northern flank of the Amazon basin, where it is separated from lower surfaces by scarps. In north-eastern Brazil, it occurs as dissected plains at 450–550 m. The best example of such plains include the Moxito plains developed on relatively weak rocks. The plains continue into Baia. Towards the coastal area, the plain has been dissected into rugged terrain with vestiges of the Velhas surface only preserved on the flat crests.

The Velhas surface has virtually replaced the Sul-Americana all over the interior of Brazil through Minas Gerais, to São Paulo and Paraná basin, penetrating right to the heads of the major Amazon tributaries like the São Francisco and Paraná. This relationship between the Sul-Americana and the Velhas surfaces is also well established in southern Brazil, where the two surfaces are developed across both basement crystalline and sedimentary rocks. The older surface, at 800–950 m is separated by a scarp 100 m high from the younger surface.

The Velhas surface is being replaced in the coastal zone of Brazil by a Quaternary surface which developed in response to the cymatogenic arching of the eastern Brazilian shield during the Plio-Pleistocene. All the regions affected by this monoclinal arching are already profoundly dissected by deep valley systems, especially in the lower courses of the W–E flowing rivers. A large section of eastern Bahia and most of Espírito Santo are already affected, but most of the interior areas, such as the Minas Gerais and the Mato Grosso are relatively unaffected. However, in these places, the surface is already in the form of terraces above the level of the present valley floors.

The axis of the monoclinal upwarping is located on the coastal area of Serra do Mar between Santos and Rio de Janeiro where the coastline was raised into a cliff up to 1 000 m high. This scarp, which can be traced further to Espírito Santo areas, has since retreated for a distance of several kilometres behind the head of Guanebara Bay near Rio de Janeiro, where it has isolated several inselbergs. Such massive inselbergs developed in granite gneisses are also common on the young surface in Espírito Santo. The retreating scarp can also be observed between Santos and São Paulo.

Conclusion

As shown in this chapter, the terrain in the humid tropics comprise mainly different types of plains at various elevations. These plains in many places are in the form of three-tier landscapes composed of wide valley floors, wide interfluvial surfaces/foot-slopes and erosional residuals. The latter are in form of hills – inselbergs, bornhardts, tors, mesas, towerkarst, conekarst, demi oranges – cuestas, ridges and mountains, which may occur singly or in chains and clusters, especially where massive intrusive rocks have been exhumed or where extruded rocks have been extensively dissected.

In some places these plains bear imprints of the Pleistocene–Holocene climatic fluctuations in the form of buried ancient alluvium or as dunes and drift material, especially across West Africa.

Opinions vary with regard to the origin of these plains, which have been ascribed to peneplanation, pediplanation or etchplanation processes. Although the first no longer commands any serious consideration, the staircase pattern of occurrence of these plains in some areas tends to lend some credence to the theory of peneplanation. However, with its little regard to the reality of deep chemical weathering, the theory is not generally applicable in explaining the complexity of landform assemblages in the humid tropics. Thus the concept of etchplanation with its focus on deep chemical weathering, the influence of rock lithology, and the fortuitous stripping of superincumbent regolith has attracted considerable attention in explaining landform development in the humid tropics. However, much research remains to be carried out before we can fully understand the characteristics and the development processes of most of the landforms in the tropics.

APPENDIX 1   A genetic landform classification

| | Constructional | | Destructional | | |
	Processes	Forms	Reduction Forms	Residual Forms	Processes
Depositional	Diastrophic	Plains and plateaux grabens, fault mountains, down-warped basins, dome mountains, folds and complex mountains	?	?	Diastrophic
	Volcanic	Lava plains and plateaux, volcanics	Calderas?	?	Volcanic
	Weathering and soil movements	Talus, rock glaciers landslide accumu-lations	Landslide scars	Exfoliation domes	Weathering soil move-ments
	Fluvial	Plains, plateaux, fans, cones, deltas, floodplains and precipitation forms	Valleys, straths and pene-plains	Monad-nocks	Fluvial
	Glacial	Moraines, drumlins, outwash plains	Troughs, cirques	Aretes, horns	Glacial
	Eolian	Loessplains, dunes	Deflation hollows	Yardangs	Eolian
	Littoral	Offshore bars, beaches, embankments	Sea cliffs, wave cut terraces	Sea stacks	Littoral
	Lacustrine	Lake plains, precipitation forms	?	?	Lacustrine
	Terraqueous	Cones, terraces	Sinks	Natural bridges (in part)	Terr-aqueous
	Impact	?	Meteor	?	Impact
	Organic	Coral reefs	?	?	Organic

Source: Howard and Spock, 1940

(right margin vertical label: **Erosional**)

APPENDIX 2   Landform classification
(R. W. Fairbridge, 1968)

Within each continent, the following geomorphic landforms are recognized:

1 Undisturbed structures (Plains and Plateaux)
   (a) Coastal Plains – pass under the sea on to the continental shelf with no break. The zone is alternatively exposed or covered by the sea.
   (b) Lowland Plains – usually conform to underlying sedimentary basins but sometimes correspond to erosion surfaces cutting across old folded foundations. May or may not be veneered by alluvium.
   (c) Upland Plains – usually discordant to the underlying structure which is not expressed in the topography except as relict hills. Occasionally found to be conformable on existing flat-lying formations, e.g. Great Plains of western U.S.A.
   (d) Plateaux – consist of horizontally bedded sediments elevated to high altitudes; or may be high interior basins with discordant foundations and borders, e.g. Altiplano of Mexico, and Bolivia, and the high plateaux of Anatolia; Iran and Tibet. They may be high discordant plains as extensively developed in East and South Africa. They may also be exhumed peneplains, e.g. Spanish Mesta.

2 Disturbed structures – Endogenic forces are dominant
   A. Youthful, Early-Maturity or Rejuvenated stage:
      (a) Fold mountains, singly or in chains, sedimentary – Jura type or Appalachian type
      (b) Fold-thrust nappe complexes – Alpine type mountains
      (c) Domes – Black hills, Ozarks in U.S.A.
      (d) Fault Blocks – Rhine Graben or Basin and Range Province
      (e) Belted Metamorphic Mountains – Laurentians of Quebec
      (f) Ancient crystalline Mountains – Wyoming and Colorado Rockies
   B. Old-age stages – Endogenic origin suppressed by exogenic modelling. All the above structure reduced to late mature or old-age landscapes.
      (a) Lowlands – degraded mountain roots, degraded metamorphic belts – Hudson River Lowlands
      (b) Lowland or Upland Hills – Disturbed structure of any category reduced to irregular topography of subdued relief, hills (less than 600 m). Intermediate between highlands and lowlands, e.g. Appalachian piedmont
      (c) Highlands – Disturbed structures of any sort deeply dissected and reduced to moderate relief, marked by more mature slopes, but with absolute relief in the mountain category ca. 1000 m, e.g. Black Forest

3 Extrusive structures
   A. Youthful, Early-mature or Revived stage
      (a) Volcanic Cones – from minor cones to high strato-volcanoes

like Fuji Yama in Japan, Egmont in New Zealand, Popocatepetl in Mexico, Mounts Hood and Shasta in northwestern U.S.A.

(b) Volcanic Domes – from small mamelons to Mauna Loa

(c) Lava Fields – has the topography of existing landsurfaces

(d) Lava Plateaus – complex regions of multiple fissure eruptions forming a regional plateau, e.g. Deccan in India, or Columbia in Oregon.

(e) Calderas – ranging from diatremes to explosion craters and explosion calderas to collapse calderas

B. Old Age (Extrusive structures) – all of the above may occur as degraded landforms following long extinction of volcanicity.

APPENDIX 3   Scale of units in geomorphology

Magni-tude	(Km²) size	Duration (yrs)	Examples	Basis
1st	10^7	500 million	1. Continental masses, ocean basins	Different categories of varied geodynamic origin: i.e. 1. geophysical, 2. geotectonics, 3. climatic
		100 million	2. Geosynclines, continental platforms	
		Tert.=∣65 m y Quat. = 2m. y	3. Morphoclimatic zones, e.g. tropical zone	
2nd	10^6		2. Scandinavian Shield, Russian platform, Hercynian belt	Subdivisions of categories 2 and 3
			3. Savanna, rainforest (major division of Linton, 1948)	
3rd	10^5		2. High Plateau, and High Anti-Atlas ('province' of Linton, 1948)	Structural units. Geomorphic effects of single climatic zones not recognizable at this scale
4th	10^4		2. Paris basin, Central Massif ('section' of Linton, 1948)	Structural units of limited regional character. Single physiographic process usually dominant
5th	10^3		2. Chalk of Picardy-Normandy ('tract' of Linton, 1948)	Tectonic and often lithologic individuality. Lower limit of isostatic compensation, single relief type
6th	10^2		Individual anticlines, synclines in folded ranges, cuestas, fault blocks ('Stow' of Linton, 1948)	Lithology is important, influences tectonic style and operation of exogenic processes. No isostatic compensation. Repetition of relief features
7th	10^4–10^5		Local relief, pediments, inselbergs, consequent gaps, etc. ('site' of Linton, 1948)	Tectonics subordinate. Exogenic processes and lithology dominant. Microclimates and biotopes influence processes
Micro-forms	Less km²		Mudflows, torrent ravines, slope facets, goletz, barchans, solifluction lobes, soil polygons, thufurs, terracettes, lappies, dreikanter, ripple marks, etc.	Exogenic processes marked by discontinuity of phenomena in place and time (requires statistical treatment). Susceptible to experimental approach

Source: Tricant, 1952; Cailleux and Tricart, 1956

References and further reading

Aitchson, P. J. and Glover, P. E. (1972) *The Land Resources of North East Nigeria*, 3, *The Land Systems*, L.R.D./D.O.S., Tolworth, Surrey, England.

Bakker, J. P. (1967) 'Weathering of granites in different climates', in *L'Evolution des Versants*, P. Macar (ed.), Congress Collegium, L'Université, Liège, 40, p. 5168.

Bakker, J. P. and Levelt, T. W. M. (1964) 'An enquiry into the problem of a polyclimatic development of peneplains and pediments (etchplains) in Europe during the Senonian and Tertiary period', *Publs. Serv. Carte Geol. Luxembourg*, 14, 27–75.

Bishop, W. W. (1966) 'Stratigraphic geomorphology: a review of some East African landforms', in *Essays in Geomorphology*, G. H. Dury (ed.), Heinemann, London, pp. 139–76.

Buckle, C. (1978) *Landforms in Africa: an introduction to geomorphology*, Longman, London.

Büdel, J. (1948) 'Das System der Klimatischen Geomorphologie (Beilrage zur Geomorphologie der Klimazonen und Vorzeit Klimate V)' *Verhandlungen Deutscher Geographentag*, München, 27, 65–100.

Büdel, J. (1957) 'Die *Doppelten Einebnungsflächen* in den feuchten Tropen', *Ziets. für Geomorph. N.F.*, 1, 201–28.

Büdel, J. (1963) 'Klima-genitische Geomorphologie', *Geographische Rundschau*, 15, 269–85.

Büdel, J. (1965) 'Die Relieftypen der Flachenspulzone: Sud-Indiens am Ostabfall Dekans gegen Madras', *Colloquium Geographie*, 8, Bonn.

Cailleux, A. and Tricart, J. (1956) 'Le problème de la classification des faits géomorphologiques', *Ann. Géog.*, 65, 162–86.

Christian, C. S. (1957) 'The concept of land units and land systems', *Proc. 9th Pacific Science Conference*, 20, 74–81.

Clarke, J. I. (1966) 'Morphometry from maps', in *Essays in Geomorphology*, G. H. Dury (ed.), Heinemann, London, pp. 235–74.

Cooke, H. J. (1973) 'A tropical karst in north-east Tanzania', *Zeits. für Geomorph. N.F.*, 17(4), 443–59.

Corbel, J. (1964) 'L'érosion terrestre étude quantitative (méthodes-téchniques-résultats)', *Ann. Géog.*, 73, 385–412.

Cotton, C. A. (1942) *Climatic Accidents in Landscape Making*, Whitcombe and Tombs, Wellington, New Zealand.

Craft, A. (1933) 'The coastal tablelands and streams of New South Wales', *Proc. Linnaean Soc. N.S.W.*, 58, 437–60.

Davis, W. M. (1899) 'The geographical cycle', *Geog. J.*, 14, 481–584.

Davis, W. M. (1900) 'Glacial erosion in the valley of the Ticino', *Appalachia*, 9, 136–56.

Davis, W. M. (1905) 'The geographical cycle in an arid climate', *J. of Geol.*, 13, 381–407.

Doornkamp, J. C., and King, C. A. M. (1971) *Numerical Analysis in Geomorphology*, Arnold, London.

Douglas, I. (1969) 'The efficiency of humid tropical denudation systems', *Trans. Inst. Br. Geog.*, 46, 1–16.

De Swardt, A. M. J. (1946) 'The recent erosion history of the Kaduna Valley near Kaduna Township', *Annual Report, Geological Survey of Nigeria*, 33–45.

De Swardt, A. M. J. (1953) 'The geology of the area around Ilesha', *Geological Survey of Nigeria, Bulletin 23*.

Du Preez, J. W. (1949) 'Origin, classification and distribution of Nigerian laterites', *Proc. 3rd International West African Conference*, Ibadan, 223–34.

Dury, G. H. (1972) 'A partial definition of the term "pediment" with field tests in humid climate areas of Southern England', *Trans. Inst. Br. Geog.*, **57**, 139–52.

Eden, M. J. (1971) 'Some aspects of weathering and landform in Guyana (formerly British Guiana)', *Zeits. für Geomorph.* N.F., **15** (2), 181–98.

Eyles, R. J. (1971) 'A classification of west Malaysia drainage basins', *Ann. Assoc. Amer. Geog.*, **61**, 460–7.

Fairbridge, R. W. (1961) 'Eustatic changes in sea level', *Physics and Chemistry of the Earth*, series 4, 99–185.

Fairbridge, R. W. (1968) 'Land mass and major landform classification', in *The Encyclopedia of Geomorphology*, R. W. Fairbridge (ed.), Reinhold, New York.

Fournier, F. (1960) *Climat et Erosion: la relation entre l'érosion du sol par l'eau et les précipitations atmosphériques*, Presses Univ., Paris.

Fripiat, J. J. and Herbillon, A. J. (1971) 'Formation and transformation of clay minerals in tropical soils', in *Soils and Tropical Weathering, Proc. Badung Symposium*, UNESCO, 1969, 15–24.

Geyl, W. F. (1961) 'Morphometric analysis and the world wide occurence of stripped erosion surfaces', *J. of Geol.*, **69**, 388–416.

Haggett, P., Chorley, R. J. and Stoddart, D. R. (1965) 'Scales standards in geographical research: a new measure of areal magnitude', *Nature*, **205**, 844–7.

Harpum, J. R. (1963) 'Evolution of granite scenery in Tanganyika', *Geol. Survey Tanganyika Record*, **10**. 39–46.

Hays, J. (1967) 'Land surfaces and laterites in the north of the Northern Territory', in *Landform Studies from Australia and New Guinea*, J. N. Jennings and J. A. Mabbutt (eds.), Cambridge Univ. Press, Cambridge.

Howard, A. D. and Spock, L. E. (1940) 'A classification of landforms', *J. of Geomorph.*, **3**, 332–45.

Jahn, A. (1974) 'Granite tors in the Sudeten Mountains', in *Progress in Geomorphology*, E. H. Brown and R. S. Waters (eds), Inst. Br. Geog., **7**.

Jeje, L. K. (1970) 'Some aspects of the geomorphology of south western Nigeria', Ph.D. thesis, Edinburgh Univ.

Jeje, L. K. (1972) 'Pediments', *Nigerian Geog. J.*, **15** (1), 79–82.

Jeje, L. K. (1974) 'Effects of rock composition and structure on landform development: the example of Idanre hills of Western Nigeria', *J. of Trop. Geog.*, **39**, 43–53.

King, L. C. (1953) 'Canons of landscape evolution', *Bull. of Geol. Ass. Amer.*, **64**, 721–52.

King, L. C. (1957) 'The uniformitarian nature of hillslopes', *Trans. Edinburgh Geol. Soc.*, **17**, 81.

King, L. C. (1962) *The Morphology of the Earth*, Oliver & Boyd, Edinburgh.

Krook, L. (1975) 'Surinam', in *The Encyclopedia of World Regional Geology Part I*: Western Hemisphere, R. W. Fairbridge (ed.), J. Wiley, London.

Langbein, W. B. and Schumm, S. A. (1958) 'Yield of sediment in relation to mean annual precipitation', *Trans. Amer. Geophysical Union*, **39**, 1076–84.

Linton, D. L. (1955) 'The problem of tors', *Geog. J.*, **121**, 470–87.

Louis, H. (1964) 'Uber rumpfflächen und talbilding in den Wechselfeuchten tropen besonders nach studien in Tanganyika', *Zeits. für Geomorph.*, **8** (Sonderheft), 43–70.

Mabbutt, J. A. (1961) 'A stripped land surface in Western Australia', *Trans. Inst. Br. Geog.*, **29**, 101–14.

Mabbutt, J. A. and Stewart, G. A. (1965) 'The application of geomorphology in resource surveys in Australia and New Guinea', *Revue de Géomorph. Dynamique*, **7–9**, 97–109.

McConnell, R. B. (1955) 'The erosion surfaces of Uganda', *Colonial Geol. Mining Resources*, **5**, 425–8.

McConnell, R. B. (1968) 'Planation surfaces in Guyana', *Geog. J.*, **134** (4), 506–20.

McConnell, R. B. (1975) 'Guyana', in R. W. Fairbridge (ed.), *op. cit.*

Miller, A. A. (1953) *The Skin of the Earth*, Methuen, London.

Ofomata, G. E. K. (1973) 'Aspects of the geomorphology of the Nsukka-Okigwe cuesta, East Central State of Nigeria', *Bull. de l'IFAN*, 35, Série A, 489–501.

Ollier, C. D. (1960) 'The inselbergs of Uganda', *Zeits. für Geomorph.* N.F., **4**, 43–52.

Pallister, J. W. (1956) 'Slope development in Buganda', *Geog. J.*, **122**, 80–7.

Pallister, J. W. (1960) 'Erosion cycles and associated surfaces of Mengo District, Buganda', *Overseas Geol. and Mineral Resources*, **8** (1), 26–36.

Peltier, L. C. (1950) 'The geographic cycle in periglacial regions as it is related to climatic geomorphology', *Ann. Assn. Amer. Geog.* **40**, 214–36.

Peltier, L. C. (1962) 'Area sampling for terrain analysis', *Professional Geographer*, **14** (2), 24–8.

Pugh, J. C. (1955) 'The geomorphology of the Northern Plateau of Nigeria', Ph.D. thesis, University of London.

Pugh, J. C. (1966) 'The landforms of low latitudes', in *Essays in Geomorphology*, G. H. Dury (ed.), Heinemann, London.

Pugh, J. C. and King, L. C. (1952) 'Outline of the geomorphology of Nigeria', *South African Geog. J.*, **34**, 30–7.

Rich, J. L. (1938) 'Recognition and significance of multiple erosion surfaces', *Bull. Gol. Soc. Amer.*, **49**, 1695–722.

Ruhe, R. V. (1956) 'Landscape evolution in the High Ituri, Belgian Congo', *Publication l'Institut National d'Etudes Agronomiques du Congo Belge*, I.N.E.A.C., S.S. 66.

Ruxton, B. P. and Berry, L. (1961) 'Notes on faceted slopes, rock fans and domes on granite in the East Central Sudan', *Am. J. of Sci.*, **259**, 194–206.

Savigear, R. A. G. (1960) 'Slopes and hills in West Africa', *Zeits. für Geomorph.*, Suppl. 156–71.

Schumm, S. A. (1965) 'Quaternary paleohydrology', in *The Quaternary of the United States*, H. E. Wright, Jr., and D. G. Frey (eds), Princeton University Press, 783–94.

Strakov, N. M. (1967) *Principles of Lithogenesis*, trans. J. P. Fitzsimmons, Oliver & Boyd, Edinburgh.

Stoddart, D. R. (1969) 'Climatic geomorphology: review and reassessment', in *Progress in Geomorphology*, Vol. 1, Arnold, London, 161–222.

Stewart, G. A. (1968) '*Land Evaluation*', Macmillan, London.

Swan, S. B. St. C. (1970) 'Land surface mapping, Johor, West Malaysia', *J. of Tropical Geog.*, **31**, 91–103.

Sweeting, M. M. (1958) 'The Karstlands of Jamaica', *Geog. J.*, **124**, 184–99.

Taylor, B. W., Baker, R. M., Loefers, C. L. and Rosayro, R. A. (1962) *Report on the Land Use Survey of the Oyo-Shaki Area*, unpublished, Ministry of Agriculture and Natural Resources, Ibadan, Nigeria.

Thomas, M. F. (1965) 'An approach to some problems of landform analysis in tropical environments', in *Essays in Geography for Austin Miller*, P. D. Wood and J. B. Whittow, (eds), University of Reading.

Thomas, M. F. (1969) 'Geomorphology and land classification in tropical Africa', in *Environment and Land Use in Africa*, M. F. Thomas and G. Whittington (eds), Methuen, London.

Thomas, M. F. (1974) *Tropical Geomorphology: a study of weathering and landform development in warm climates*, Macmillan, London.

Townshend, J. R. G. (1970) 'Geology, form and slope processes and their relation to the occurrence of laterite', *Geog. J.*, **136**, 392–9.

Tricart, J. (1952) 'La géomorphologie et la notio d'echelle', *Revue Geomorph. Dyn.*, **3**, 213–18.

Tricart, J. (1972) *The Landforms of the Humid Tropics, Forest and Savannas*, trans. C. K. de Jonge, Longman, London.

Tricart, J. and Cailleux, A. (1965) *'Introduction à la géomorphologie climatique'*, S.E.D.E.S., Paris.

Wayland, E. J. (1934) 'Peneplains and some other erosional platforms', *Bull. Geol. Survey of Uganda, Annual Report Notes*, 1, 74, 366.

Weaver, G. D. (1965) 'What is landform?', *Professional Geographer*, **17**, 11–13.

Wigwe, G. A. (1966) 'Drainage composition and valley forms in parts of Northern and Western Nigeria', Ph.D. thesis, University of Ibadan.

Wright, R. L. (1963) 'Deep weathering and erosion surfaces in the Daly River Basin, Northern Territory', *J. Geol. Soc. of Australia*, **10**, 151–65.

Young, A. (1963) 'Soil movement on slopes', *Nature*, **200** (4902), 129–30.

Young, A. (1969) 'Natural resource survey in Malawi: some considerations of the regional method in environmental description', in M. F. Thomas, and G. Whittington, (eds), *op. cit.*

Young, A. (1970) *Slopes*, Oliver & Boyd, Edinburgh.

7 Rivers and river patterns

There are several possible connotations of the word 'river'. Geologically, the word generally refers to the main trunk (segment) of a drainage system. However, following the quantitative revolution of the post-Hortonian era, and the accompanying problems of defining unambiguously the 'trunk stream' of a drainage basin, geomorphologists and geologists alike have come to define the word 'river' more broadly as a stream of water bearing the waste of the land from higher to lower ground, and as a rule to the sea; or as the trunk stream and all its branches in a river system (Amer. Geol. Institute, 1962). Thus by implication, the classical/hierarchical distinction between 'rill', 'rivulet', 'brook', 'creek' and 'river' is fast disappearing from the literature as words and phrases such as 'stream', 'stream segment', 'stream order', 'river system', 'drainage basin' and so on become well established, especially in quantitative drainage studies. Accordingly, the word 'river' is here used in the same sense that the less discriminatory term 'stream' is often used with little or no relation to size, and so covers all water-scoured courses or valleys through which water flows, however infrequently. These courses, in the humid tropical environment, always carry water during at least one season of the year (in the savanna region), but throughout the year in the rainforest and other heavy rainfall (monsoonal) areas.

After hills and mountains, which are in many cases fashioned by rivers, and perhaps also vegetation, river courses are among the most conspicuous features of the landscape. Rivers and river activities have, therefore, been among the first to be described by perceptive observers. Among these was Playfair, who, in 1802, wrote as follows:

> Every river appears to consist of a main trunk, fed from a variety of branches, each running in a valley proportioned to its size and all of them together forming a system of valleys, communicating with one another, and have such a nice adjustment of their declivities, that none of them join the principal valley, either on too high or too low a level; a circumstance which would be infinitely improbable, if each of these valleys were not the work of the stream that flows in it.

Playfair wrote with a view to convincing scientists of his time that rivers make the valleys in which they are found, which was not easy at a time when the Plutonists of that era had converted several scientists to their own cataclysmic view of landform, especially valley formation. And although subsequent observations (for example, on misfit streams) have invalidated parts of Playfair's observations, they still form the groundwork of the science of streams, especially fluvial geomorphology. Playfair's concept of the trunk stream in fact survived the early days of the quantitative revolution (cf. Horton, 1932, 1945), and fell into obscurity only after the works of Strahler (1952) and his Columbian colleagues on river-basin morphometry.

In Chapter 4 we considered the removal processes in the humid tropics, including surface flow. The result of that particular process is the excavation of a channel, also called river valley, the characteristics of which form the topics for the present chapter. These characteristics are described under the following headings: initiation, including humid tropical river hydrology; the thalweg or longitudinal cross-section; and the transverse or valley cross-section, using examples of humid tropical rivers.

The river valley and its initiation

The river valley has at least three dimensions defined respectively as length, width and depth, which together define it. Both the size and the shape (form) of valleys vary widely from one part of the earth's surface to another, reflecting among other factors the environmental factors of climate, geology, including structural-geological history, topography, and the nature of flow through the valley.

The main cause of rivers is rainfall: rivers are, therefore, sometimes *ephemeral* – that is, they carry water only during and immediately after a rain; they may be *intermittent* – they flow during part of the year (the wet-season streams of the humid tropics); or they may be *perennial* and flow all the year round, as in the rainforest and other continuously wet areas.

In many rivers there is usually a direct relationship between rainfall and river flow or discharge (Table 7.1a, b). The deficit between rainfall and runoff is often attributable to evaporation plus evapotranspiration and seepage or infiltration, part of which (see page 000) eventually contributes to runoff (base flow). These relationships are best expressed in the concept of the hydrological cycle, in which the water that falls as precipitation (rain, snow, hail or sleet) may be (1) temporarily retained on the surface (as glaciers, snow banks, lakes or ponds); (2) evaporated and/or transpired by plants back into space; (3) retained in plants or in the soil; or (4) run off into streams, both directly and through seepage (through flow). Eventually, the water is returned into space via several paths, where it condenses and falls as rain (Figure 7.1). The aspect of the hydrological cycle of particular interest to geomorphologists generally and to this chapter in particular is that in open channels, which carves the various forms and pattern described below.

There are several possible sources of the water in channels or rivers. They include unconcentrated (sheet) flow and sub-surface flow, but the most important sources are the karst areas, basaltic lava country, and the perhaps more common 'outflow from coalescing lines of sub-surface water movement . . . in deep regolith material in the humid tropics'

TABLE 7.1a Summary of rainfall–runoff relation for the river catchment of the Nigerian basement complex by hydrologic regions

Region	Group of river catchments	Rainfall (p, ins.)			Runoff (R, ins.)			
		Six-year mean total	Six-year mean	% p of total	Six-year mean total	Six-year mean	% R of total	R as % of p
	A₁ Western State	941.34	56.37	19.98	211.73	12.46	13.00	22.10
	A₂ Kwara State	722.83	49.50	17.51	256.72	17.11	17.85	34.64
	B₁ Benue-Plateau	307.64	51.50	18.40	225.19	20.56	21.45	39.62
	B₂ North-Central	628.47	48.34	17.14	211.15	16.24	16.95	33.60
	B₃ North-Western	328.31	33.74	11.96	151.52	15.15	15.81	44.90
	B₄ Kano	443.66	42.33	15.01	157.50	14.32	14.94	53.83

Source: Okechuwu, 1974

TABLE 7.1b   Runoff from land-use surfaces within small catchments (total rainfall = 343 mm)

Characteristics of land-use surface types	Runoff percentage of rainfall
Relatively undisturbed grass surface, with tree cover	29.6
Maize farmland with 5% slope	26.7
Fallow land with short bush	20.7
Maize farmland with less than 2% slope	16.3
Forest	0.00

Source: Akintola, 1974

(Douglas, 1977; Faniran, 1980). Another possible source of streams according to Douglas (1977) is the mass of rock debris (coreboulders) of many forested granitic terrains in the humid tropical regions, but this can be safely combined with the second (already mentioned) sources, especially that of deeply weathered regoliths. In the case of coreboulders, what has happened seems to be partial or differential weathering within corestone remnants (see Chapter 3).

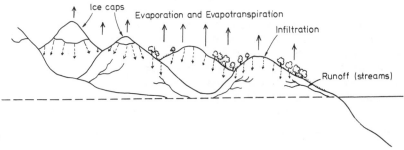

Fig. 7.1 The hydrological cycle (modified from Douglas, 1977)

Let us now return to the form (size and shape) of the river channel. Among other factors, variations in form are largely a response to variation in the input of water into the surface and sub-surface sources of runoff or components of the hydrological cycle. The relative importance of these two sources, which have been formulated into models of channel initiation, has also been contested. Whereas some people think that the Hortonian (overland) model is not strictly applicable in the forested humid tropics, others think that the second model (through-flow model) is only an extension and complication of Horton's overland flow model.

According to the *Hortonian model of channel initiation*, if water is poured on a bare, unconsolidated, sloping surface, rills will develop as soon as flow becomes sufficiently turbulent to erode the loose material.

However, for turbulent flow to occur, the infiltration capacity of the soil should have been exceeded sufficiently either by the amount or the rate of precipitation. It is reasonable to expect this to happen further down the hillside (valley-side) slope, and to expect a number of flow environments, with the depth of flow increasing from the hill-crest to the valley bottom, and with it the power of runoff to initiate a channel. Horton (1945) thus recognized three zones as (1) belt of no erosion, (2) belt of active erosion, and (3) belt of deposition (Figure 7.2).

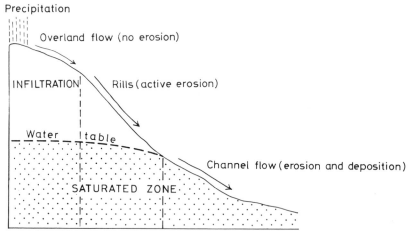

Fig. 7.2 The Hortonian model of channel initiation

Although the rills and gullies developed along hillside slopes are channels of some sort, they are not the type that make the rivers which usually occur at the base of such slopes. Also, these channels carry water well after the overland flow component has dried out, while many river sources are associated with springs, pipes or similar sub-surface sources. These are the sources which *through-flow model of channel initiation* (Figure 7.3) emphasize. Here, the spring head of source and its location

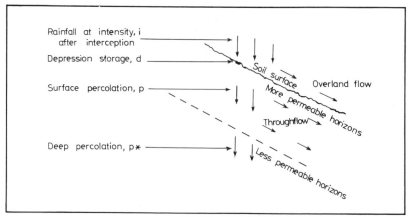

Fig. 7.3 The through-flow model of channel initiation

relative to the other heads become very crucial, especially in relation to the various measures of drainage composition, particularly drainage density. Moreover, the rate of retreat of these heads determines the ultimate size of the basins, if not their shapes. This idea is similar to that

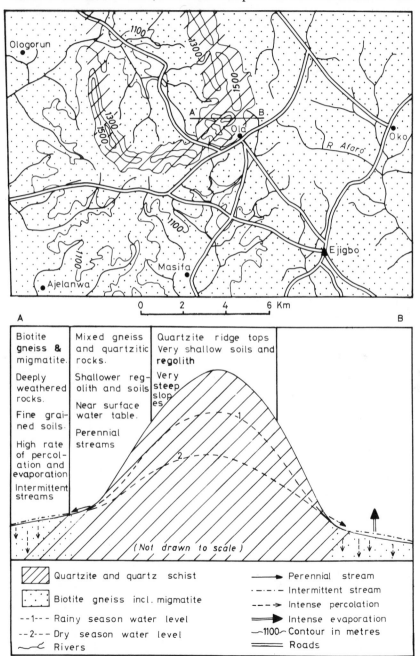

Fig. 7.4 Springline, Ejigbo area, Oyo State, Nigeria

195

implied in the stream piracy (capture) principle, which is widely discussed in traditional geomorphology texts on the drainage network of many areas. Finally, the through-flow model appears to fit many situations in the humid tropical world, especially those covered with deep regoliths. In these areas, springs are widespread, some of which not only form the sources of many large rivers, but also flow all year round near their heads, even in seasonal rainfall savanna areas such as the Ejigbo area, Oyo State, Nigeria (Faniran, 1980; Figure 7.4).

Both Horton's overland flow model and the through-flow model of runoff and channel initiation imply the assumption, although to a varying extent, of a widespread, if not a basin-wide, source of channel water. In the case of Horton's overland flow, the contributory area is definitely the whole basin, whereas in the through-flow model, the weathered mantles, pipes and other aquifers are the major sources. By contrast, recent research suggests that the area of storm runoff may be more restricted and variable from basin to basin.

According to the *partial area model of runoff and channel initiation*, as through flow and surface runoff converge in natural declivities, some parts of the catchment area become so wet as to initiate runoff. A number

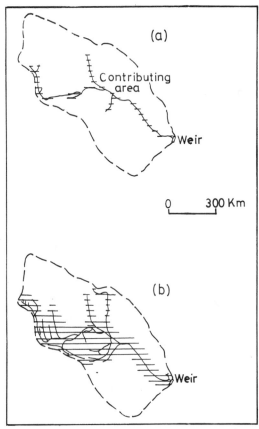

Fig. 7.5 Partial area model of runoff and channel initiation after different storms

of such aquifers may exist within a basin and feed the river at different times, depending on local conditions. However, an integrated drainage system may develop (Figure 7.5). The partial area model also recognizes possibilities of expansion and contraction of the functioning drainage net, including possibilities of variations in drainage density from one storm to another.

The partial area model, however, need not necessarily be seen as a separate model, and can be regarded as an extension of the through-flow or sub-surface flow model. Here, the sub-surface sources are the aquifers which may be rain-fed, fed by prior streams, or from other sources, including weathering basins (Faniran, 1980; Faniran and Omorinbola, 1980). Such areas are easily saturated during storms, and so initiate runoff which develops into channels. They do not, however, necessarily get saturated at the same time or to the same extent, thus explaining their variable contributions to channel flow. A very important factor here, apart from the aquifer factor, is the nature of tropical storms, especially the convectional types, which cover limited areas and tend to occur at random over space, including drainage basins. Thus, depending upon the part of the basin having rainfall at a particular time, the contributory area will vary (see Gerson, 1974, p. 5).

The hydrology and hydraulics of humid tropical rivers

Hydrology is a relatively young discipline in the humid tropics, and so suffers like other topics from the lack of definitive statements regarding regional parameters and their behaviours. Thus, until recently information about the rivers of the region was limited to their description in relation to geological structure – such as dendritic, trellised and so on (Figure 7.6, Table 7.2), while areas were described qualitatively as 'poorly drained', 'well-drained', 'marshy' and so forth. The mechanics of how river channels and networks form as well as the internal functions and relations of rivers, which form the main inputs of present-day hydrology, were treated in general, vague and qualitative terms. Consequently, water resources development projects, especially in the tropics, were based on what is often described as a 'pragmatic' approach, involving dams and not based on any hydrological information – they were examples of planning without facts.

However, this situation has changed for the better in many countries, especially since the establishment by the United Nations of the International Hydrological Decade. Many countries have now amassed sufficient hydrological data to facilitate the description of regional hydrology in these countries in particular, and the humid tropics in general (Figure 7.7).

An example of this kind of work is available for West Africa, where Rodier (1961) and Ledger (1964), among others, have described hydrological regions with specific characteristics. According to Rodier (1961), two major hydrological regions – equatorial and tropical – are

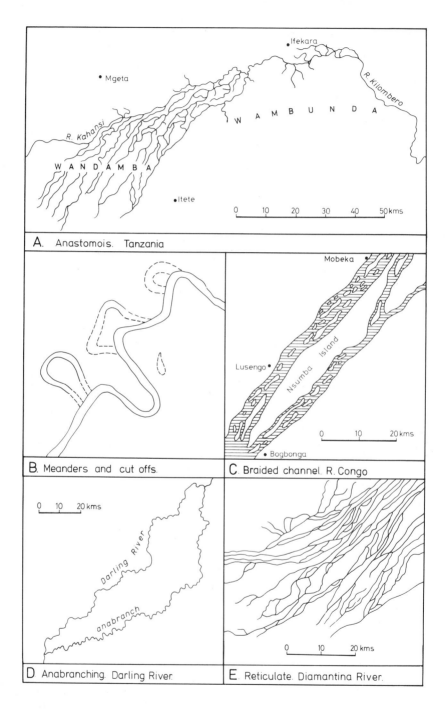

A. Anastomois. Tanzania

B. Meanders and cut offs.

C. Braided channel. R. Congo

D. Anabranching. Darling River.

E. Reticulate. Diamantina River.

Fig. 7.6 River channel patterns

TABLE 7.2   Descriptions and characteristics of basic patterns

River pattern	Geomorphic significance
Dendritic	Horizontal sediments or bevelled, uniformly resistant crystalline rocks. Gentle regional slope at present or at time of drainage inception.
Parallel	Generally indicates moderate to steep slopes, but also found in areas of parallel, elongate landforms.
Trellis	Dipping or folded sedimentary, volcanic or low-grade metasedimentary rocks: areas of parallel fractures: exposed lake or sea floors ribbed by beach ridges. Type pattern is regarded here as one in which small tributaries are essentially same size on opposite sides of long parallel subsequent streams.
Rectangular	Joints and/or faults at right angles. Lacks orderly repetitive quality of trellis pattern: streams and divides lack regional continuity.
Radial	Volcanoes, domes and erosion residuals. A complex of radial patterns in a volcanic field might be called multiradial.
Annular	Structural domes and basins, distrimes and possibly stocks.
Multibasinal	Hummocky surficial deposits: differentially scoured or deflated bedrock: areas of recent volcanism, limestone solution and permafrost. This descriptive terms is suggested for all multiple-depression patterns whose exact origins are unknown.
Contorted	Contorted, coarsely layered metamorphic rocks. Dikes veins, and migmatized bands provide the resistant layers in some areas. Pattern differs from recurved trellis in lack of regional orderliness, discontinuity of ridges and valleys, and generally smaller scale.

Source: Modified from Ritter, 1977

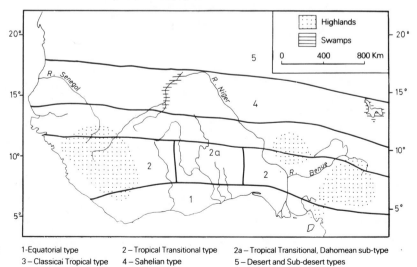

1-Equatorial type   2 – Tropical Transitional type   2a – Tropical Transitional, Dahomean sub-type
3 – Classical Tropical type   4 – Sahelian type   5 – Desert and Sub-desert types

Fig. 7.7 Hydrological regions, West Africa

199

3.	Range: 5–16	Range: 2–8	Range: 2–3.5	Wet season: July–October, 70–74% of total runoff concentrated in period from mid-October. Low water from December to June.	Varies greatly according to slope. Annual flood ranges from 10 to 80, may have and ten-year, from 40 to 120	Usually no dry season flow, through some rivers may have small flows of up to 0.10
4(2)	Range: 1–3	Range: 0.2–0.5	Range: 5–10	Very short wet season July–September; and long dry season with 8–9 months with no flow	Very low, probably of the order of 5–10 for ten-years' floods	No dry season flow
5	Runoff is limited to fairly small catchments in mountainous areas. May reach 5–10 for basins of 5 000 sq km	Sporadic occurrence of runoff	Not applicable	Flow limited to one or two floods per year resulting from periodic storms of the tornado type, usually in August and September	Can be quite high in mountainous areas due to absence of vegetation; 100–200 for catchments of 3 000–5 000 sq km	No dry-season flow

Source: Modified from Ledger, 1964

TABLE 7.3 Regional hydrological characteristics of West African rivers

Region	Coefficient of runoff (%)	Mean annual run-off in litres/sec/sq km	Variation of annual runoff	Seasonal flow pattern	Flood magnitudes in litres/sec/sq km	Minimum flows in litres/sec/sq/km
1.	Range: 15–30	Range: 5–20	Range: 1–4	Two periods of high flow: (1) (April–August (2) October–November. Severe dry season from December to March	Range: 40–70	0.30–2.0
2.	Range: 25–50	Range: 10–35	Range: 1–2	Wet season: May–October, 40–50% of total runoff concentrated in September and October. Significant dry-season flow	Annual: range from 50–90 Ten-year, from 70 to 120. Values doubled in areas of accentuated relief	Range 0.50–2.0
2a(i)	Range: 6–10	Range: 2–4	Range: 10–20	Wet season: June–November, approx. 60% of total runoff concentrated in September and October. Long dry season: from December to May.	Varies from 15–50 for annual to 40–100 for ten-year flood	Usually none, but some Ivory Coast rivers have from 0.10–0.50

identifiable in West Africa. The equatorial regimes in the southern parts show two separate periods of high water (for example, N'zi at Zienda, Figure 7.8a), while tropical regimes have a single high-water season. The tropical regimes, perhaps because of their areal coverage, are more complex and so are distinguishable into sub-types; namely: (1) the classical tropical regime consisting of a season of high water from July to October and a low-water period from December to June; (2) a broad transitional zone, south of the classical tropical regime, characterized by a longer period of high water and a much shorter period of low-water season; and (3) the sahelian type, with a shorter period of high water than low-water (cf. Figure 7.8; Table 7.3).

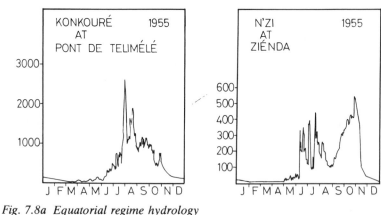

Fig. 7.8a Equatorial regime hydrology

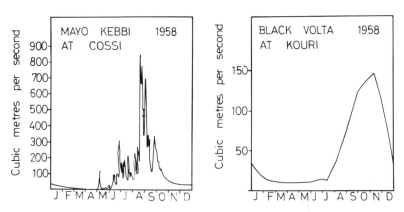

Fig. 7.8b Savanna regime hydrology

The concept of regionalization of hydrological behaviours was confirmed by a recent study in the Upper Owenna basin of Ondo State, Nigeria, where the discharge regimes of the third-order basins studied show four flow characteristics: (1) perennial flow, characterized by high discharge volumes; (2) perennial flow with low discharge volumes; (3) seasonal flow, characterized by low discharge volumes during the wet season; and (4) seasonal flow with high discharge volumes during the wet season (Ogunkoya, 1980). The factors responsible for the areal variations include climate (rainfall, evaporation and evapotranspiration); geology, including superficial deposits; vegetation and topography (cf. Ledger, 1964, pp. 75–82).

The relevance of the above hydrological and hydrometeorological parameters of the drainage basin in geomorphology has been brought into focus since the dawn of quantification marked especially by the work of Horton (1932, 1945). Since then, morphometric and other statistical studies have established various levels of correlation on the one hand among the hydrological and hydrometeorological variables, and on the other between these and basin (geomorphological) variables. Thus, statistical analyses have shown, among other things, that valley-side

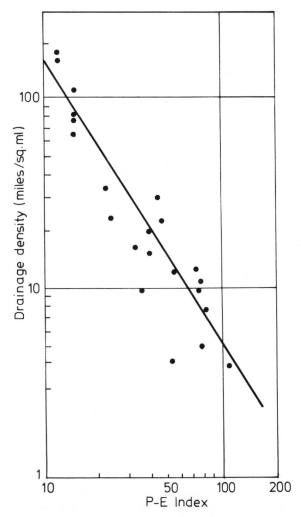

Fig. 7.9 P.E. index morphometry (after Melton, 1957)

slopes, drainage density, basin area and stream length are statistically dependent on hydrometeorological (climatic) variables alone or on some combination of these and other physiographic – vegetation, infiltration capacity, soil resistance – variables (Melton, 1957; Chorley, 1957). For example, mean monthly values of precipitation and temperature, themselves important determinants of the hydrological characteristics of an area, included in the calculation of the precipitation effectiveness (P.E.) index (Thornthwaite, 1931), have been shown to be related in some fashion to several morphometric factors (Figure 7.9). Figure 7.9 also shows the mutual relationship existing between and among these variables: certain morphometric components can exist as both dependent and independent variables.

Geomorphologists are particularly interested in the frequency and magnitude of flow events, because each has an important bearing on the various fluvio-geomorphological processes and the resultant forms within the drainage basin. Moreover, the flow of water and sediment in an open channel is conditioned by a number of forces – namely: surface tension, the components of the water's weight acting in the direction of the channel slope; shear stresses developed at the solid and free-face boundaries; internal inertia forces due to the turbulent nature of the flow; normal pressure at the walls and bed; and the movement of sediment (Sellin, 1969; Douglas, 1977); the mutual interaction of which accounts for the complexity or otherwise of channel flow, and the resultant variations in channel form.

One important aspect of channel flow relates to the mutual relationship between channel shape, velocity distribution, discharge and work. The actions of frictional forces along the banks and bed of the channel and at the air-water (free) interface and of internal inertia forces cause an uneven distribution of water velocity in any given cross-section (Figure 7.10). Velocity tends to be greatest where frictional forces are least. Engineers tried to account for this by using a formula involving the cross-sectional area of channel flow, the slope of the channel and the wetted perimeter, in the calculation of velocity. This is the Chezy formula:

$$V = C\sqrt{RS} \quad (7.1), \text{ where}$$
$$V = \text{Velocity}$$
$$C = \text{Constant (Chezy's)}$$
$$R = \text{hydraulic radius}$$
$$S = \sin$$

Chezy's C is derived from a formula

$$C = 1.5 \frac{Rt}{n} \ldots (7.2), \text{ in which C is a}$$

function of the hydraulic radius and a roughness coefficient n.

Table 7.4 shows possible values of n, according to Manning's formula, for natural channels. An alternative to the estimation of Manning's n is the examination of the relative roughness of a channel, which is the ratio of depth to the size of the roughness elements (Wolman, 1955). Yet another

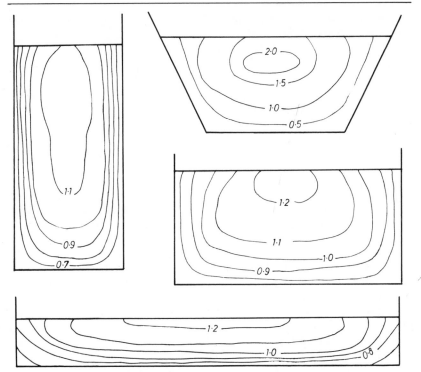

Fig. 7.10 Velocity distributions in channels

TABLE 7.4   Values of roughness coefficient n for natural streams

Description of stream	Normal n
On a plain:	
Clean straight channel, full stage, no riffles or deep pools	0.030
Same as above but with more stones and weeds	0.035
Clean winding channel, some pools and shoals	0.040
Sluggish reaches, weedy, deep pools	0.070
Mountain streams:	
No vegetation, steep banks, bottom of gravel cobbles and a few boulders	0.040
No vegetation, steep banks, bottoms of cobbles and large boulders	0.050
Floodplains:	
Pasture, no brush, short grass	0.030
Pasture, no brush, high grass	0.035
Brush, scattered to dense	0.050–0.10
Trees, dense to cleared, with stumps	0.150–0.04

Source: Morisawa, 1968

205

measure, the Reynold's number (Re), estimates resistance from the velocity and hydraulic radius:

$$Re = \frac{VR}{VK} \dots (7.3), \text{ where}$$

Re = Reynold's number
V = Velocity
R = hydraulic radius
VK = kinematic viscosity (μ/p or v),
   μ = dynamic viscosity and p = density;
μ/p = $1.21 \times 10^{-5\text{ft}^2}$/sec
(See Chorley, 1971, p. 126.)

 The point in all these computations is the proper understanding of the functioning of the basin hydrology with a view, in geomorphology, to relating it to the forms existing along river channels. The calculations also help to highlight regional variations. Thus, if it can be established that the hydrological and hydraulic characteristics of humid tropical rivers are distinct in some ways, the effect, all other things being equal, should be reflected in the forms present in such channels. The problem, however, is that information on the rivers of humid tropical regions is too limited for any reasonable and definitive inferences and deductions to be made. And as earlier observed (Chapter 4), attempts made at inference have been largely misleading, as in the following from Twidale (1976, p. 273):

The flow of the rivers is remarkably even, for much rainfall is intercepted by vegetation and soil and only slowly infiltrates into the rivers. Extremes of flood and drought, typical of the arid regions for instance, are lacking; there is little disequilibrium between the river and its interfluves ... though precipitation is high and there is commonly a dense network of permanent rivers and streams which have incised deep valleys in the selvas regions, running water is not a dominating component of the morphometric system in the selvas regions. Indeed degradation of the tropical humid lowlands is apparently among the slowest known ... the slopes are well clothed and protected by vegetation; the landsurface is in equilibrium with the climate.

Several of these generalizations, including the remarkably even flow of the rivers, the protective role of vegetation and the landsurface–climate equilibrium postulate (see Chapter 4) have now been controverted.

 Moreover, the observed differences in hydrology may not be sufficiently important to affect the resultant forms, while the problem of the principle of 'equifinality', already referred to earlier, may also be decisive. In fact this principle appears to apply best to fluvial processes, whatever their nature and magnitude, which tend to produce similar form (most probable states). The influence of regional hydrology may only be expected to show observable landforms in detail, rather than the general types which are similar in all areas where the fluvial processes operate. Leopold and Wolman stated this thus:

From the consistency with which rivers of all sizes (and in all climates) increase in size downstream it can be inferred that the physical laws governing the formation of the channel of a great river are the same as those operating in a small one [quoted by Dury, 1970, p. 197].

Relief, structure, soil and vegetation are additional complications in regional fluvial geomorphology. For example, Ledger (1961) observed for the Rima basin in Sokoto State, Nigeria, marked differences in hydrological behaviour between the portion underlain by basement complex and sedimentary rocks, even though the climate over the entire basin is uniform. He wrote as follows:

there is a marked difference between the hydrology of the upper catchment and that of the lower valley. The rivers of the upper basin have characteristically irregular hydrographs showing a rapid alternation of high and low flow.... Average discharge increases downstream and is at a maximum where the rivers leave the crystalline rocks and flow on the sedimentary plain.

On leaving the pre-Cambrian rocks, the tributaries ... flow on to the broad, flat plain ... where, no longer restricted in narrow valleys by resistant rocks, they have developed wide flood plains ... become very wide and shallow ... channel capacities are reduced, and a considerable proportion of the peak flow enters the flood plain through which it flows as a slow-moving, shallow sheet of water ... the floodplain acts as a huge storage lake, rapidly reducing the irregularities of the hydrographs ... so that the peaks are much lower ... maximum flood heights at Sokoto are some 4–6 weeks later than those upstream ... (and there is a progressive downstream reduction in the volume of water flowing through the lower valley due to evaporation and seepage losses) [pp. 479–80, and see Figure 7.11].

Also, writing about the significance of the relief factor, Ledger (1964) noted how the major rivers in West Africa, especially those originating in mountainous areas of higher rainfall and steeper gradients, deposit large volumes of sediment on entering the plains in their middle and lower reaches. The rivers thus become wide, aggraded, braided, shallow, slow-moving sheets of water subject to enormous losses by seepage and evaporation.

Several workers have also studied river pattern in relation to geology, soils and relief. Doornkamp and King (1971) reported higher drainage densities in hilly catchments than elsewhere in Uganda, although generally low drainage densities (range 0.5–1.5 km/km^2) have been reported over the Nigerian basement both in the rainforest and the savanna areas (Wigwe, 1966; Jeje, 1970; Thorp, 1970). The low drainage densities are thought probably to reflect the widespread occurrence of deep regoliths with infiltration capacities high enough to absorb a large proportion of the precipitation (Thomas, 1974, p. 141).

By contrast, a rapid rise in drainage density with increasing relief has been observed. Apart from the Uganda study (Doornkamp and King,

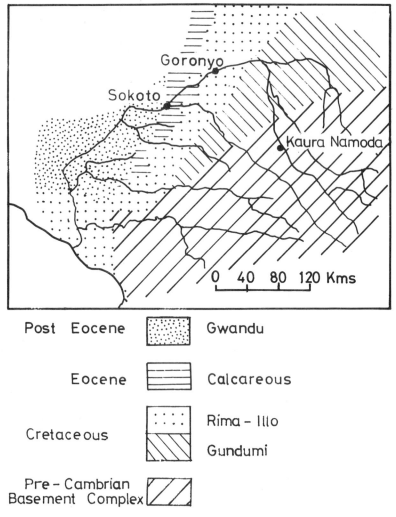

Fig. 7.11a Geology of the Rima basin, Sokoto State, Nigeria

1971), Eyles (1971) reported drainage densities of up to 12.5 km/km² in mountainous catchments of west Malaysia. The drainage density over the whole of the Malaysian peninsula is also found to be generally higher than on the basement area of the African continent, a situation which is attributable to the higher relief, thinner regoliths and higher annual rainfall in the former than the latter area. Similar observations in relation to relief and slope can also be inferred from Peltier's (1962) pioneer study.

Finally, vegetation cover exerts strong influences on basin hydrology through its impact on interception of rainfall, evapotranspiration and soil permeability, among other factors, and the combined effect of these processes on stream flow. Nevertheless, the possible influence of

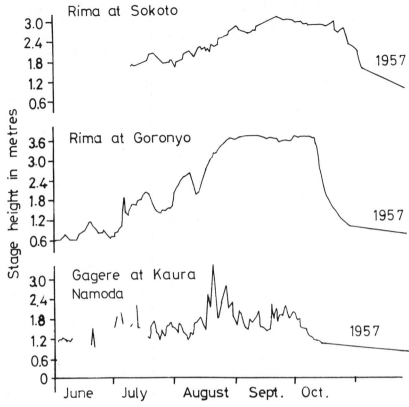

Fig. 7.11b Hydrography of the Rima basin, Sokoto State, Nigeria

the unique meteorological and hydrometeorological aspects of humid tropical lands may not be completely denied, and may be discernible from detailed study of, for example, drainage-basin morphometry.

River patterns

There are two main approaches to the study of river patterns – the traditional descriptive approach and the quantitative analytical approach. Whereas the former is well established in the literature, the latter is a more recent development. Also, whereas examples of most river patterns have been described from several countries and climatic environments, the available information, in quantitative terms, is as yet rather skeletal, scattered and inconclusive. We know, for example, that rivers meander and braid, which phenomena are related to energy distributions among other parameters within a given channel; but we are not in a position to say whether the frequency, magnitude and other attributes of river patterns show any regional trend(s). The same is true of the descriptive/genetic patterns such as dendritic drainage patterns, which are generally believed to be structurally controlled. There is need to study these in relation, especially, to climatic variables in order to establish their several and combined influences. These types of study are yet to be undertaken; the following material on river patterns is therefore essentially of a general nature except that the examples used relate to humid tropical environments.

River patterns, whether in the traditional qualitative or the quantitative sense, are best considered in two parts:
1 in terms of its general layout – network analysis; and
2 in terms of the channel characteristics and features – the long and the transverse profiles.

The drainage network

The description of the layout of drainage nets is also best done in the language of traditional geomorphology. Here, a number of descriptive and genetic terms are in currency, namely: *consequent, subsequent, obsequent, resequent* and *insequent*. As the genetic terms and forms can be rather ambiguous, the descriptive terms are more popular. The best-known examples are the *sub-parallel, dendritic, trellised, annular, centripetal, rectangular, antecedent* and *superimposed* patterns (Figure 7.12).

The sub-parallel pattern

This comprises a series of streams running approximately parallel to one another. They are especially common in areas of uniformly dipping rocks such as are found in the dip slope of a cuesta, gently-sloping monoclines and recently emerged coasts. In fact, sub-parallel patterns are essentially initial drainage patterns on 'young' terrains. The best examples are coastal streams, especially straight coasts such as are characteristics of West and East Africa. Streams flowing down scarp faces also assume sub-parallel forms – for instance, in the Rift Valley zone of East Africa. The alignment of joints and faults can also lead to the development of sub-parallel patterns (cf. the river system developed in parts of the Idanre hills of south-western Nigeria, Jeje, 1974, among others, Figure 7.13).

Dendritic patterns

These are ubiquitous over the earth's surface, associated with areas of uniform lithology, gently-dipping strata and generally low relief. The ubiquitous nature of this pattern is indicated by the rivers of West Africa most of which, minus the coastal (sub-parallel) creeks, illustrate dendritic pattern *par excellence*.

Trellised patterns

These are common in cuesta topographies where the main (dip-slope) streams run broadly at right angles across alternatively resistant and non-resistant rock outcrops. Consequently, the best illustrations come from folded and faulted terrains. Similarly, radial-annular, centripetal, rectangular, antecedent and superimposed patterns are directly connected with one geological structure or another.

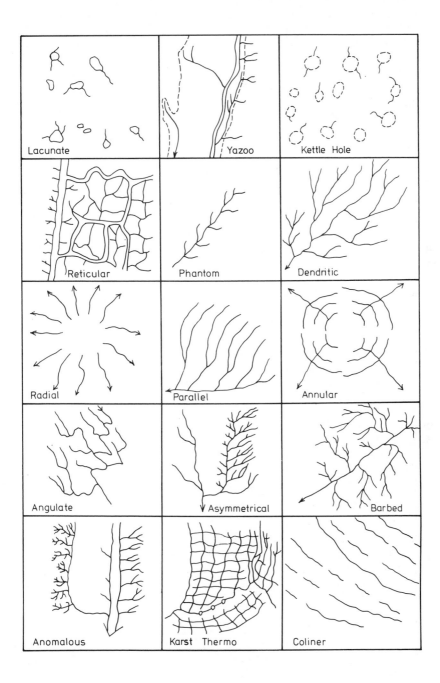

Fig. 7.12a Drainage network patterns

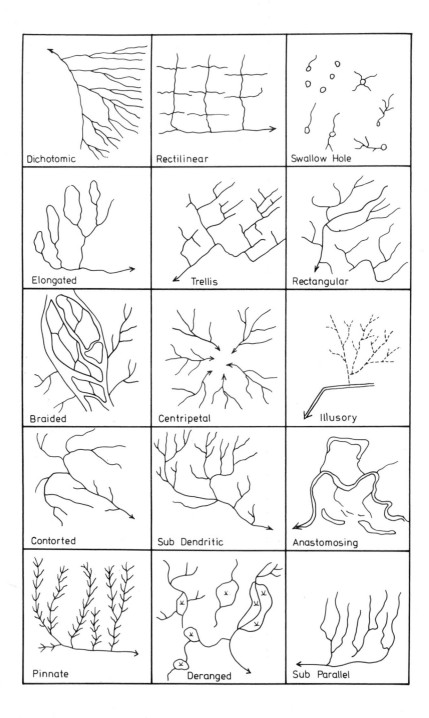

Fig. 7.12b Drainage network patterns

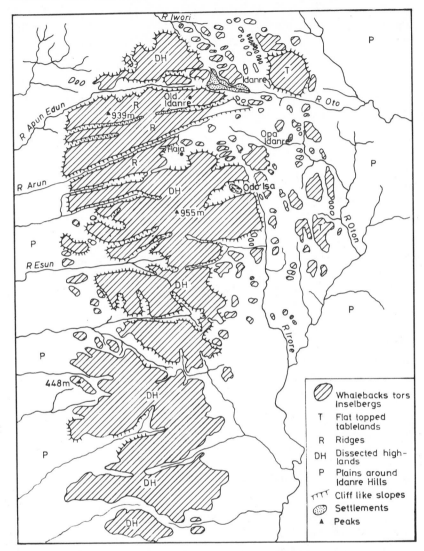

Fig. 7.13 Sub-parallel drainage patterns, Idanre hills, Nigeria

However, as shown by the African continent, there seem to be some areas of major difference. The intensity of the drainage net tends broadly to reflect climate; drainage density appears to be higher in the wetter than in the drier areas; rivers in the wet areas also tend to have higher bifurcation ratios than those in drier areas; and the rivers of identical size in the wet areas generally have higher discharges than those in drier areas.

The river course

The river channel has three dimensions – length, width and depth – which can be used to describe it. However, in geomorphological terms, only the first two parameters are worth considering since the depth is often subsumed under the general shape, especially of the transverse profile. The long profile refers to the length parameter, which is also the most important, especially in terms of the variety of associated landforms.

The river's long profile

A careful study of the long profile of rivers reveals a variety of forms, which can be described under at least three major categories of general shape, erosional features and depositional forms. The characteristic features of the general shape of the long profile of rivers include concavity and convexity, graded (equilibrium) and ungraded forms; forms such as meanders, braids and so on also describe the general shape of the long profile. The erosional features of geomorphological note include waterfalls, rapids, cataracts, knickpoints, potholes, striae and polished bedrocks, eroded banks and slip-off slopes and gorges; while the major depositional features include midstream and other bars, floodplains, levees, deltas, alluvial fans and terraces.

Profile shape and geometry

The best description of the shape of the river long profile is its slope or gradient, which is the graphic representation of the ratio of the fall of the channel to its length over a given reach. It is the configuration of the channel bed or bottom in longitudinal view. This, in most cases, and particularly in humid regions and for established (mature) stream channels, is concave – upwards towards the sky.

Several explanations have been offered for the typical shape of thalwegs, whether of newly cut badland gullies or of large rivers such as the Nile, the Congo or the Amazon (Figure 7.14). Even rivers which flow

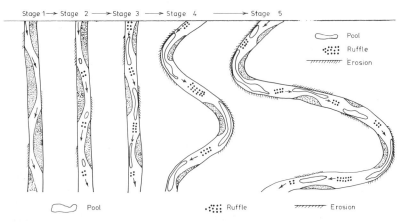

Fig. 7.14 Straight and meandering thalwegs

through arid lands in their lower reaches (for example, the Nile, the Rio Grande) do not depart much from this shape. There are several possible explanations, including (1) variations (increase) in discharge from source to mouth; (2) variations (decrease) in the calibre of sediment load downstream; (3) variations in the nature of the fluvial processes operating at different parts of the river course – erosion in the upper part and deposition in the lower part; and (4) the tendency for river courses and their controlling factors to assume the most probable state – namely, a state in which energy loss is minimized. Faniran (1981) describes the situation thus:

> The truth is that concavity, like many other features of the drainage basin, is more likely to be the result of a number of interdependent factors, including capacity, competence, discharge, the amount and calibre of the river load, the bed and bank material, the topographic relief, and the length, depth, width and gradient of the river itself.

The concave upwards-towards-the-sky shape of long profiles has also been described in terms of the graded profile, profile of equilibrium or stable channels. The idea of stability is inherited from the pioneer work of Gilbert (1877), which proposed that a graded channel is that where the load of a given degree of communition is as great as the stream is capable of carrying, so that the entire energy of the descending water is consumed in the transportation of the water and its load, and there is nothing left for corrasion. Faniran (1981) further comments as follows:

> It is this generalized shape of the river's long profile that is indicative of the general goal of energy movements at the earth's surface, namely: to remove obstacles on its paths and carve a profile (path) which minimizes energy consumption but maximizes its movement with the minimum of interruption.

Nevertheless, there are few, if any, ideally completely graded river profiles the world over, even in homolithologic terrains. The situation with respect to the African continent has been explained in terms of periodic and other endogenic processes, as of hard rocks which lie across the river courses. The features which complicate the general concave upwards-towards-the-sky (smooth) curve of river beds include waterfalls, cataracts, rapids and knickpoints, which are also among the most prominent erosional features along river courses. Douglas (1977, p. 137) describes the situation as follows:

> marked variations of gradient occur where streams flow over rocks of differing resistance to erosion, as on the River Nile where the Shabulka Gorge and five cataracts below Khartoum mark the points where the river leaves the sandstones and crosses resistant crystalline rocks. Major waterfalls such as the Victoria Falls on the Zambezi . . . are caused by belts of igneous rock. . . . Not all variations in declivity of the longitudinal profile may be accounted for by contrasts in resistance to rocks over which a river flows. . . . The longitudinal profile [may give]

way to a steeper, irregular profile . . . as a result of rejuvenation of the drainage system by a fall in sea level.

Whereas the shape of the river channel or course refers generally to the vertical or pictorial plan, the river patterns or geometry refer to the trace of the plan, as shown, for instance, on planimetric maps and aerial photographs or other forms of remotely sensed imagery. Several patterns have been observed which are yet to be satisfactorily classified. The following are a few of the patterns that have been discussed in the literature: straight, irregular, sinuous (meandering), braided, anastomosing, reticulate, anabranching and deltaic-distributary. They are described briefly here under three major headings of straight, braided and meandering river patterns, even though the boundaries between them are sometimes arbitrary and indistinct (Ritter, 1977). The distinction between straight and meandering rivers is based on a property called sinuosity, defined as the ratio of stream length (measured along the centre of the channel) to valley length (measured along the axis of the valley) (Schumm, 1963); the transition between a straight and meandering stream is placed at an arbitrary sinuosity value of 1.5. Similarly, a braided channel is taken as characterized by the division of the river into two or more channels.

The braiding index (B.I.) is based upon the sum of the length of islands or bars in any given reach (l), and the length of the reach measured midway between the banks (m) and expressed as B.I. = 2 l/m. A value of B.I. higher than 1.5 is an indication of a braided channel. Measurements can be made in the field or from large-scale topographic maps or aerial photographs. Many attempts have been made to quantify the overall meander pattern in different rivers, but the best-known appears to be by Speight (1965). His spectral analysis of the meander plan of rivers involves the measurement of a number (n) of changes of angle direction of the river measured at specified, standard distances along the river course in order to determine meander intensity X(K) where

$$X(K) = \frac{1}{n} [Cx\,(o) + \sum \{\, Cx\,(1)\,(1 + Cos\,\frac{\pi\,1}{m}\,)\,Cos\,\frac{\pi\,kl}{m}\,\}]$$

l = distance of sampled points apart
Cx = autocorrelation between successive angles[*]
m = number of frequency bands
n = number of direction angles between regularly spaced points

Straight channels

These are very rare in nature, especially over a significant distance. In fact, it has been shown from empirical evidence that it is rare to find river beds or channels which are straight for distances exceeding ten times their width. In Figure 7.15a, for example, although the river valley, marked by

---

[*]Autocorrelation is a method of comparing a repeated sequence of certain events with itself at successive positions. The goodness-of-fit is an index of their similarity (see J. C. Davis, *Statistics and data analysis in geology*, Wiley, New York, 1973, pp. 232–44).

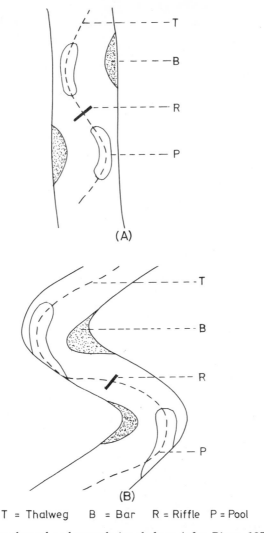

T = Thalweg   B = Bar   R = Riffle   P = Pool

Fig. 7.15a Straight channel and meandering thalweg (after Ritter, 1978)

its banks, appears straight over several kilometres, the thalweg migrates back and forth across the valley bottom. Furthermore, in many alluvial channels with a poorly sorted bedload, the channel floor is usually irregular, marked by alternating shallow zones (riffles) and deeps (pools). Also present along such channels are associated bars; the pools are directly opposite the alternate bars and the riffles about midway between two successive pools (Figure 7.15b). In short, straight river channels display many of the channel features associated with meandering channels (see below). A straight channel does not therefore necessarily imply a uniform stream bed or a straight thalweg, while a number of its features – bars, riffles, pools – are closely analogous to those found in meanders.

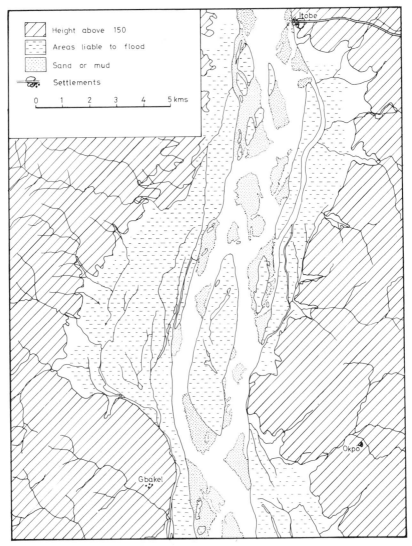

Fig. 7.15b River Niger below Lokoja: braids and bars

Braided channels

Braiding is generally associated with straight and some meandering channel reaches, and so further illustrates the complexity that may be expected in the study of river patterns. A basic aspect of the formation of braids is the division of a single channel into a network of anastomosing branches and the growth and stabilization of the intervening bars or islands (Figure 7.16). They vary from simple types in which a central bar divides a stream into two to complex types (the anastomosing pattern), in which several channels divide and join again in a reticulate pattern (Figure 7.17).

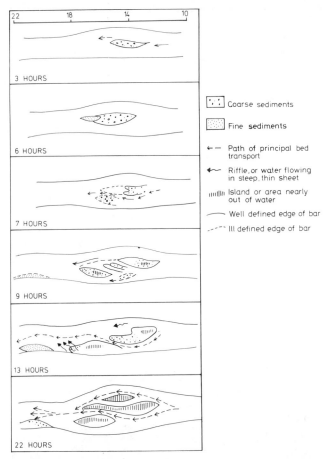

Fig. 7.16 Development of braids in a flume channel (after Leopold, Wolman and Miller, 1964)

Detailed studies of braided systems in both artificial (flume) and natural (river) environments (Leopold and Wolman, 1957; Fahnestock, 1961; Smith, 1970, 1974; Church, 1972; Rust, 1972) showed that divided reaches have different channel properties compared to their adjacent undivided counterparts. Braided zones are usually steeper and shallower; total width is greater, although each channel may be narrower than the undivided channel. However, changes in channel positions and in the total number of channels are more rapid. The braided patterns result from the simultaneous interaction of several factors, including erodible banks leading to the supply of abundant and possibly heterogeneous sediment load; rapid and frequent changes in discharge, and a river channel in which the operating processes can 'respond' appropriately to the external controls on the input variables. Braiding is now understood to represent an example of process-response mechanisms in which rivers drop their load when incompetence and/or incapacity occurs, so as to increase their gradients and flow capacity to carry the deposited load again.

Apart from channel divisions within a restricted channel area, rivers divide in many other ways, including the anabranching, anastomosing, deltaic-distributary and reticulate patterns. In the anabranching pattern the rivers divide and rejoin after many kilometres. The world-famous example is the Darling river in New South Wales, Australia, but examples abound in the drier areas of the humid tropics, including the Sudan and other countries of sub-arid tropical Africa. The anastomosing pattern is common in the same general areas as the anabranches – for example, the Diamantina river in Queensland (Australia). The Inland Niger Delta is a good example of a deltaic distributary pattern. They all, like the braid, illustrate process-response mechanisms in natural channels.

Meandering channels

By far the most common river form is the meandering pattern. Apart from forming distinct patterns in their own right, meander characteristics and forms are also present in straight as well as braided river courses. Thus, along an anabranch may be found varying degrees of meandering (Figure 7.16).

Before the recent developments in fluvial geomorphology, especially in the realm of quantification and morphometric analysis, scientists accepted such theories as random diversions of flow by slumped

Plate 17 The incised meander of Orogun river, University of Ibadan Botanical Garden (A. Faniran)

boulders or fallen trees and other obstacles or other vagaries of nature; old age; size (for example, a large river); and rotation of the earth as the major causes of meandering. More recently, other factors such as energy loss maximization or minimization, bank erosion, sediment load, helicoidal flow, and maladjustment between the hydraulic properties of the flow and channel shape have been used, some of them very successfully, to explain the origin of meanders. It has been shown, especially through flume experiments, that meandering is a more stable geometry than straight, non-meandering channel pattern is (Langbein and Leopold, 1966) or, as claimed by Yang (1971), that meandering is the only possible stable, unbraided, natural channel form, relating to the properties of the streams themselves, including water in the channel, river load, bed and bank material, and the energy present within the system.

As noted earlier, meandering reaches contain the same physical components observed in straight channels (pools, riffles, bars). As water is forced against the outer bank of a meander, its slightly elevated level at that point gives the flow a circulatory motion. The water moves along the surface towards the undercut bank and along the bottom toward the point bar. The resulting helical flow pattern reaches its greatest velocity slightly downstream from the axes of the meander bends at the position of the pools; the velocity decreases gradually downstream until the flow approaches or reaches the next riffle and the lateral circulation disappears, only to pick up again at the approach to the next riffle. It is probably this mechanism that operates in all examples of meanders so far studied, which gives meanders in rivers of all sizes similar dimensional and consistent geometric and hydraulic relations, as shown in Table 7.5.

Like other topics in geomorphology, studies of meandering rivers in the humid tropical regions are limited and have not been comparative. On the contrary, they have been studied with a view to illustrating and exemplifying known general principles and established 'laws'. For

TABLE 7.5 Empirical relationships between parameters that define meander geometry

Dependent relationship	Source
Wavelength	
$= 6.6w^{0.99}$	Inglis (1949)
$= 10.9w^{1.01}$	Leopold and Wolman (1957)
$= 4.7r_m^{1.01}$	Leopold and Wolman (1957)
$= 30\,Qbf^{0.5}$	Dury (1965)
Amplitude	
$A = 18.6w^{0.99}$	Inglis (1949)
$A = 10.9w^{1.04}$	Inglis (1949)
$A = 2.7w^{1.1}$	Leopold and Wolman (1957)
A = wavelength (ft)	A = amplitude (ft)
w = width (ft) at bank-full stage	Qbf = bank-full discharge (cfs)
r_m = radius of curvature (ft)	

Source: Ritter, 1977, p. 237

221

example, in Speight's (1965, 1967) study of the meanders of the Anagabunga river in central Papua, he found that

1 most peaks were very stable in successive spectra, allowing considerable confidence in the reality of the peaks, some of which might otherwise have been classified fortuitous oscillations or 'noise';
2 a number (three) of long (low-frequency) wave lengths dominate each spectrum;
3 there was a sharp drop in meander intensity at a wave length shorter than the shortest of the long wave lengths;
4 the mountain courses have a meander spectrum very similar to those of reaches on the alluvial plain, indicating that the tendency to orderly, multiple wave-length periodicity in the direction of stream flow was not suppressed even when the river was incised in bedrock and relief is prominent;
5 peak wave lengths were more stable in the downstream direction than any other parameter, including channel width;
6 the sinuosity index varies between 1.8 and 2.3, but the channel itself is marked by several features typical of other fluvial regions, including active meanders, chutte cutoffs, scrolls, point bars and ox-bow lakes (Figure 7.17).

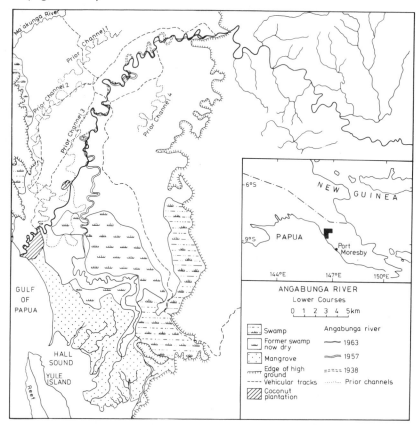

Fig. 7.17 Angabunga river, Papua

However, it should be observed that, although the processes of braiding and meandering have been described as distinct types, they often occur along the river. The Ganga river, for example, is known to have fully developed meanders between Allahabad and Benares, while it braids prominently downstream of the junction with the Gogra, Gandak and Sone rivers (Chitale, 1970). Straight stretches also exist along rivers with prominent meanders and braids. Thus the Damodar river, near Durgapur barrage, has a straight, wide shallow and steep channel and carries a heavy load of sediment (Douglas, 1977). In short, the natural river channel shows a succession of river patterns, often as a continuum.

Finally, it is necessary to observe, with respect to the observed relationships between river patterns and river discharge, or between one river pattern and another, that extreme events are capable of offsetting the system. This is particularly true of humid tropical regions, which record exceptional storms and floods associated with hurricanes and related violent tropical weather conditions. Examples include those of:

1 January 1974 in the rivers of western Cape York Peninsula of Queensland, which cut coastal sand bars and developed a complex series of new channels;

2 Cyclone Brenda in 1968, which caused the Dumbea river in New Caledonia to erode not only the banks facing the minor channel but the floodplain itself, creating new channels and redepositing the eroded material further downstream;

3 the major floods – for example, December 1926 and January 1971 – in west Malaysia, which produced rapid channel changes through the temporary damming of tributaries by debris and the development of cutoffs and new courses through old levee and terrace deposits; and

4 the Ogunpa floods in Ibadan, which have modified the channel in several ways. Douglas (1977, p. 163) summarized the situation as follows:

While in general most work of changing the landscape is done by events recurring as the mean annual flood ... individual drainage basins may have suffered from a particular extreme event, whose traces have not yet been effaced by the more regularly operating, but lesser magnitude geomorphic event. The long term trends ... in channels are thus likely to be interrupted over time ... by catastrophic events which do more work than all the flows of several hundred or even thousand years before.

Erosional features

Rivers produce a number of features as a result of their corrosive, corrasive, abrasive, cavitational and other hydraulic activities. The most important of these features, which have already been mentioned above, are waterfalls, cataracts and rapids; riffles and pools; potholes; and striae and polished rocks. Gorges and other ravine-like forms, as well as eroded banks and slip-off slopes relate to the river cross-section and so will be described later. Those pertaining to, found along or associated with the river long profile are described below, grouped on basis of size and location into two.

Waterfalls, cataracts, rapids and knickpoints

These are by far the most conspicuous and so the most familiar erosional fluvial landforms. They are quite impressive, and provide benefits to man as sites for tourist attraction as well as for hydroelectric power generation. Their major causes include eustatic changes, uplifts and the influence of horizontal, alternately hard and soft formations across river courses. Eustatic changes, in particular, produce knickpoints with recognized geological and geomorphological implications. They have, for instance, been used in dating geologic events and in calculating rate of continental erosion.

Although related in their causative factors, the appearance of the landforms differs to varying degrees. A waterfall is a sharp break in the long profile of a river, where the water drops for a vertical distance in one fall (Figure 7.18a). By contrast, a cataract is a step-like succession of

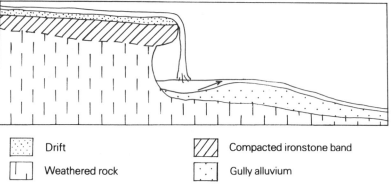

	Drift		Compacted ironstone band
	Weathered rock		Gully alluvium

Fig. 7.18a Waterfalls

waterfalls, while rapids are sections of the long profile in which river flow is broken by short, vertical drops which seldom affect the entire width of the channel at any part of the river cross-section. Whereas a cataract or waterfall is rarely drowned by floods of whatever magnitude, rapids may be inundated by high flows. A knickpoint may be a waterfall, cataract or rapid with established eustatic events. All of them are commonly studied together since they tend to occur in juxtaposition without definite geographical orientation. They also have generally common causes. We may therefore use one to represent the others, as is done for waterfalls in what follows.

One of the most comprehensive studies of waterfalls is that by Schwarzbach (1967), who, among other suggestions, proposed a system of classification based on genetic principles; he distinguished between those that have developed independently of the erosive activities of the

TABLE 7.6   Schwarzbach's genetic classification of waterfalls

A. DESTRUCTIVE FALLS (with more or less distinct headward erosion)

   I *CONSEQUENT FALLS* (Falls caused by the sudden blockage of a valley or by any other deviation imposed on the talweg)
   a. Falls at the downstream edge of the blockage (landslide, lavastream, etc.)
   b. Rapids over larger boulders
   c. Falls at a point where a river rejoins its old course beyond the deviation, often on the old valley side.
   d. Falls located at a new junction point with the main valley following a blockage.
   e. Falls following the cutting off of a meander.

  II *SUBSEQUENT FALLS*
   1. Horizontal or gently dipping (hard and soft strata). The majority are secondarily subsequent with headward erosion, e.g. Great Niagara. Gullfoss: Waterfalls steps.
   2. Excavation of a steeply dipping boundary between hard rocks upstream and soft rocks downstream. 'Primary' subsequent falls with little if any headward erosion. The falls are often stationary.

      (a) the sharp boundary is volcanic in origin (vertical barrier falls);
   (b) the boundary under attack is a fault;
   (c) the boundary is erosional in origin (vertical barrier falls).

   3. A joint zone is eroded, and headward erosion is bound to the joints.

B. CONSTRUCTIVE FALLS (downstream movement and no headward erosion). The break is usually the result of $CaCO_3$ precipitation.
   1. CaCO precipitation at the mouth of a secondary valley or on a slope.
   2. Overflow of waters penned in a valley lake dammed directly as the result of travertine deposition.

Source: Modified from Douglas, 1977, p. 139

river from those that have developed as a result of such activity. A further distinction was made between destructive and constructive waterfalls (Table 7.6).

The highest reported waterfalls occur in parts of the humid tropics – for example, the Angel Falls in Venezuela (*c.* 1000 m high). Others include the Victoria Falls in Zimbabwe (over 120 m) and the Iguazu Falls in Brazil (*c.* 100 m). Prominent waterfalls also occur in Surinam, Guyana

225

Plate 18 A waterfall on quartzite near Erin-Ijesa, Oyo State, Nigeria (L. K. Jeje)

and parts of tropical and sub-tropical Africa, especially on the uplifted edges of old land surfaces. Many of these are much higher and bigger than the world-famous Niagara Falls (about 60 m high).

The waterfalls in humid tropical environments have, however, not been studied to the same extent and detail as those in extra-tropical areas, such as the Niagara or the Yosemite Falls in North America. Thus, while rates of recession have been obtained for the Niagara and other falls in extra-tropical areas, such information is hard to come by for tropical rivers. Among the reasons advanced, apart from the proverbial problem of the humid tropical environment, is the possibility that water action is less efficient in humid tropical areas. Tricart (1959), for instance, claimed that only one pebble (over 5 cm long) was found in over 200 swirlholes in the Felou Falls on the Senegal river. It is also claimed that pebbles, where they occur in tropical rivers (for instance, the Sungai Dong in Pehang, Malaysia), are transported over relatively short distances before they are reduced in size by the processes of chemical weathering, abrasion and attrition, among others (Michel, 1973).

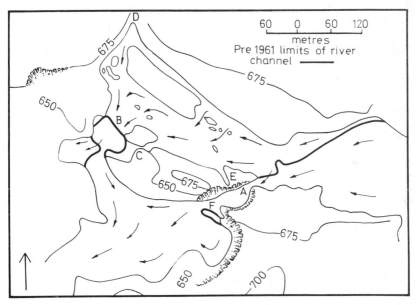

Fig. 7.18b Murchison Falls (after Ware, 1967)

One important factor of waterfall development is the geological structure of the area concerned. This is fairly well illustrated by the Murchison Falls (Figure 7.18b). Douglas (1977) commented thus on the relationship between structure and discharge during a phase in the development of the Murchison Falls in Victoria, Uganda:

> Before 1961, the whole of the discharge of the Victoria Nile passed through a joint-controlled 6 m gap [see A in Fig. 7.18b]. The line of the joint slot continued north-eastwards across the former line of the river and caused the river to take a sharp turn . . . west. The water gathered momentum on entering the slot to descend the modest 42 m of the fall.

> At the end of 1961 the discharge of the Nile increased suddenly. . . . Higher water levels . . . eventually led to the re-adoption of the former course of the river, over the joint step and to around a former dry cascade . . . a lip some 10 m lower than the old joint slot. The Murchison Falls now comprises three separate cascades [A, B, C, on Fig. 7.18b]. . . . These changes . . . illustrate the interaction of structure and hydrology in the development of streams [pp. 141–42].

Further evidence adduced in support of the ineffectiveness of the process of waterfall development in humid tropical regions is the probable effect of the case-hardening of bed and bank rocks (hardpan, duricrust and so on) (Tricart, 1959). Zonneveld (1969), in particular, described the iron and manganese layers (surface incrustations) along sections of the Corantijn river in Surinam as additional evidence in support of the slow rate of waterfall development and the antiquity of waterfalls, the origin of many of which has been put at the Tertiary period (Schwarzback, 1967; Büdel, 1957).

Finally, waterfalls and related features, whatever their causes, have developed *plunge pools* prominently at their bases, the sizes of which vary with the size and force of the falling water above them. In some cases, the depth of the plunge pool may be close to the height of the waterfall itself. In such cases, the basal pool easily extends backwards to undercut the main feature, which may later collapse to effect the retreat of the waterfall. This process usually leaves behind a narrow, steep-sided valley or chasm, which represents the headward progression of the waterfall. The Howick and Victoria falls, consisting of a broad-floored valley above the narrow gorge-like chasm, have been quoted as illustrations of this process, which, if true, suggest that waterfalls in humid tropical areas may not be as stable and permanent as is being suggested in the literature.

Ripples, potholes, striae and polished beds

Unlike the features just described, these are among the smallest erosional features along a river channel. Ripples refer to the undulating bed of, or the alternating deeps and rises (shallows) along, river channels. These features are also sometimes called riffles and pools, dunes and antidunes. They are particularly prominent in alluvial channels, although similar phenomena occur in channels cut on bedrock (Leopold *et al.*, 1964).

Ripples are produced primarily by the wave-like (helicoidal, helical) flow pattern of streams and vary with the nature of the river regime. Under a low flow regime, resistance to flow is great and the rate of sediment transportation reduced. The bedform, therefore, consists of mainly ripples or dunes. The contrast holds under the upper flow regime, resulting in smooth or plane bedform. The transitional flow regime tends to produce a more erratic system of dunes and antidunes.

Whereas riffles and pools are essentially erosional features in crystalline rock areas, ripples, and most especially the dunes and antidunes, are restricted to sedimentary environments and so are both erosional and depositional forms (cf. Fig. 7.19; cf. also Simons *et al.*, 1965; Langbein and Leopold, 1966).

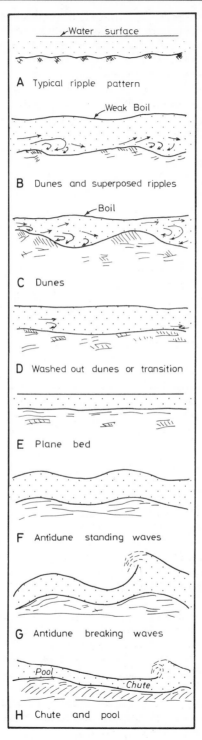

A Typical ripple pattern

B Dunes and superposed ripples

C Dunes

D Washed out dunes or transition

E Plane bed

F Antidune standing waves

G Antidune breaking waves

H Chute and pool

Fig. 7.19
Dunes and antidunes

Potholes and polished (striated) beds occur almost invariably in massive (crystalline) rocks. Potholes develop most readily, though not invariably, in the weak parts of the river bedrocks, which are exploited by the corrosive, abrasive, evorsive and hydraulic processes of flowing streams. As to be expected, the resulting forms vary widely in shape and size. Some potholes along the Tugela river (Natal) are known to be several metres deep (King, 1953), but most of the potholes on the bed of the Ogun river in south-western Nigeria are generally shallow and small, some of them indistinguishable from the associated grinding pits. Usually the floors and sides of the potholes are smoothed (polished) by the abraiding rock pebbles, many of which are perfectly well rounded. Polished and striated surfaces are widespread along rivers cut in bedrock, especially on flats, waterfall and rapids.

Depositional features

Fluvially produced landforms, particularly those along river channels, are accomplished partly by the erosional capability inherent in moving sediment: laden water, and, perhaps more importantly, by the deposition of debris along the river course. Consequently, relatively few fluvial features are purely erosional with little or no sediment associated with them. However, many fluvial landforms owe their origin to a combination of both erosion and deposition, with the pure cases being probably the end-members of a continuum.

Like their erosional counterparts, depositional fluvial landforms are many and varied, including the more areally restricted forms such as bars, dunes and antidunes as well as very extensive and complex forms such as floodplains and deltas.

Bars, dunes and antidunes

Bars are generally mentioned in connection with coastal landforms (see Chapter 12), especially in shallow, epi-continental or shelf waters. The bars found along river courses (midstream and bankside bars) are formed by similar processes. Midstream bars are similar to some ripples and/or dunes (see above), except that they are generally much larger and less regularly spaced. Also, not all midstream bars are made up of sediments: many are made up of hard rock (for example, the rock bars). Where rock bars occur alternatively with pools, as in the bedrock rivers of basement complex areas, a situation close to the riffle-and-pool phenomenon of alluvial channels is present. Rock bars are, however, not relevant here.

Alluvial bars, dunes and antidunes, where established, are invariably the initial stages of more complex forms such as braids and floodplains. Similarly, bankside or point bars are associated with meanders, found in the inside (low-energy) portions directly opposite the slip-off slopes. The bars are sometimes called floodplain scrolls, perhaps because they combine in the process of the evolution of the typical floodplain (Figure 5.2a).

Floodplains

These, as observed in Chapter 5, are among the most conspicuous of fluvial landforms; they are present along most major rivers and their tributaries. Although difficult to define topographically and geologically, floodplains are not difficult to recognize, especially where they are well developed. They are composed of relatively flat alluvial terrains, subject to periodic flooding. According to Leopold and Wolman (1957), most tropographic floodplains are subject to flooding nearly every year or every other year, with a recurrence interval of 1.5.

The origins, formation and morphological characteristics of floodplains have already been examined in Chapter 5. However, perhaps the most important aspect of floodplains is their role in the working or operation of the entire river system as an open natural system. Ritter

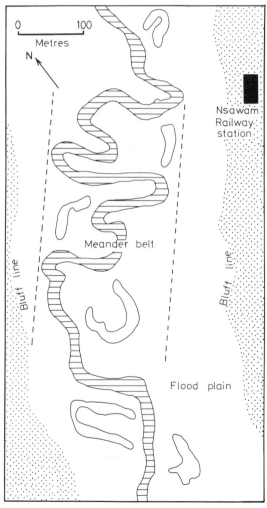

Fig. 7.20 Typical floodplain, Nsawam, Ghana

231

(1977, p. 259) summarized this role thus:

> a floodplain plays a very necessary role in the overall adjustment of a river system. It not only exerts an influence on the hydrology of a basin (lag, etc.) but also serves as a temporary storage bin for sediment eroded from the watershed.

The most closely studied floodplains are in the extra-tropical areas – for example, the rivers of the USA and of Australia (Fisk, 1944, 1947; Leopold and Wolman, 1957; Hickin, 1970). However, impressive and extensive floodplains abound in humid tropical areas which await study more detailed than Reiner and Mabbutt's (1968) study of the Sepik river in New Guinea. Examples are described in Chapter 5 (see also Figure 7.20).

Alluvial fans

These are fluvial deposits associated mainly with piedmont topography, especially where streams issue out of mountains to lowland, but also where a river runs into a terrain that drastically reduces its velocity – such as where a tributary joins a slower, lower-gradient trunk stream (Figures 7.21a and b). The morphology and the formation of this feature have been examined in Chapter 5.

Attempts have been made to interpret fans in terms of the equilibrium principle, but no success comparable to that achieved in respect of floodplains has yet been achieved. In fact, the views expressed range from complete denial (Beaty, 1970; Lathman, 1973) to total agreement (Hooke, 1968; Lustig, 1965). There is no doubt that this issue has rightly aroused much interest, which should be encouraged. It is possible that alluvial fans, like braids, meanders and floodplains, represent mechanisms adopted by rivers in their attempt to accommodate the vagaries of the basin environment; in which case they represent a tendency towards, if not actual, equilibrium. However, much information is required, as shown in the following observation by Ritter (1977, p. 289):

> It is clear from these examples that we do not truly understand what fans tell us about their conditions of equilibrium. . . . Perhaps, like many geologic features, fans can form in a number of ways under various combinations of their controlling factors. Each fan should, therefore, be considered in its own individual setting. It will probably have relict components retaining the effects of a past climate, areas of erosion and deposition, young and old parts, and a morphology that demonstrates a statistical balance. The interpretation of what all those things mean is a real challenge to any geomorphologist.

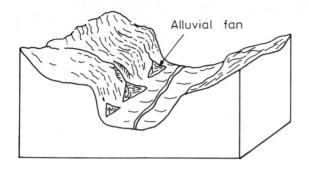

Alluvial fan

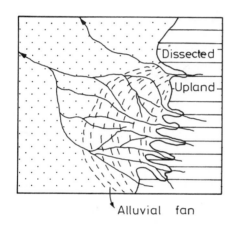

Dissected

Upland

Alluvial fan

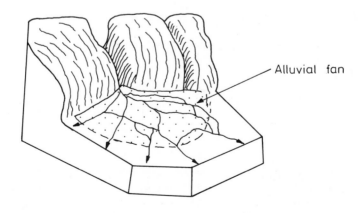

Alluvial fan

Fig. 7.21a Alluvial fans

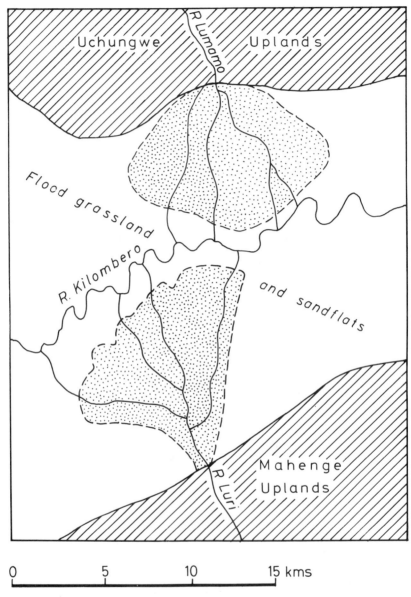

Fig. 7.21b Alluvial fans along the Kilombero valley, Tanzania (after Jatzold)

Terraces

Terraces are abandoned floodplains that were formed when the river
flowed at a higher level than at present. Consequently, unlike floodplains,
alluvial fans, deltas, and other depositional features along river channels,
the surface of the terrace is not directly related to the modern hydrology

234

of the basin in that it is not inundated as frequently as an active floodplain, for example. The details of formation as well as examples are discussed in Chapter 5.

Lakes

These are water bodies caused either by a barrier across a river channel, or by other geological agents which form depressions on the earth's surface. We are here interested in lakes as part of a river system, either with inlets and outlets (as in the case of Lake Kutubu in New Guinea and Lake Victoria in central East Africa), or without outlets (such as Lake Chad in Central Africa). Absence of an outlet to a lake in humid climates is rather unusual: it is either because the lake is a recent formation or because it has an underground outlet, as in the case of Lake Tebera, again in New Guinea.

Lakes along river courses may result from fluvial action, landslides, or organic activity, including dams built by man. Apart from the scouring (potholing) activity of flowing streams, floodplain evolution involves many types of lake formation. In addition to ox-bow lakes (see Chapter 5) and lakes in other rivers channel remnants and swales between meander scrolls, lakes form when levees block the outlet of tributary streams. Along the Pehang river in west Malaysia, great quantities of sediment deposited on levees during floods block small lowland tributaries, creating lakes such as the Tasek Bera and Tasek Chini. Douglas (1977) also described floodplain lakes in areas with no tributary inflows – for example in the Ka valley of New Guinea. These are formed by the combined processes of silt/clay deposition and peak development, a process also common to lakes resulting from organic activity. Landslides and other forms of mass wasting also block rivers to form lakes, if only temporarily. However, man-made lakes are more permanent.

Deltas

Most of the deposits along river sources are temporary phenomena: they will ultimately be carried further downstream and deposited at the stream mouth – that is, at sea level. In other words, deltas are the locations where the sediment being transported by rivers ultimately comes to rest; they, therefore, like the other features described so far, represent bona fide components of the fluvial system. The morphology of the various types of deltas is described in Chapter 5.

Deltas are particularly prominent features of the rivers of the humid tropics, and so suggest in very clear terms that these rivers, like the others in extra-tropical areas, erode their beds and banks and carry the resulting sediment, to be deposited, whether temporarily along river channels or ultimately at the river mouth at sea level. Even then, deltas are not permanent features. Like the other depositional features along the river course – braids, lakes, alluvial fans, levees and so on – they either get colonized and made part of dry land, or are drowned. Nevertheless, these featues provide one of the most valuable sources of information about not only the history of the landscape but, perhaps more importantly, the fluvial system as a whole.

The river's cross-profile

Like the long profile, the form and associated features of the river cross profile are functions of the flow characteristics: the quantity and character of the river load; the nature of the initial channel; and the nature of the bed and bank material, including the basin biotic (plant and animal) life. Although the river cross profile refers, ideally, to the entire river valley – that is, including the valley-side slopes – the region or area of interest here is more restricted, limited to the wetted perimeter or the area affected by the rather frequent floods. This is the portion of the river basin described as the channel, as opposed to the valley-side slope (Chapter 9), and often characterized by its depth, roughness and slope, as well as by flow parameters of velocity, discharge and resistance. The most often used parameter is the width–depth ratio.

The ideal cross-section of a channel in homogeneous bed and bank material is sinusoidal in shape; most natural channels are, however, far from this ideal shape: they are generally trapezoidal in straight reaches and asymmetric at curves and bends (Figure 7.22).

It is not always easy to separate the channel from the valley-side slope, especially in high-relief, fine-textured terrains. Here, the valleys result from two interrelated processes of river-bed incision and denudation of valley-side slopes. On the basis of the ratio of incision to valley-side widening, river valleys have been classified into gorges, V-shaped valleys, canyons and flat-floored valleys among others. Among flat-floored valleys are flood-plains and terraces, which have been described as examples of depositional features along the river long profile.

Gorges, canyons, ravines and gullies result essentially from rapid down-cutting by rivers. A gorge by definition is a vertical, steep-sided narrow and deep valley cut almost invariably in hard rock. By contrast, canyons tend to be V-shaped; whereas gullies develop almost exclusively in unconsolidated (soil) material. Ravines are smaller than the others and are most common in dry (desert) environments.

The most common cause of gorges is back-cutting of waterfall walls by rivers. That examples abound in the humid tropical regions is a testimony that humid tropical rivers can be effective as erosive agents. In Africa, gorges characterize many rivers, particularly in the vicinity of waterfalls. Among these is the world-famous Victoria Falls gorges on the Zambezi river, which occur in a series over a distance of about 100 km downstream of the falls. The following description of the Kariba gorge (the best-known of the Zambezi gorges), is quoted from Wellington (1955, pp. 396–97):

At the mouth of the Songwe tributary ... below the falls, the depth

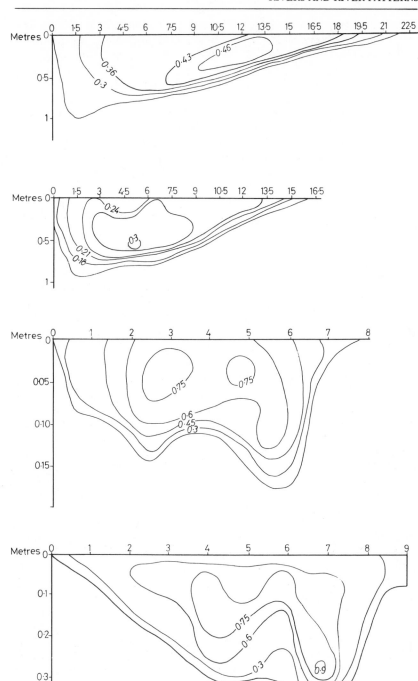

Fig. 7.22 Natural channel geometry

[of the gorge] is 460 feet (140 m), at Chimamba the Karamba confluence 800 feet (244 m). . . . The river bed in the gorge is broken in many places by low falls, the largest reported being 10 miles (16 km) from the Victoria Falls, where . . . there is a fall of about 60 feet (19.2 m). The impressiveness of the Chimamba derives apparently from the narrowness of the river channel. When one stands on the brink of the lower cataract and sees the whole volume of the Great Zambezi converging into a simple pass only 50 or 60 feet (15 or 18 m) in width, shuddering and then plunging for 20 feet (6 m) . . . that seems in its impact visibly to tear the grim basaltic rocks asunder, one learns better . . . what force there is in the river and one wonders no longer at the profondity of the gorge.

Another example of spectacular valley (channel) form in humid tropical regions is the gully. Like gorges, gullies develop as a result of accelerated erosion, mainly in disturbed watersheds. In fact, the examples described so far are from areas of high population density such as those of south-eastern Nigeria (Ofomata, 1965; Floyd, 1965; Ebisemiju, 1976), Ibadan (Faniran and Areola, 1974) and Zaria (Ologe, 1971, 1972), to list only a few. In these locations, they are caused by a combination of climatic, geomorphic and biotic factors which favour accelerated removal of bed and bank material. Being in unconsolidated materials, gullies are like canyons and ravines, but unlike gorges, they have V-shaped valleys, more typical of valleys in humid regions where both down-cutting and valley-side denudational processes operate.

Yet another set of impressive valley (channel) forms are the intrenched (incised) and rejuvenated valleys. Where a number of these occur in succession within a river valley, terraces occur. Intrenchment and rejuvenation may involve great depths, or, as in the case of the river valleys in the typical inselberg country, only a few decimetres. Studies conducted over large parts of the Nigeria's basement complex show that this phenomenon, although comprising just about 5 per cent of the total surface area which is dominated by extensive low-relief surface, often called pediment, is distinct and mappable (Burke, 1967).

Incised river valleys are particularly prominent in meandering rivers with wide floodplains. With rejuvenation caused by uplift, sea-level changes or climatic change, the rivers cut into their beds to form features similar to ravines in arid regions. The actual form varies according to whether the rivers cut down vertically (as, for example, entrenched meanders), or whether lateral shifting is involved, in which case undercutting or slip-off slopes (for example, ingrown meanders) result (Figure 7.23). Two or more episodes of intrenchment will produce terraces and/or terracettes. A good example is the Jatau, a tributary of the Gbako in the Niger State of Nigeria, which is incised for 15 m east of Kataregi railway station. Similar phenomena can result from structural or meteoro-hydrological reasons. A V-shaped valley cut into horizontal sedimentary strata of unequal hardness tends to have a step-like profile, as in the Grand Canyon in the USA, the middle course of the Zambezi, or the Tibetan courses of the Brahmaputra in India.

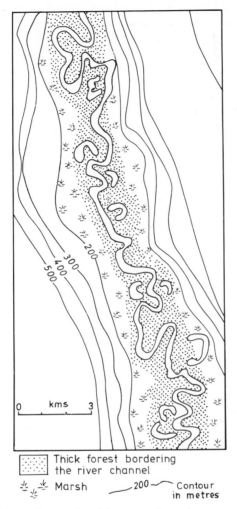

Thick forest bordering
the river channel

Marsh   —200— Contour
in metres

Fig. 7.23 Incised, superimposed and ingrown channels

Conclusion

One important aspect of river-pattern studies relates to the light that they
throw on the relationship between fluvial processes and forms: that is,
river-channel patterns reflect specific adjustments to fluvial variables.
Since the river system is a dynamic one, involving continuous work by
geomorphological and other natural processes, it follows that each
pattern and form would remain stable within certain finite limits of the
controlling factors called thresholds. When such limits are passed, the
system readjusts to the new limits to establish a new pattern or form.
Consequently, stream patterns can only represent, at best, 'ephemeral'

fluvial properties, the degree of ephemerality depending on the rate of change in the controlling factors.

A few examples of this close relationship between process and form should help. One good example is that of change from straight to meandering and back to straight channel pattern. This change results from a number of events along the river channel. The first of these events relates to the mechanics of river erosion and deposition, especially the formation of bars, pools and riffles. These, especially the bars, cause secondary flow patterns which encourage the processes of bank erosion and lateral accretion and the exaggeration of the meander patterns. The same mechanics that produce meandering also cause chuttes and cutoffs that straighten the river course again, or produce a system of multiloops or other complexly woven patterns in the continuous process of adaptation of form to process.

Another example is the transition between meandering and braided patterns. The critical factor here is the channel slope and its effect on discharge of both water and sediment. However, slope is not always an independent variable, since it also gets adjusted, as a dependent variable, to changes in the sediment discharge system. Yet another example relates to the channel cross-profile: when an increase in precipitation, and so of discharge, could lead to changes in the quantity and calibre of the river sediment. Although this may not lead to marked changes in the slope and the width/depth ratio, especially if the channel is widened and deepened at the same time and rate, should either channel widening or deepening be prevented for any reason, both the slope and the width/depth ratio will be affected in such a way as to maintain the system. One will compensate for the other in a way that the attempts of the river to establish and maintain the most efficient conditions for transporting water and sediment are not jeopardized. Similarly, depositional forms – bars, braids, alluvial fans, lakes and so on – as well as the other irregularities along river courses, are temporary phenomena in the course of process–form adjustments: the ultimate goal of a stream is to remove all obstacles along its course and carve out for itself a smooth, concave upwards, self-regulatory profile, or curve of equilibrium, which is the most effecient (probable) state for performing its major function, within the basin hydrological-cycle system, of transporting both water and sediment from the mountains to the adjoining seas and oceans.

In this scheme, the role of the humid tropical environment seems to relate more to degree than to kind. In other words, the operations of the fluvial systems are similar irrespective of the climatic regime. However, where, as in the humid tropics, the rainfall regime is unique in some respects, these are reflected in the hydrology to which the physical properties of the channel adjust to establish the patterns observed in nature.

References and further reading

Akintola, F. O. (1974) 'The parameters of infiltration equations of urban landuse surfaces', Ph.D. thesis, Univ. of Ibadan.

Alavi, N. A. (1979) 'Water storage and abstraction from basement complex drainage basins of Oyo and Kwara states', Ph.D. thesis, Univ. of Ibadan.

American Geological Institute (1962) *Dictionary of Geological Terms*, Dolphin Books, Doubleday, New York.

Beaty, C. B. (1970) 'Age and estimated rate of accumulation of an alluvial fan, White Mountains, California, USA', *Amer. J. Sci.*, **268**, 50–77.

Blake, D. H. and Paijmans, K. (1973) 'Landform types of eastern Papua and their associated characteristics', *Land Res. Ser. CSIRO, Australia*, **32**, 11–21.

Büdel, J. (1957) 'Die *Doppelten Einebnungsflächen* in den feuchten Tropen', *Zeits. für Geomorph.*, N.F. **1**, 201–88.

Burke, K. C. (1967) '*The scenery of Ibadan*', mimeo., Dept of Geology, Univ. of Ibadan.

Burke, K. C. and Durotoye, B. (1971) 'Geomorphology and superficial deposits related to late Quaternary climatic variation in southwestern Nigeria', *Zeits für Geomorph.*, N.F. 15, 320–44.

Chitale, S. V. (1970) 'River channel patterns', *J. Hydr. Div. Amer. Soc/Civ. Engrs. 96 (NYI)*, 201–21.

Chorley, R. J. (1957) 'Climate and morphometry', *J. Geol.*, **65**, 628–38.

Chow, V. T. (1959) *Open-channel Hydraulics*, McGraw-Hill, New York.

Church, M. (1972) 'Baffin Island sand-dunes: a study of arctic fluvial processes', *Canada Geol. Surv. Bull. 216.*

Coleman, J. M. (1966) 'Brahmaputra River: channel processes and sedimentation', *Sedim. Geol.*, **3**, 129–239.

Doornkamp, J. C. and King, C. A. M. (1971) *Numerical Analysis in Geomorphology: an introduction*, Arnold, London.

Douglas, I. (1977) *Humid Landforms*, MIT Press, Cambridge, Mass.

Dury, G. H. (1970) 'General theory of meandering valleys and underfit streams', in G. H. Dury, (ed.) *Rivers and River Terraces*, Macmillan, London.

Ebisemiju, S. F. (1976) 'The structure of the interrelationships of drainage basin characteristics', Ph.D. thesis, Univ. of Ibadan.

Eyles, R. J. (1971) 'A classification of West Malaysian drainage basins', *Ann. A.A.G.*, **61**, 460–7.

Fahnestock, R. K. (1961) 'Competence of a glacial stream', *U.S. Geol. Surv., Prof. Paper 424 B.*, 211–13.

Faniran, A. (1980) 'The shallow water reserve in basement complex areas in southwestern Nigeria', *24th I.G.C.*, Tokyo, Japan. Abstract, Vol. 1.

Faniran, A. (1982) *African Landforms*, Heinemann, Ibadan (in press).

Faniran, A. and Areola, (1974) 'Landform examples from Nigeria, Vol. 2, a gully', *Nig. Geog. J.*, **17** (1), 57–60.

Faniran, A. and Omorinbola, E. O. (1980) 'Trend surface analysis and practical implications of weathering depths in basement complex rocks of Nigeria', *J. Mining Geol.*, **23**, 113–26.

Fisk, H. N. (1944) *Geological Investigation of the Alluvial Valley of the Lower Mississippi River*, Vicksburg, Miss.

Fisk, H. N. (1947) *Fine-grain Alluvial Deposits and the Effects on Mississippi River Activity*, 2 vols. U.S. Army Corps Engrs., U.S. Waterways Exp. Stn.

Floyd, B. (1965) 'Soil erosion and deterioration in Eastern Nigeria', *Nig. Geogr. J.*, **8**, 33–44.

Geddes, A. (1960) 'The alluvial morphology of the Indo-Gangetic Plain: its mapping and geographical significance', *Trans. Inst. Br. Geog.*, **28**, 253–76.

Gerson, R. (1974) 'Karst processes of eastern Upper Galilee, north Israel', *Jour. Hydr.*, **21**, 91–152.

Gilbert, G. K. (1877) *Report on the Geology of the Henry Mountains*, U.S. Dept. Interior, Washington, D.C.

Hickin, E. J. (1970) 'The terraces of the Lower Colo and Hawkesbury drainage basins, New South Wales', *Aust. Geog.*, **2**, 278–87.

Hooke, R. (1968) 'Steady-state relationships on arid region alluvial fans in enclosed basins', *Ann. J. Sci.*, **266**, 609–29.

Horton, R. E. (1932) 'Drainage basin characteristics', *Trans. Am. Geoph. Un.*, **21**, 522–41.

Horton, R. E. (1945) 'Erosional development of streams and their drainage basins: hydrophysical approach to quantitative morphology', *Bull. Geol. Soc. Am.*, **56**, 275–370.

Howard, A. D., Fairbridge, R. W. and Quinn, J. H. (1968) 'Terraces, fluvial: introduction', in *Encyclopedia of Geomorphology*, R. W. Fairbridge (ed.), Reinhold, New York, 117–23.

Inglis, C. C. (1949) 'The behaviour and control of rivers and canals', *Res. Publns. Poona, India No. 13*, 2 vols.

Jeje, L. K. (1970) 'Some aspects of the geomorphology of south-western Nigeria', Ph.D. thesis, Univ. of Edinburgh.

Jeje, L. K. (1974) 'Relief and drainage development in Idanre Hills of Western Nigeria', *Nig. Geog. J.*, **17** (2), 83–92.

King, L. C. (1953) 'Canons of landscape evolution', *Bull. Geol. Soc. Am.*, **64**, 721–53.

Lathman, L. H. (1973) 'Calcium carbonate cementation of alluvial fans in southern Nevada', *Ibid.*, **84**, 3013–28.

Langbein, W. B. and Leopold, L. B. (1966) 'River meanders – theory of minimum variance', *U.S.G.S. Prof. Paper 422 H.*

Ledger, D. C. (1961) 'Recent hydrological change in Rima Basin, Northern Nigeria', *Geog. J.*, **127**, 477–86.

Ledger, D. C. (1964) 'Some hydrological characteristics of West African Rivers', *Trans. Inst. Br. Geog.*, **35**, 73–90.

Leopold, L. B. and Wolman, M. G. (1957) 'River channel patterns: braided, meandering and straight', *U.S.G.C. Prof. Paper 282 B.*

Leopold, L. B., Wolman, M. G. and Miller, J. P. (1964) *Fluvial Processes in Geomorphology*, Freeman, San Francisco.

Lustig, L. K. (1965) 'Clastic Sedimentation in Deep Sprigs Valley, California', *U.S.G.S. Prof. Paper 352 F.*

Melton, M. A. (1957) 'Analysis of the relations among elements of climate, surface properties and geomorphology', *Project NR 389–042*. Tech. Rept. II, *Columbia Univ. Dept. of Geol.*, ONR Geogr. Branch, New York.

Michel, P. (1973) 'Les bassins des fleuves Sénégal et Gambie: étude géomorphologique', *Mem. ORSTROM. 63.*

Morisawa, M. (1978) *Streams, their Dynamics and Morphology*, McGraw-Hill, New York.

Ofomata, G. E. K. (1965) 'Factors of soil erosion in the Enugu area of Nigeria', *Nig. Geog. J.*, **8** (1), 45–59.

Ogunkoya, O. O. (1980) 'Hydrology of third-order basins in the upper Owana drainage basin of south-western Nigeria', Ph.D. thesis, Univ. of Ife, Ile-Ife.

Okechukwu, G. K. (1974) 'Fluvial geomorphic interrelationships in some river catchments in the Nigerian Cambrian basement complex', Ph.D. thesis, Univ. of Ibadan.

Ologe, K. O. (1971) 'Gully development in the Zaria area, Northern Nigeria', Ph.D. thesis, Univ. of Ibadan.

Ologe, K. O. (1972) 'Gullies in the Zaria area: a preliminary study of headscarp erosion', *Savanna*, **1** (1), 55–66.

Peltier, L. (1950) 'The geographical cycle in peniglacial regions as it is related to climatic geomorphology', *Ann. A.A.G.*, **40**, 214–36.

Peltier, L. C. (1962) 'Areal sampling for terrain analysis', *Prof. Geog.*, **14**, 2, 24–8.

Playfair, J. (1802) *Illustrations of the Huttonian Theory of the Earth*, Edinburgh.

Reiner, E. and Mabbutt, J. A. (1968) 'Geomorphology of the Wewak Lower Sepik Area', *Land Res. Ser. C.S.I.R.O. Aust.*, **22**, 61–7.

Ritter, D. F. (1977) *Process Geomorphology*, W. M. C. Brown, Dubuque, Iowa.

Rodier, J. (1961) *Régimes hydrologiques de l'Afrique Noire à l'ouest du Congo*, Paris.

Schumm, S. A. (1963) 'The sinuosity of alluvial rivers on the Great Plains', *Bull. Geol. Soc. Amer.*, **74**, 1089–1100.

Rust, B. R. (1972) 'Structure and process in a braided river', *Sedimentology*, **18**, 221–45.

Schwarzbach, M. (1967) 'Islandische Wasserfalle und eine genetische suplemetik des Wasserfalles iberhaupt', *Zeits. für Geom.* N.F. 11, 377–417.

Scott, A. J. and Fisher, W. L. (1969) 'Delta systems and deltaic deposition', in *Delta Systems in the Exploration for Oil and Gas*, W. Fisher, L. Brown, A. Brown, A. Scott and J. McGowan, (eds), Univ. of Texas, Austin.

Sellin, R. H. J. (1969) *Geology Illustrated*, Freeman, San Francisco.

Simons, A. B., Richardson, E. V. and Mordin, C. F. (1965) 'Sedimentary structures generated by flow in alluvial channels', in *River Morphology*, S. A. Schumm (ed.), Benchmark Papers in Geology, Cowden, Hutchinson Ross, pp. 303–15.

Smith, N. D. (1970) 'The braided stream depositional environment: comparison of the Platte River with some Silurian clastic rocks, north central Appalachia', *Bull. Geol. Soc. Am.*, **81**, 2993–3014.

Smith, N. D. (1974) 'Sedimentology and bar formation in the upper Kicking Horse river, a braided outwash stream', *J. Geol.*, 205–23.

Speight, J. G. (1965) 'Meander spectra of the Angabunga river', *J. Hydrol.*, **3**, 1–15.

Speight, J. G. (1967) 'Spectral anaysis of meanders of some Australasian rivers', in *Landform Studies from Australia and New Guinea*, J. N. Jennings and J. A. Mabbutt (eds), A.N.U. Press, Canberra.

Strahler, A. N. (1952) 'Dynamic basis of geomorphology', *Bull. Geol. Soc. Amer.*, **63**, 923–38.

Thomas, M. F. (1974) *Tropical Geomorphology: a study of weathering and landform development in warm climates*, Macmillan, London.

Thornthwaite, C. W. (1931) 'The climates of North America according to a new classification', *Geog. Rev.*, **21**, 633–55.

Thorp. M. B. (1970) 'Landforms', in *Zaria and its Region*, M. J. Mortimore (ed.), Dept. of Geography, Ahmadu Bello Univ., Zaria. Occ. Paper, vol. 4.

Tricart, J. (1959) 'Observations sur le façonnement des rapides des rivières intertropicales', *Bull. Sec. Geog. Comite Trav. Hist. et Science*, 1958, 289–313.

Twidale, C. R. (1976) *Analysis of Landforms*, J. Wiley, New York.

Wellington, J. H. (1955) *Southern Africa: a geographical study*, Vol. 1, Cambridge University Press.

Wigwe, G. A. (1966) 'Drainage composition and valley forms in parts of northern and western Nigeria', Ph.D. thesis, Univ. of Ibadan.

Wolman, M. G. (1955) 'The natural channel of Brandywine creek, Pennsylvania', *U.S. Geol. Survey Prof. Paper*, 271.

Wolman, M. G. and Miller, J. P. (1960) 'Magnitude and frequency of forces in geomorphic processes', *J. Geol.*, **68**, 54–74.
Yang, C. T. (1971*a*) 'On river meanders', *J. Hydrol.*, **13**, 231–53.
Yang, C. T. (1971*b*) 'Formation of riffles and pools', *Water Resources Res.*, **7**, 1567–74.
Zonneveld, J. I. S. (1969) 'Soela's in de Corantijin (Suriname)', *K. ned. Aardr. Genoots Geogr. Tijd.*, **13**, 45–55.

8 Duricrust and landscape

The word 'duricrust' has been accepted in the literature as a generic term for all surficial crusts which have resulted from the processes of rock weathering and soil formation. This usage is best exemplified by Dury (1969), and Faniran (1970a), both of whom used the chemical composition of these materials to classify them into types (see Chapter 3). The generic nature of the term is also well illustrated by the range of included materials and the number of local names (cf. Gentilli, 1968; Goudie, 1973). Given this situation, it is not surprising that various and diverse views have been expressed on the nature of the relationship between duricrusts and landscape or topography (cf. Newbold, 1846; Lake, 1891; Simpson, 1912; Prider, 1966; Faniran, 1969a; Goudie, 1973). Unless it is possible successfully to disaggregate duricrust types and establish those with and those without definite and recognizable connections, through detailed local studies such as has been undertaken in parts of Australia (Hays, 1967; Faniran, 1968, 1969a; among others), it will be impossible to resolve the duricrust–landscape controversy.

The prospects of achieving this type of success lie in the related concepts of deep weathering and duricrusting, both of which seek to relate duricrusts principally to the process of deep chemical weathering which operates almost exclusively in the humid tropical regions (Chapter 3). According to this viewpoint, the process of rock weathering proceeds to very advanced stages in certain parts of the humid tropics where favourable climatic (rainfall and temperature) factors – topography, rock type, vegetation and time period over which the process operates uninterrupted – exist. It is the nature of the variations in the controlling factors of rock type, climate, vegetation and topography, among others, which accounts for the various types of duricrust. The most important types are laterites or duricrusts rich in iron and aluminium hydroxides; ferricretes (also called iron stones, hardpans and ferricrusts) or ferruginous crusts; calcretes or calcareous crusts; and siliceous crusts. It is important to observe here that these materials are invariably associated with the indurated zones of deep-weathering profiles, of the process of which they are part. That is to say that the crusts have formed *in situ* during the process of weathering/pedogenesis and so are genetically related to them. Where this genetic connection is lacking – that is, where the crusts are not directly connected with *in situ* deep-weathering profiles – the observed relationship, if any, between duricrust and topography should not be confused with that (if any) observed for *in situ* (primary) duricrusts. The typical deep-weathering profile comprises at least two indurated, mottled and pallid zones, overlying ordinarily weathered, weathering or unweathered country rock (Faniran, 1968, 1969a, 1970a, 1974a; cf Chapter 3; see also Table 8.1).

Basic concepts

There are at least two aspects of this topic of duricrust topography relations. First, it involves a consideration of the topographic factor in duricrust formation; secondly, it implies the significance of duricrusts in landform evolution. The second issue is partly discussed in the chapter on slopes (Chapter 9), and so will not be discussed at length here except as it relates to more general issues. This is also the aspect which has been widely discussed in the literature (cf. Ollier, 1969; Goudie, 1973; Thomas, 1974). By contrast, the topographic factor in deep weathering has only been included in the broad and often generalized discussion of the factors of deep weathering rather than formed into substantive areas of study. It is further discussed in what follows.

A review of the literature on the relations between duricrust and topography shows that at least three viewpoints or concepts have been advanced and amplified. Two of these concepts seem to recognize some form of special relationships (cf. Woolnough, 1927; Playford, 1954; Prider, 1966; Faniran, 1969a); the third view appears to controvert or deny any form of special relationships between duricrusts and topography (Hallsworth and Costin, 1953; Maignien, 1966).

Woolnough's peneplain or 'low relief' concept of duricrusting is well known. It was expressed in the literature (cf. Newbold, 1844, 1846; Lake, 1891; Oldham, 1893; Dixey, 1920; Davis, 1920) long before it was formally presented in the form of a hypothesis, which it assumes in subsequent works. Similarly, but perhaps to a lesser extent, Playford's (1954) contrary hypothesis, that duricrusts formed not on, but subsequent to, the uplift and dissection of 'peneplains' or low-relief surfaces, continues to find support in certain quarters (cf. Gordon et al., 1958, pp. 142–43; Trendall, 1959, 1961, 1962; King, 1962, discussed by Dury, 1966, p. 305; De Swardt, 1964; De Swardt and Trendall, 1968; Prider, 1966, pp. 444–46). De Chatelat (1938) and Maignien (1966, p. 66) have also reported laterites in areas of high relief.

The third concept is here represented by the assertion by Hallsworth and Costin (1953) that laterites occur not only as cappings on isolated hills, but also at the bottoms of valleys and basins and at all positions on the slopes. Although this position has not been formally defended in the literature, it is contained in a number of casual remarks and field observations, especially by those who use such terms as *laterite* for any red (Krasnozem) or yellowish soils, including some stone lines. This type of unbridled usage of the term *laterite*, and by implication *duricrust*, has in fact been attacked (cf. Corbett, 1965; Faniran, 1969b; Faniran and Areola, 1977). The suggestion has been put forward that terms such as *laterite* be restricted to *in situ*, rather than to transported and other secondary materials.

The issue of transportation cannot of course be taken too far. In fact, virtually all weathered materials at the earth's surface must have

TABLE 8.1 Deep weathering in Nigeria

Sample location	Rock type	DEPTH (m)			
		Total	Indurated	Mottled	Pallid zone
Rayfield, Jos plateau	Basement complex	–	9.6±0.9	19.9±3.6	5.5±2.2
Barakin Ladi, Jos plateau	Basement complex	–	38.6±12.3	8.4±2.7	9.4±3.1
Pasakai, Jos plateau	Granite, B.C.	10.3±9.9	–	–	–
Dan Mongu, Jos plateau	Alluvium	31.2±10.3	–	–	–
S.W. Nigeria	Sedimentary	66.9±19.8	32.3±6.9	120.4±27.1	60.0±36.0
S.W. Nigeria	Basement complex	13.9±3.0	–	–	–
Nigeria	Basement complex	9.8±4.5	5.2±1.9	26.8±6.2	26.0±7.2
Nigeria	Sedimentary	–	82.1±11.3	243.7±34.4	249.8±34.5

Source: Faniran, 1974

247

been moved to some extent. The point to make, however, relates to the distance involved and the extent of disruption suffered by the original profile. As long as the material (duricrust) overlies a mottled and/or pallid zone on top of ordinarily weathered, weathering or unweathered rock, primary (*in situ*) situation can be assumed. Luckily, it is not very difficult to distinguish between primary and secondary (re-worked, transported and so on) duricrusts, a series of which may be encountered along a single valley-side slope, as indicated in the following observation by De Swardt (1964) among others; he has traced the existence of a characteristic suite of ferricreted surfaces through most of West and parts of East Africa. Goudie (1973) went on further thus:

> The upper surface was formed by general lowering of the landscape over a long period of humid tropical conditions and a thick laterite formed. Then, there was a change to more arid conditions that led to the cessation of laterite formation and the incision of drainage. Waste material from the old laterite formed a new ferruginous alluvial horizon, and as conditions became more humid, primary lateritization took place on the lower surface as well. [p. 42; cf. Figure 8.1].

We shall concern ourselves here mainly with primary duricrusts, which are usually indurated and so possess certain special characteristics (see below) that influence the associated topographical forms. In the meantime, we shall return to our three concepts or postulates concerning duricrust–topography relations.

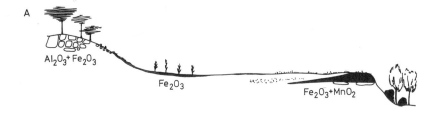

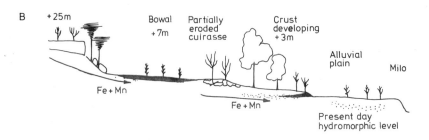

Fig. 8.1 Duricrust and slope (after Maignien, 1966)

248

The low-relief concept of duricrusting

In terms of the volume of relevant literature as well as the extent of time coverage, Woolnough's peneplain (pediplain/etchplain), otherwise described as low-relief, concept of duricrusting appears to be by far the most popular of the three concepts. The concept has been widely illustrated from almost every continent and for most duricrust types. Goudie (1973, p. 41) put the idea thus:

> Ferricretes, silcretes and calcretes (the three types of duricrust studied) are very frequently associated in the landscape with extensive flat-topped tablelands and plateaux of characteristic morphology.

However, duricrust/relief relation has been most extensively studied in Australia. Faniran (1968) described the situation as follows:

> Within Australia, duricrusts have been observed and described on relatively flat plateau and ridge-tops as early as the first days of serious geological reconnaissance (cf. for instance Darwin, 1844; Von Sommer, 1848, p. 52; Gregory, 1861, p. 480; Daintree, 1872, p. 272; Wilkinson, 1877, p. 80; Jack, 1886; Tennison-Woods, 1888, pp. 290–302; Jutson, 1914; Jackson, 1920; for earliest observations. Cf. also for instance Jackson, 1957; Mulcahy, 1959, 1960, 1961, 1967; Mulcahy and Hingston, 1961; Jessup, 1960a, 1961b, for more recent reports) [p. 133].

Much of the time span of the above references belong, in geomorphological history, to the period when Davis's influence was very strong – hence the reference to peneplains. However, the basic idea of low-relief influence does not change even with the introduction of alternative concepts of pediplanation (King, 1953), etchplanation (Thomas, 1966; Faniran, 1974*b* among others) (cf. Chapter 6). In fact, King (1962, p. 177), quoted by Goudie (1973), pp. 41–42), recognized duricrusts associated with extensive flat-topped tablelands and plateau, using them to buttress his concept of 'scarp retreat in the reduction of broad landscape from one base level to another'.

One basic problem in the application of the low-relief concept of duricrusting is to explain plausibly the occurrence of duricrusts at various elevations (levels) in the landscape. For instance, in parts of Australia studied by one of the present co-authors, duricrusts were recorded from levels varying from sea level to plateau surfaces above 1 000 m. A similar situation exists in Nigeria, where duricrusts cap the Coastal Plain Sands close to sea level, in the Lagos area, as well as hills on the Jos plateau at elevations over 1 300 m (Faniran, 1969, 1970*b*; Ologe and Shepherd, 1972).

Several hypotheses have been advanced to explain this situation, among which are the multi-erosion-surface hypothesis (Du Preez, 1949, 1954, *a,b*); the flexuring hypothesis (King, 1962); differential–post-duricrust diastrophism hypothesis (Faniran, 1968, 1969*a*); and the multi-base (water) level hypothesis.

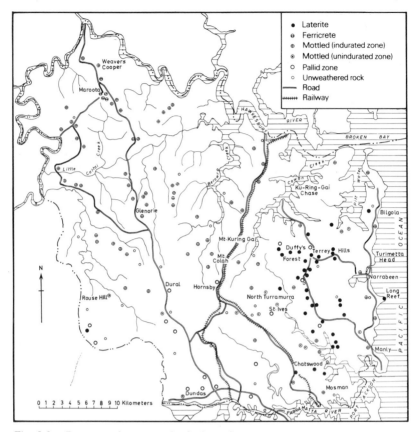

Fig. 8.2a Deep weathering in the Sydney district

The multi-erosion-surface hypothesis states in essence that each duricrusted surface represents a peneplain or other erosion surface remnant, separated from the one adjacent to it by a period of uplift which initiated a new phase of planation. Accordingly, it should be possible to reconstruct the geomorphological history of an area by mapping and possibly dating the duricrusted surfaces in that area. The situation is similar to that expressed in Figure 8.2*a*, except that in real terms the number of surfaces that can be so mapped is likely to be many more. It is also basic to the work of several geomorphologists such as Davis, King and Pugh, who have produced maps showing erosion surfaces for different areas (cf. Chapter 6).

The idea of multi-erosion surface, however, does not enjoy much popularity, since it is capable of producing too many and too complex systems of erosion surfaces and episodes than can be accommodated, explained or applied in serious geological study. In fact, Mulcahy (1964, p. 55) has warned against accepting residual duricrusts as evidence of peneplains (also pediplains and etchplains) without careful examination and study of the nature and the geomorphological relations of such duricrusts.

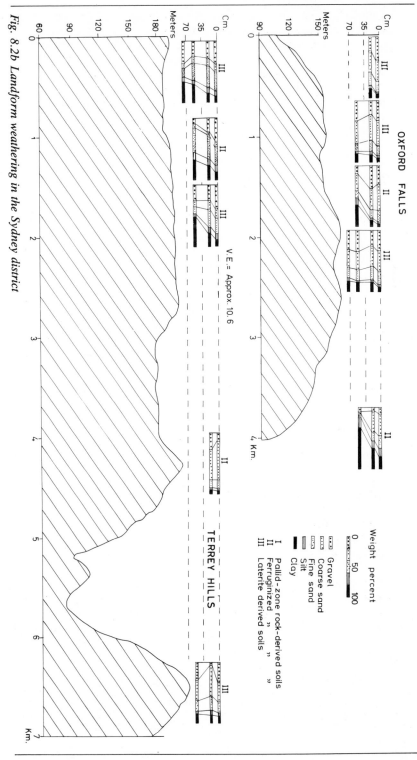

Fig. 8.2b Landform weathering in the Sydney district

Similarly, the flexuring hypothesis raises serious geological problems, and, more seriously, fails to exemplify itself in many areas. Several studies in Australia confirm that it does not apply in that continent; it is also difficult to follow such flexures over large areas, even in Africa where the idea seems to have originated.

Conversely, the concept of uplifted and possibly also faulted duricrusted surfaces appears to have received more favourable responses. It has been widely applied in some cases with slight or even significant modifications, by geomorphologists and geologists alike (cf. Newbold, 1844, 1845; Blanford, 1859; Woolnough, 1927; Hays, 1961, 1967; Wright, 1963; Langford-Smith and Dury, 1966; and Maignien, 1966). The hypothesis was specifically tested by one of the co-authors in a part of Australia where fossil duricrusts are widespread (see Faniran, 1969a).

Two types of objective approach were made in the attempt to quantify the nature and extent of the connection between duricrusts and topography, including slopes, in the area. Firstly, maps showing the distribution of deep weathering and duricrusts (Figure 8.2a) and that of landforms (Figure 8.2b) were prepared and compared, with a view to establishing the extent of coincidence in the two distributions. Moreover, topographical cross-sections were also used. The following observations were made:

1 Duricrusts and duricrust profiles almost invariably occur on tops of hills and plateaux while their truncated forms (underlying mottled, pallid and country-rock zones) tend to outcrop mostly along the valley sides of dissected country.

2 Duricrusts (laterites) are particularly widely preserved on flat-topped hills and interfluves, especially in the main interfluve region separating the major local drainage basins.

3 Duricrusts (laterites) have been mostly destroyed within the largest river system (the Hawkesbury–Broken Bay), but are still preserved at least in part within the other two basins. In terms of area coverage, the Hawkesbury–Broken Bay basin covers 70 per cent, the Parramatta–Port Jackson 20 per cent, and the Pacific Ocean 10 per cent of the area in reference.

4 Duricrusts (laterites) and their associated underlying materials occur at several levels from sea level to the highest elevation of about 214 m at Terrey Hills.

5 The general attitude of the duricrusts and associated profiles seems to be that of materials changing, along valley-side slopes of dissected terrains, from duricrust (laterite) on hilltops through mottled to pallid zone and/or unweathered country rock. Similar catenary schemes have been described in South Australia by Stephens (1946), in Queensland by Gunn (1967) and in Nigeria by Faniran (1974b, 1978) and Faniran and Areola, (1977).

The second approach was a little more detailed and stringent. It involved the collection of data on local valley-side slopes and the subjection of the collected data to statistical tests. The procedures employed are standard and are described in Faniran (1969a, pp. 54–56).

The null hypotheses tested seek to establish whether there are significant differences in the slope populations of the various samples chosen to reflect the possible connections between local topography and the occurrence of duricrusts and duricrust profiles. Slope could not be measured from the little-dissected and mostly low-relief hilltops, plateau-tops and interfluves where most of the duricrusts occur, because local slopes were too gentle for the methods used (cf. Faniran, 1970*b*). Nevertheless, the results of statistical tests indicate that there are strong connections between duricrusts and slopes. Valley-side slopes in areas where duricrusts (laterites) are still preserved, and where they were probably formed originally, are significantly different from those of areas where duricrusts (laterites) do not occur. Slopes are generally gentler in the former than in the latter areas.

The implications of the above conclusion are at least two. It is possible that duricrusting occurred over restricted areas of good drainage – that is, in the interfluve regions, hilltops and plateau-tops – following the uplift and dissection of the original low-relief surface, as has been proposed by, among others, Playford (1954), Gordon *et al*., (1958) and Trendall (1962) of the high-relief school (see below); or that duricrust affected a low-relief surface, which was subsequently uplifted to different heights and dissected.

The totality of the evidence in the Sydney area of Australia suggests that the latter alternative holds better than the former. It appears that the entire area, or a large part of it, was deeply weathered and that the present-day duricrusts on tops of hills, plateaux and interfluve regions are evidence of differential post-duricrust uplift, dissection and truncation. Several duricrust-free areas still show evidence of previous duricrusting in truncated duricrust profiles, and the rate of destruction was obviously affected by the valley-side slope. The corollary that duricrusts tend to maintain gentler valley-side slopes than the underlying materials does not hold, especially for the shale areas.

The situation in the Sydney area cannot, of course, be transferred wholesale to other areas, even when the term 'duricrust' is restricted in meaning to material produced by the process of deep weathering, which necessarily implies broad area coverage. It is, for instance, not strictly applicable in Nigeria, where duricrusting seems to have affected surfaces of different elevations which cannot be linked with known episodes of diastrophism. Also, the differences observed in duricrust types tend to reflect parent rock rather than age, especially for the primary types. Although it is possible to distinguish the re-worked and thus secondary duricrusts along valley-side slopes, most of the duricrust cap rocks are similar in physical and petrographical terms, especially when formed from identical rock types. It is therefore likely that they were all formed during a single episode but at different levels, guided by the factors of local ground water and topography (Faniran, 1970*a*).

The high-relief concept of duricrusting

Like the low-relief hypothesis, the high-relief concept has been advanced to explain observed duricrust–topography relations in parts of Australia.

The greatest proponent of the idea is Prider (1948, 1966), who has been supported by one of his students (Playford, 1954). Both studied the laterites of Western Australia, the distribution and topographic relations of which are very similar to those of the Sydney region. Nevertheless, both Prider and Playford insist that duricrusting post-dated the uplift and dissection of the original 'peneplain' surface, rather than predate them. Prider (1966) described the situation as follows:

> I first noted the occurrence of laterite on sloping ... mesas in the course of mapping with students at Lawnwood in 1945 ... the laterite-capped mesas slope down to a central valley. ... Playford pointed out how, at Moonyoonooka, laterite was formed ... on irregular surface sloping down toward the present stream valleys, indicating that the drainage system found today was already established at the time the laterite was forming. The same characteristics hold throughout the north-central part of the Perth Basin ... along the Darling Scarp and further inland .... Playford (1954) states, 'even if one believes that a large part of Western Australia was peneplained in Tertiary times, the period of lateritization must have occurred after uplift and following dissection of the old peneplain ...'. I find myself in complete agreement with Playford (pp. 445–446).

The arguments advanced for the high-relief hypothesis revolve mainly on the question of the significance of the sesquioxide-rich indurated and mottled zones. The thickness of these zones in some profiles (see below) is such that large-scale fluctuations are called for, which are possible only in high-relief regions. Another related argument relates to the concept advanced by Trendall (1962), among others, which requires the removal of great thicknesses of weathered rock in order to precipitate the high concentrations of sesquioxides in the indurated and mottled zones.

However, groundwater fluctuation and 'winnowing' are not the only processes, and in fact are not among the most popular processes, of duricrusting. The occurrence of truncated profiles in dissected terrains and along the slopes of duricrust-capped hills and plateaux contradict the high-relief or post-dissection duricrusting idea. Accordingly, this concept has not been applied widely. Prider's conclusion that 'the Great Plateau [of Western Australia] surface was already considerably dissected ... prior to the formation of the laterite profile in Pliocene times *after* the epeirogenic uplift of the Western Australian Shield' (p. 450) remains an isolated view, difficult to substantiate on either practical (field) or theoretical grounds. It is significant also to observe in relation to geological age that most authors put deep weathering and duricrusting in Australia, as elsewhere, to at least the Miocene, in some cases earlier (Exon *et al.*, 1970).

The complex, no-relation concept of duricrusting

The assertion, or suggestion, that duricrusts have no definite relationship with topography makes sense as long as the terminology of duricrusts is not streamlined. Perhaps the greatest culprit here is the term 'laterite', which remains among the most controversial in the scientific literature. A review of the literature on this material shows that the term has been largely abused, so much so that many scientists think it should be discarded from the scientific literature (Hamming, 1968). Since several types of material have been described as laterite, Hallsworth and Costin (1953) are therefore justified to report them, as they do in many places, 'at all positions on the slopes', as well as on hilltops and valley bottoms.

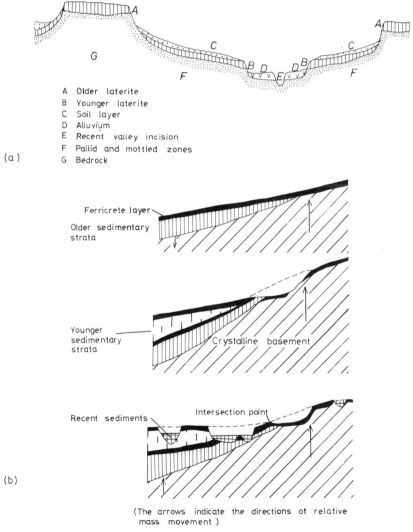

(a)

A Older laterite
B Younger laterite
C Soil layer
D Alluvium
E Recent valley incision
F Pallid and mottled zones
G Bedrock

Ferricrete layer
Older sedimentary strata

Younger sedimentary strata

Crystalline basement

Recent sediments

Intersection point

(b)

(The arrows indicate the directions of relative mass movement)

Fig. 8.3 Complication of duricrust/topography relations

Similar observations have also been reported by one of the co-authors both in Australia (Faniran, 1968) and in Nigeria (Faniran, 1974*b*). The difference, however, lies in the conclusions drawn from such observations. It is the view of the present authors that duricrusts *stricto sensu* occur at definite positions on the landscape, as described above.

Another possible source of confusion relates to post-duricrust events, which can so modify the original duricrust–topography relations to the extent of obscuring them. Goudie (1973, pp. 42–52) described some examples (Figure 8.3), which are very illustrative. What Goudie, however, failed to do is provide information on the underlying materials to confirm whether the materials in reference are '*in situ*' (primary) duricrusts with associated mottled and/or pallid zones. Goudie also failed to provide information on the scale, if any, of his illustration which would give an idea of the angle of slope, ruggedness of the surface and so on. It is our submission here that the situations described by Goudie are not all identical and so cannot be treated together. With sufficient discrimination of materials and the associated topography, the nature of their mutual relationships is certain to emerge.

In conclusion, therefore, it appears that of the three concepts being used to describe the nature of duricrust–topography relations, especially in terms of the topographic requirements of duricrusting, the low-relief concept is not only the most popular in terms of the volume of literature, but also the best documented. Although the occulary, cartographical and statistical tests reported above may not necessarily establish causal relationships between process and form, the correspondence shown appears sufficiently convincing for now. There are very strong indications that (1) duricrusting affected broad areas of low relief; (2) the areas may or may not be a single continuous surface; and (3) post-duricrust events – uplifts, down-faulting, sub-aerial (fluvial) action and so on – have greatly modified the original duricrusted terrain to produce the present-day situation of duricrusts on hilltops, plateau-tops and other interfluves and their truncated forms in dissected and fluvially eroded terrains.

The geomorphic effects of duricrusts

Geomorphologists are interested more in the way in which rocks (including duricrusts) affect geomorphological processes than in the type of topography on which particular rock formations were formed. Accordingly, and as hinted earlier, many geomorphologists have studied the way in which duricrusts, where they occur in the landscape, influence the associated geomorphological processes and forms. This appreciation started with the earliest explorers and travellers (see above) who took pains to describe the striking impact of duricrusts on landscapes generally and landforms, vegetation and landuse in particular.

The geomorphological attitude of duricrusts relate to certain characteristics of the material – namely, their physical characteristics of profile thickness, texture, structure, hardness and so on; chemical characteristics; and mineralogical characteristics, some of which have already been discussed in Chapter 3.

Some physical and chemical properties of duricrusts

The above discussion underscores the importance of the profile characteristics in the identification of the typical duricrust profile. In like terms, the profile characteristics of the duricrust, especially in terms of depth, are potentially a very important factor in the geomorphic significance of duricrusts of whatever type.

As shown in Tables 8.2a and b the thickness of the indurated zone or hardpan varies between duricrust types. Generally speaking, the indurated zone is thinnest in calcrete profiles and thickest in the laterite/ferricrete duricrusts. Accordingly, all things been equal, laterites, alcretes, ferricretes and silcretes are likely to persist longer in the landscape and so exert the strongest geomorphic effects. However, since the geomorphological environments to which these materials are exposed are not similar, there can be no easy comparison as such.

Another important set of physical attributes of duricrust of geomorphic relevance are the texture, structure and hardness characteristics, which together determine its strength or resistance to stress. Unfortunately, there is hardly any reliable data on any of these. Quantitative studies have produced some generalizations such as have been made by Goudie, among others, thus:

1 Silcretes with low porosities, high quartz content and low absorption capacities (often less than 1 per cent), are likely to be more resistant to both chemical and physical weathering than other types of duricrusts.

TABLE 8.2a Maximum thickness of laterite and ferricrete profiles (metres)

Source	Location	Total	Indurated zone
Mikhaylov (1964)	Conakry (Guinea)	120	–
Connah & Hubble (1960)	Queensland	87.5	–
Lake (1891)	India	46	–
Wright (1963)	Northern Territory Australia	46	6.09
Harden & Bateson (1963)		88	7.62
Humbert (1948)	New Guinea	37	1.52
Wolfenden (1961)	Sarawak	34	5.48–701
Santos-Ynigo (1952)	Philippines	30	–
Cooper (1936)	Ghana	18	–
Vann (1963)	Amapa (Brazil)	18	–
Sillans (1958)	Central Africa Rep.	–	10.0
Mabbutt (1965)	Central Australia	–	9.14
Hill (1955)	Jamaica	–	7.62
Maud (1968)	Natal, S. Africa	–	4.57
Dowling (1966)	Nigeria	–	3.05
Nel et al. (1939)	Transvaal, S. Africa	–	1.82

Source: Goudie, 1973

TABLE 8.2b   Maximum thickness of calcrete profiles

Source	Location	Thickness (m)
Goudie (1973)	Molopo valley, S. Africa	60
Davidson (1968)	Millstream, W. Australia	47
Brown (1956)	Liana & Stacado, USA	45
Von Backstrom (1953)	Lichtenburg, S. Africa	38
Govt. of Kenya (1962)	Uaso Nyiro, Kenya	35
Siragusa (1964)	Argentina	20
Dutcher & Thomas (1966)	Gefara, Tunisia	15
Johnson (1967)	Ghannel Islands USA	13
Goudie (1973)	Wajir, Kenya	12
Firman (1967)	Australia	10
Gautier (1922)	Algeria	10
Nagregaal (1969)	Spain	10
Liddle (1964)	Venezuela	10
Monroe (1966)	Puerto Rico	10
Sanders (1968)	Wiluna (W. Australia)	10
Bretz & Horberg (1949)	New Mexico	10
Goudie (1973)	Kalahari	10
Sanchez & Artes (1967)	Spain	20–30
Goudie (1973)	Cyprus	20
Ruellan (1967)	Morocco	10
Gaucher (1948)	Algeria	5–10

Source: Goudie, 1973

2 Calcretes with water absorption capacities generally between 4 and 10 per cent would prove less resistant to, say, salt weathering (in the desert) than silcrete; calcretes, however, have a durability comparable to many other sedimentary rocks.

3 Laterites can be quite porous, with absorption capacities of up to 15 per cent; they are, therefore, probably less resistant than other silcretes or calcretes.

4 Ferricretes have higher bulk specific densities, owing to their iron content, than calcretes, silcretes or laterites (Table 8.3a–e).

5 Indurated duricrusts are generally non-plastic with Plasticity Index (P.I.) of less than 3; exceptions are the nodular types (bauxites, laterites and some silcretes) with P.I.s of 20–30 and Lower Liquid Limit (LLL) of 50 or more.

6 Indurated duricrusts can be very hard, with Moh's values of as much as 7 (for porcellanitic silcretes) and 5 for some calcretes.

7 Shear-strength values provided by Eyles (1967) for Malaysian laterites suggest considerable degree of strength. The 25 samples studied had a mean value of 21.35 Nam^{-2}, with some exceeding 42 Nam^{-3}.

8 High infiltration capacities are also characteristics of some duricrusts, especially weathered laterites. Eyles (1967) reported a mean value of 12.5 cm with a range of 0.86 cm h^{-1} and 42 cm h^{-1}. With such high values, surface runoff will seldom, if ever, occur, while high shear-strength values, where present, will offer protection against slumping and related physico-denudational (mass-wasting) processes.

TABLE 8.3a Calcrete chemistry (percentage)

		$CaCo_3$	$MgCo_3$	SiO_2	Al_2O_3	Fe_2O_3
1. Australia	Mean	87.97	6.28	6.49	1.58	2.26
	Range	58.36–99.40	1.00–45.41	0.56–17.86	0.26–3.81	0.06–7.10
2. India	Mean	61.02	–	17.92	3.13	3.62
	Range	27.07–81.00	3.19–49.09	0.04–5.97	1.65–15.97	
3. S. Africa	Mean	79.13	8.72	11.83	2.38	1.51
	Range	–	–	3.34–46.06	0.23–7.40	0.23–740
4. N. Africa	Mean	74.81	–	11.33	0.70	0.95
	Range	27.64–95.59	–	1.38–69.38	0.20–1.98	0.40–2.78

Source: Goudie, 1973

TABLE 8.3b Silcrete chemistry (percentage)

	SiO_2	Fe_2O_3	MgO	CaO	TiO_2	Al_2O_3
South Africa						
Mean	93.75	1.65	0.46	0.45	1.62	1.17
Range	86.32–97.14	0.11–5.04	–	–	0.23–2.77	0.09–3.28
Australia						
Mean	95.63	2.66	0.06	0.29	0.57	1.63
Range	58.70–99.50	0.03–31.60	–	0.06–1	0.10–1.80	0.10–18.40

Source: Goudie, 1973

TABLE 8.3c   Ferricrete chemistry (percentage)

	a	b	c	d	e	f	g	h
SiO_2	2.70	5.40	19.83	0.62	–	17.08	16.40	35.60
Al_2O_3	9.60	44.30	12.47	10.54	31.37	20.83	23.50	18.80
Fe_2O_3	74.00	23.10	47.80	14.43	19.12	40.18	45.00	33.90
TiO_2	0.68	1.40	0.86	3.92	1.12	1.72	1.80	1.40
H_2O+	12.40	24.50	10.49	9.60	9.10	11.05	13.30	10.1

a–W. Africa   b–W. Africa   c–W. Africa   d–Guyana   e–W. Africa   f–India
g–W. Africa   h–W. Africa

Source: Goudie, 1973

Table 8.3c summarizes the available data for silcrete and calcrete.

By contrast, many primary (intact) duricrusts have very low infiltrations. In fact Akintola (1974) recorded no infiltrations whatsoever after several tests on the duricrusted (hardpan) surfaces in the Ibadan built-up area. This is related to the degree of induration and hardness attained by the materials as well as to their textural and structural characteristics.

Despite all these elements of strength, several duricrusts also have elements of weakness in the void, rind, and some type of cement structures, not to talk of the pipes. As observed elsewhere (Faniran, 1968, 1971b), these features are exploited by the weathering process in the destruction of the duricrusts.

The mineralogy and chemistry of duricrusts have been largely documented (cf. Goudie, 1973). As shown in Table 8.3a calcretes are dominantly composed of calcium carbonate ($CaCO_2$), with a world mean value of 79.28 per cent, and a 97 per cent $CaCO_2$ purity reached by some South African examples. Other components include silica (SiO_2) (12.3 per cent), alumina (Al_2O_3) (2.12 per cent), and ferric oxide (Fe_2O_3) (2.03 per cent).

By contrast, silcretes show a great dominance of silica in their makeup, being probably the purest of duricrusts, with over 90 per cent silica (SiO_2) content (Table 8.3b). Fourteen South African samples analyzed by Visser (1964) gave a mean SiO_2 content of 95.63 per cent. All other constituents are relatively minor, with Al_2O_3, Fe_2O_3 and TiO_2 being the most significant of the minor elements (Table 8.3b).

The chemical and the constituent materials in ferricretes and alcretes are discussed in Chapter 3, and will not be repeated here. However, it is sufficient to observe that a petrographical study of the Sydney duricrusts gave the major mineralogical components as gibbsite ($Al_2O_3 . 3H_2O$), goethite (FeO), haematite (α-Fe_2O_3), maghemite (γ-Fe_2O_3), kaolin and quartz (Tables 8.3 d & e). These are among the hardest minerals in their natural forms. It is therefore not surprising that many primary duricrusts are relatively immune to the effects of the weather and so persist for long periods in the landscape, to the extent of fossilizing such surfaces or landscapes.

261

TABLE 8.3d Laterite chemistry – Sydney district (percentage)

Type	SiO_2	Fe_2O_3	Al_2O_3	TiO_2	MnO_2	Others	Parent rock
1. Bulk	19.30	24.90	55.30	0.20	0.06	0.24	Fine-grained Sandstone
2. Birds'-eye	5.70	59.80	34.10	0.40	Tr.	–	Sandstone
3. Matrix	25.70	25.10	49.20	0.25	Tr.	–	Sandstone
4. Bulk	39.81	23.80	35.30	0.10	0.11	0.89	Corse-grained Sandstone
5. Pisolite	49.50	24.20	26.00	0.01	Tr.	0.29	Sandstone
6. Pisolite	35.80	35.00	29.00	0.07	Tr.	–	Sandstone
7. Pisolite	38.60	26.20	34.70	0.20	Tr.	0.30	Sandstone
8. Matrix	37.20	55.80	6.90	0.52	Tr.	–	Sandstone
9. Bulk	72.10	9.30	17.90	0.50	–	–	Cemented gravel

Source: Faniran, 1968

TABLE 8.3e   Laterite mineralogy (X-ray defraction) (percentage) – Sydney district

Sample

Bulk	Sibbsite	Goethite	Haematite	Kaolin	Quartz
1	30	15	5	25	25
2	5	30	5	25	35
3	40	20	5	10	10
4	35	5	20	30	10
AV.	27.5	20	8.75	22.5	20
Pisolite					
1	20	10	25	25	20
2	20	5	35	30	10
3	15	10	70	0	5
AV.	18.3	8.3	43.3	18.3	8.3
Matrix					
1	50	30	10	5	5
2	35	15	0	25	25
3	5	25	5	25	25
Total	25.5	16.5	18.0	20.0	18.5

Source: Faniran, 1970*b*

Duricrust and landform evolution

The characteristics of duricrusts described above help to explain the following landforms, among others, commonly found in humid tropical environments of the present day as well as fossil ones: caprock escarpments and associated hillside slopes; mesas; relief inversions; pseudo-karsts and pipes, slumps and debris-strewn pediments and plains (gibbers); reefs and lunates; flat irons; patterned ground; playas or pans (depressions); and pseudoanticlines. We shall describe a few of them.

Caprock escarpments and mesas

These are among the best illustrations of the geomorphological role of duricrusts. They are present in areas where post-duricrust removal processes have reduced the areal extent of the original cover. Very fine examples have been described from the Nupe sandstones, false-bedded sandstone and Jos plateau areas of Nigeria (Ofomata, 1967; Ologe and Shepherd, 1972) in the Buganda region (Pallister, 1956) and large parts of Australia, among other places. Writing about the Australian examples, Ollier and Tuddenham (1962, p. 98) described how

The hard band acts as a datum for slope development. . . . All outlying hills have pretty much the same sort of shape so long as they retain a duricrust capping. . . . When the duricrust is eventually removed, however, the lower slopes are rapidly reduced.

Goudie (1973, pp. 44–45) has also described how, in New Mexico, the prominent Mescalero ridge is capped by a resistant caliche; on the Liano Estacado the resistant caprock is responsible for the development of escarpments which go by the name of 'Caprock Escarpments' or 'The Breaks'; in southern Africa, Pliocene calcrete forms the prominent capping to the Kalk plateau of Namibia and to the canyon sides of the Kuiseb and Swakop rivers; and in the Pampas of Argentina where 'Tosca' (calcrete) forms steep bluffs to some of the rivers of the great Pampean plain, and abrupt sides to little plateaux. Finally, Ologe and Shepherd, (1972, p. 83) wrote about the Jos plateau thus:

> Laterite-capped mesas are a common feature of certain parts of the Jos Plateau. These landforms are developed in the deeply-weathered Older Basalts which appear to have covered most of the plateau surface sometime in the late Palaeocene or early Miocene. . . . These mesas tend to occur along main watersheds and are the remnants of what was once a much more extensive duricrusted surface which has been dissected by fluvial action. They are found both as isolated hills and as small groups . . . are low, flat-topped hills, seldom higher than 250′ (76.2 m) above the local base level. Apart from the sharp break of slope at the edge of their flat tops, they have a smooth (concave) profile.

Other locations, in Nigeria, where laterite-capped mesas occur include Kano (the Dala and Dutsi), between Zaria and Kaduna, on the Nsukka–Okigwi uplands of Anambra State, in the Gongola basin in the Niger–Benue trough, around Bida, and in the Sokoto basin (see Chapter 6).

Plate 19 Duricrust-capped hill near Bida, Niger State, Nigeria (A. Faniran)

Inverted relief and related features

Duricrusts are known to have played quite major roles in relief inversion. This is particularly true of duricrust types formed in depressions. For example, calcrete often increases in thickness towards depressions, as in parts of southern Africa and Brazil. Subsequent erosion tends to affect the non-duricrust areas and leave the former depressions as hills. Maignien (1966, p. 72) made the following observations regarding parts of West Africa:

> Incrusted zones have protected the zones in which they [calcretes] developed while erosion has lowered the surrounding zones. Old crusts occupy the highest positions, and are gradually reduced by lateral retreat of slopes.

Goudie (1973, p. 46) also described examples of relief inversion from Arabia, where 'a crust up to 3 m thick formed in the valleys and was subsequently left standing above the non-indurated sands of the area'.

Pseudo-karsts and related features

The typical duricrust profile is characterized by, among other things, a thick zone of decomposed rock consisting, in an ideal situation, of mottled zone, a pallid zone and a zone of weathered and/or weathering country rock. This underlying material varies in thickness from place to place, and can measure up to 100 m. Since the overlying duricrust (indurated zone) may have spaces (voids, pipes and so on) through which water reaches the underlying decomposed and unsolidified rock, percolating water does work on the material through underground springs and other forms of flow. Where the underlying material is exposed, gullies develop which can cut back into the slope and lead to serious sapping. The gullies and valleys may widen and coalesce to give swampy, ever-widening depressions. Woolnough (1927) described an Australian example thus:

> The hard capping, resting upon a considerable thickness of thoroughly decomposed rock material, produces an extremely weak structure. As soon as the agents of denudation pierce the crust, erosion proceeds rapidly . . . extensive valleys are carved out and . . . residuals of characteristic shapes are developed.

A major process in this type of duricrust destruction consists of pseudo-karst and/or piping. This process was observed as far back as 1909 by Gautier (p. 60). Aubert (1963) also described how the attack of the lower friable horizons beneath a ferricrete cap has, in the Fouta Djallon area, led to the formation of caverns and caves; while Chevalier (1949) reported the formation of hollows by piping in duricrust areas. In Guyana, sinkholes of about 800 × 400 m surrounded by undermined laterite scarps have been described by McBeath and Barron (1954). Other reports include:

1 that by Thomas (1968, 1974) in the Sula mountains of Sierra Leone, who also thought the Lake Sonfon depression to be a result of the wholesale flushing out of the underlying clay over a wide area as a consequence of the encroaching head-waters of the Sende river;

2 Dixey's (1920, p. 219) claim, also in Sierra Leone, that streams are known suddenly to appear and disappear much as they do in limestone districts, giving Hart's Cave (32 m × 11 m × 4 m) developed beneath a laterite and a swallow hole (Devil Hole) 8 m deep and over 2 m wide with passages leading away from its base as examples.

3 rounded to elliptical depressions rimmed by steps of lateritic crusts in the Buganda plateau (Goudie, 1973, p. 47);

4 the pseudo-karstic forms of buttes and tablelands developed in Western Australia from rock hollows formed about 5 m beneath the crust through 'blow holes' to natural bridges (Jutson, 1934); and

5 the oval shaped pseudo-karstic hollow (210 m × 126 m × 1.5–5 m) in Buganda (Patz, 1965).

However, de Chatelat's (1938) study (reported by Goudie, 1973) remains one of the most detailed of these phenomena. The study area is Guinea, where he (de Chatelat) described caves beneath a laterite layer which had been eroded by underground water into the friable sub-crust layer. The Foulas (local inhabitants) call the associated lakes and depressions 'vendous'; they are characterized by the following features among others;

1 schist bedrock overlain by a lateritic cuirasse;

2 location at the head of amphitheatres on gentle slopes; and

3 depth of about 1.5 m.

Duricrust 'karst' is also well developed in western India, where Goudie studied the Panchgani tableland features south of Poona, and wrote as follows (1973, p. 48):

> On the surface of the crust there are numerous shallow depressions up to 140 m in diameter and 3 m in depth. On the edges of the Tableland lateral slumping and undermining has led to the development of scallops of similar diameter but with depths up to 10 m. The biggest scallops were associated with springs or seepage from beneath the ferruginous crust.

If pseudo-karsts develop in laterites and ferricretes, they are even better developed in calcretes, where they are associated as well with patterned ground phenomena. This is due mainly to the well-known lime content and relative solubility of $CaCO_2$, the major constituents of calcretes. Accordingly, calcrete areas have a whole suite of landforms more characteristic of limestone terrains, including unusual drainage patterns, sub-surface flow and closed depression.

Sinkholes develop best where, as observed earlier for laterites, thick masses of indurated calcrete overlie less consolidated material. Examples already described include:

1 A portion of Namibia west of Kalkrand which shows a swallow-hole pattern of internal drainage, in marked contrast to a more normal, dendritic pattern in the area (Figure 8.5a).

2 The High Plains of the USA where Levings (1961, p. 68) observed characteristic lacunate drainage pattern as a result of the solution of the Ogallala caprock (calcrete) and the underlying calcareous layers.

3 The Troodos mountains of south-east Cyprus with discontinuous drainage systems.

4 The pans of southern Africa and the 'playas' of North America, especially those in calcrete regions (Judson, 1950; Nicholson and Clebsch, 1961; Havens, 1966; Figure 8.5d). The best examples from North America occur in Texas and New Mexico, where Goudie (1973, p. 55) mapped closed depressions, most of which are less than 12 m deep, but some of which, in the Liano Estacado, are at least 16 m across and 50 m deep.

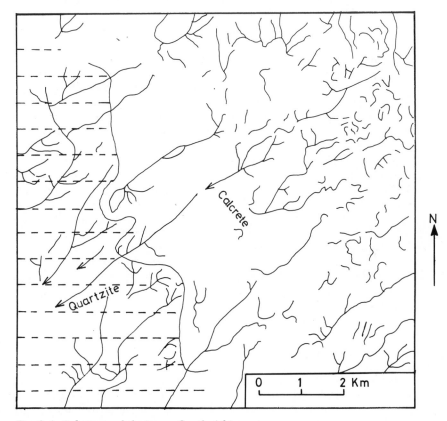

Fig. 8.4 Calcrete and drainage, South Africa

The southern African examples have been described from the Boschpan–Water pan area north of Vryburg and parts of Botswana near Khuis. The pans here (southern Africa) are localized, associated mainly with Dwyka tillite and shales, but also with calcrete.

Flat-irons and other slope features

Two of the most interesting slope forms of duricrusted terrains are flat-ions and debris-strewn surfaces or gibbers. Flat-irons are composed of remnants of old slope facets, while gibbers result from the deposited duricrusted fragments. Both occur on hill-slopes, particularly pediments.

The best examples of flat-irons have been described from Cyprus, West Africa and North Africa. In Cyprus, south-east of Nicosia, flat-irons have been observed to occur on both sides of valleys; they are still being shaped by seepage and other denudational processes. Sapping beneath the crust seemed to have attacked the rear steep portions of the flat-iron facing the main valley slope.

The duricrust type in Cyprus is of course calcrete, just like those of North Africa. Similar feature have been observed in the ferricrete/laterite areas of West Africa. Flat-irons have been described, for example, by Brammer (1956) from Upper Volta, with heights in excess of 60 m, separated from the escarpment by a gap of more than 1 km and developed on deeply weathered schists or weathered debris on the piedmonts of hill masses (Goudie, 1973, p. 50).

As observed above (Chapter 5), the debris produced by the breakdown of duricrusts sometimes litter the slopes to produce debris-strewn slopes. These are particularly prominent in arid and semi-arid regions where the stony desert tableland soils (in Australia) frequently consist of silcrete fragments produced by the destruction of silcrete-capped mesas. In West Africa, lateritic debris is known to litter many slopes. De Chatelat (1938) described the disordered manner in which the fragments occur on the slopes (lateritic pavements) as 'chaos lateritiques'. The progressive breakdown and sorting of these fragments on slopes have also been linked with the development of some pediments (Dury, 1966), while Pallister (1956) thinks that lack of dimunition in the size of the fragments after they have been reduced to the size of a pea may be a factor in the maintenance of essentially straight, as opposed to the more normal concave, pediment profiles in parts of Uganda. Perhaps the most interesting examples here are those described by Eden (1971) in Guyana as 'residual laterites of mottled appearance on inselberg slopes near Shea' (p. 191).

Other minor duricrust features

In certain cases, the duricrust cover is rather limited in areal extent and so cannot develop the type of features described so far. This is particularly true of duricrusts formed under localized environments, including some deposited and re-worked duricrusts. Thus a ferricrete/laterite formed by spring flow or water seepage into an area of constricted drainage or localized depressions will cover a very small area, depending upon the size and shape of the receptacle. Such materials, if hardened (indurated), may persist in the landscape after the surrounding material has been destroyed, as small isolated 'reef' (Schofield, 1957) or lunate bands (Williams, 1969). Finally, differences in duricrust types may be reflected in slope forms at the local scale, including the free-face or scarp slopes of many mesa terrains and the rounded profiles of the underlying materials.

References and further reading

Akintola, F. O. (1974) 'The parameters of infiltration equation of urban landuse surfaces', Ph.D thesis, Univ. of Ibadan.

Aubert, G. (1963) 'Soil with ferruginous or ferrallitic crusts of tropical regions', *Soil Sci.*, **95**(4), 235–42.

Blanford, W. T. (1859) 'Note on the laterite of Orissa', *Mem. Geol. Surv. India*, **1**, 80–94.

Brammer, H. (1956) 'A note on former pediment remnants in Haute Volta', *Geogr. J.*, **122**, 526–7.

Bretz, J. H. and Horberg, L. (1949) 'Caliche in S.E. New Mexico', *J. Geol.*, **57**, 491–511.

Brown, C. N. (1956) 'The origin of caliche in the northeast Llano Estacado, Texas', *J. Geol.*, **64** (1), 1–15.

Casewell, J. in *Field Studies in Libya*, S. G. Willimott and J. I. Clarke (eds), Res. Paper No. 4, Dept. of Geography, Univ. of Durham.

Chevalier, A. (1949) 'Points de vue nouveaux sur les sols d'Afrique tropicale sur leur dégradation et leur conservation', *Bull. Agric. du Congo Belge*, **40**, 1057–92.

Connah, T. H. and Hubble, G. D. (1962) 'Laterites', in *The Geology of Queensland*, D. Hill and A. K. Denmead (eds), Nelbourne, pp. 373–86.

Cooper, W. G. G. (1936) 'The bauxite deposits of the Gold Coast', *Bull. Gold Coast Geol. Surv.*, 7.

Corbett, J. R. (1965) 'Mineralogy as a guide to soil chronology and provenance in the Inverell area', Ph.D thesis, Univ. of Sydney.

Daintree, Re (1872) 'Desert sandstone', in 'Notes on the Geology of the Colony of Queensland', *Quart. J. Geol. Soc. Lond.*, **28**, 275–78.

Darwin, C. (1844) referenced by Tennison-Woods (1888) see below.

Davidson, W. A. (1968) 'Millstream hydrological investigation', *Dept. Mines, Western Australia Annual Report*, No. 57, 61.

Davis, W. M. (1920) 'Physiographic relations of laterite', *Geol. Mag.*, **57**, 429–31.

De Chatelat, E. (1938) 'Le modèle latérique de l'ouest de la Guinée française', *Rev. Géogr. Phys. Geol. Dyn.*, **11** (1), 5, 120.

De Swardt, A. M. J. (1964) 'Lateritisation and landscape development in parts of Equatorial Africa', *Zeit für Geomorph.*, N.F. 8, 313–33.

De Swardt, A. M. J. and Trendall, A. R. (1969) 'The physiographic development of Uganda', *Overseas Geol. Min. Resource*, **10**, 241–88.

Dixey, A. F. (1920) 'Notes on lateritisation in Sierra Leone', *Geol. Mag.*, **26**, 783–6.

Dowling, J. W. F. (1966) 'The mode of occurrence of laterites in N. Nigeria and their appearance in aerial photography', *Engrg. Geol.*, **1** (3), 221–33.

Dunstan, B. (1920) 'Geological notes on the Gloncury–Camooweal–Burketown–Boulia Area, N.W. Queensland', *Old Geol. Surv. Publ.*, No. 265.

Du Preez, J. W. (1949) 'Laterite: a general discussion with a description of Nigerian occurrence', *Bull. Agri. Congo Belge*, **50**, 53–60.

Du Preez, J. W. (1954a) 'Notes on the occurrence of oolites in Nigerian laterites', *19th Internl. Geol. Congr.*, **21**, 163–9.

Du Preez, J. W. (1954*b*) 'Origin, classification and distribution of Nigerian laterites', *Proc. 3rd Internl. W. Afr. Conf.*, (1949), 223–34.

Dury, G. H. (1966) 'Duricrusted residuals on the Barrier and Cobar pediplains of N.S.W.', *J. Geol. Soc. Aust.*, **13**, 299–307.

Dury, G. H. (1969) 'Rational descriptive classification of duricrusts', *Earth Sci. J.*, (Waikato), **3** (3), 77–86.

Dutcher, L. C. and Thomas, H. S. (1966) 'The occurrences chemical quality and use of underground water in the Tabulbah area of Tunisia', *U.S.G.S. Prof. Paper 1157 E.*

Eden, M. J. (1972) 'Some aspects of weathering and landforms in Guyana, formerly British Guiana', *Zeits für Geomorph.*, N.F. 15 (12), 181–98.

Exon, N. F., Langford-Smith, T. and McDougall, I. (1970) 'The age and geomorphic correlations of deep weathered profiles, silcrete and basalt in the Roma-Amby region of Queensland', *J. Geol. Soc. Aust.*, **17** (1), 21–30.

Eyles, R. J. (1967) 'Laterite at Kerdau Panang, Malaya', *J. Trop. Geog.*, **25**, 18–23.

Faniran, A. (1968) 'A deeply weathered surface and its destruction', Ph.D thesis, Univ. of Sydney.

Faniran, A. (1969*a*) 'Duricrust, relief and slope populations in the Sydney district, N.S.W.', *Nig. Geog. J.*, **12**, 53–62.

Faniran, A. (1969*b*) 'On the use of the term "laterite" ', *Prof. Geog.*, **21** (4), 231.

Faniran, A. (1970*a*) 'Landform examples from Nigeria, No. 2, the deep-weathering (duricrust) profile', *Nig. Geogr. J.*, **13** (1), 87–8.

Faniran, A. (1970*b*) 'The Sydney duricrust: note on terminology and nomenclatures', *Earth Sci. J.* (Waikato), **4** (2), 117–28.

Faniran, A. (1971*a*) 'Implications of deep weathering for the location of natural resources', *Nig. Geog. J.*, **14** (1).

Faniran, A. (1971*b*) 'The parent material of Sydney laterites', *J. Geol. Soc. Aust.*, **18** (1), 159–62.

Faniran, A. (1971*c*) 'Maghemite in the Sydney duricrust', *Amer. Mineral*, **55**, 925–33.

Faniran, A. (1974*a*) 'The extent, profile and significance of deep weathering in Nigeria', *J. Trop. Geog.*, **38** (1), 19–30.

Faniran, A. (1974*b*) 'Nearest-neighbour analysis of inter-inselberg distance: a case study of the inselbergs of South Western Nigeria', *Zeits für Geomorph.*, *Suppl. 8d.*, **20**, 150–69.

Faniran, A. (1978) 'Deep weathering duricrusts and soils in the humid tropics', *Savanna*, **7** (1), 61–5.

Faniran, A. and Areola, O. (1977) *Essentials of Soil Study*, Heinemann, London.

Firman, J. B. (1967) 'Stratigraphy of late Genozoic deposits in S. Australia', *Trans. Geol. Soc. S. Australia*, **91**, 165–78.

Flach, K. W., Nettleton, W. D., Gile, L. H. and Cady, J. G. (1961) 'Pedocementation: indurated by silica carbonates and sesquioxides in the Quaternary', *Soil Sci.*, **107c** (6), 442–53.

Gaucher, G. (1948) 'Les sols rubifriés et les sols à croutes du Bas Chelif et des basses plains oranaises (Regions d'Inkermann, de Relizane, de Perregaux, et de Saint-Denis-du-sig)', *Comptes Rendus, Academie des Sciences, Paris*, **225**, 130–5.

Gaucher, G. (1958) 'Les conditions géologiques de la pédogenèse Nord-Africaine', *Bull. Service Geol. Algérienne*, **20** (n.s.), 55–94.

Gautier, E. F. (1909) *Madagascar-Essai de Géographie*, Paris.

Gentilli, J. (1961) 'Quaternary climates of the Australian region', *Ann. New York Acad. Sci.*, **95**, 463–501.

Gentilli, J. (1968) 'Duricrust', in *The Encyclopedia of Geomorphology*, R. W. Fairbridge (ed.), Reinhold, New York.

Gordon, (Jr.), M., Tracey, (Jr.), J. I. and Ellis, W. M. (1958) 'Geology of the Arkansas bauxite region', *U.G.S.S. Prof. Paper 299.*

Goudie, A. (1973) *Duricrusts in Tropical and Subtropical Landscapes.* Clarendon, Oxford.

Govt. of Kenya (1962) *An Investigation into the Water Resources of the Ewaso Ngiro Basin, Kenya.* Dry. Branch Report No. 4, Nairobi.

Gregory, F. T. (1861) 'On the geology of a part of Western Australia', *Quart, J. Geol. Soc., Lond.,* 475–83.

Gunn, R. N. (1967) 'A soil catena on denuded laterite profiles in Queensland', *CSIRO, Aust. Div. Land Resour. Reg. Surv., Ann. Report 1965–66,* p. 17.

Hallsworth, E. G. and Costin, A. B. (1953) 'Studies in pedogenesis in New South Wales, IV: The ironstone soils', *J. Soil Sci.,* **4,** 24–46.

Hamming, E. (1968) 'On laterites and latosols', *Prof. Geog.,* **20** (4), 238–41.

Harden, G. and Bateson, J. N. (1963) 'A geochemical approach to the problem of bauxite genesis in British Guiana', *Econ. Geol.,* **58,** 1301–08.

Harrison, J. B. (1910) 'The residual earths of British Guiana commonly termed laterite', *Geol. Mag., N.S. Dec. V.,* Vol. 7, 120–3.

Havens, J. S. (1966) 'Recharge studies on the High Plains in northern Lea County, New Mexico', *U.S.A. Geol. Survey, Water Supply Paper,* 1819–F.

Hays, J. (1961) 'Laterite and land surfaces in the Northern Territory, Australia, north of the MacDonell Ranges', in *Geochronology and land surfaces in relation to soils in Australia,* Symposium, Aust. Acad. Sci. A.F., Adelaide.

Hays, J. (1967) 'Land surfaces and laterites in the north of the Northern Territory', in *Landform Studies from Australia and New Guinea,* J. N. Jennings and J. A. Mabbutt (eds), A.N.U. Press, Canberra, pp. 182–210.

Hill, V. G. (1955) 'The mineralogy and genesis of the bauxite deposits of Jamaica, B.W.I.', *Amer. Miner.,* **40,** 676–89.

Humbert, R. D. (1948) 'The genesis of laterite', *Soil Sci.,* **65,** 281–90.

Jack, R. L. (1886) *Handbook of Queensland Geology,* Brisbane.

Jackson, E. A. (1957) 'Soil features in arid regions with particular reference to Australia', *Aust. Inst. Agr. Sci. J.,* **53,** 196–208.

Jessup, R. W. (1960*a*) 'The laterite soils of the southeast portion of the Australian arid zone', *J. Soil Sci.,* **11,** 106–13.

Jessup, R. W. (1960*b*) 'The stony tableland soils of the southeastern portion of the Australian arid zone and their evolutionary history', *J. Soil Sci.,* **11,** 188–96.

Jessup, R. W. (1961*a*) 'Evolution of the two youngest (Quaternary) soil layers in the southeast portion of the Australian arid zone', *J. Soil Sci.,* **12,** 52–63.

Jessup, R. W. (1961*b*) 'A Tertiary–Quaternary pedological chronology for the south east portion of the Australian arid zone', *J. Soil Sci.,* **12,** 199–213.

Johnson, D. L. (1967) 'Caliche on the Channel Islands', *Mineral Information, California Div. Mines and Geol.,* **20** (12), 151–8.

Judson, S. (1950) 'Depressions of the northern portion of the southern High Plains of eastern New Mexico', *Bull. Geol. Soc. Amer.,* **61,** 253–74.

Jutson, R. T. (1934) 'The pluysiography (geomorphology) of Western Australia', *Geol. Surv. W. Aust., Perth, Bull.,* **95.**

King, L. C. (1953) 'Canons of landscape evolution', *Bull. Geol. Soc. Amer.,* **64,** 721–52.

King, L. C. (1962) *The morphology of the Earth,* Oliver & Boyd, Edinburgh.

Lake, P. (1891) 'The geology of south Malabar between the Beypore and Ponnani Rivers', *Mem. Geol. Surv. India,* **24** (3), 210–46.

Langford-Smith, T. and Dury, G. H. (1965) 'Distribution, character and attitude of the duricrust in Northwest New South Wales and the adjacent areas of Queensland', *Amer. J. Sci.,* **263,** 170–90.

Liddle, R. A. (1964) *The Geology of Venezuela and Trinidad*, 2nd edn., Ithaca, N.Y.

Levings, W. S. (1951) 'Late Cenozoic erosional history of the Raton Mesa Region', *Colorado Sch. Mines Quarterly*, **46** (3).

Mabbutt, J. A. (1965) 'The weathered land surface of Australia', *Zeits für Geomorph.*, 82–114.

Maignien, R. (1966) 'Review of research on laterites', UNESCO (Paris) *Natural Resour. Res.*, **4**.

Maud, R. R. (1968) 'Further observations on the laterites of coastal Natal, S. Africa', *Trans. 9th Intern. Congress of Soil Sci.*, Adelaide, **4**, 151–8.

McBeath, D. M. and Barron, C. N. (1954) 'Report on the lateritic ore deposit at Iron and Wamara Mountains, Berbice', *Rep. Geol. Surv. Dept. Br. Guiana*, 1953, 87–100.

Mikhaylov, B. M. (1964) 'Part played by vegetation in the lateritization of mountainous areas of the Liberian Shield', *Doklady Akad. SSR*, **157**, 23–4.

Monroe, W. H. (1966) 'Formation of tropical karst topography by limestone solution and precipitation', *Carr. J. Sci.*, **6** (12), 1–7.

Mulcahy, M. J. (1959) 'Topographical relations of laterite near York, W. Australia', *J. Roy. Soc. W. Aust.*, **42**, 44–8.

Mulcahy, M. J. (1960) 'Laterite and lateritic soils in south-western Australia', *J. Soil Sci.*, **11** (2), 206–65.

Mulcahy, M. J. (1961) 'Soil distribution in relation to landscape development', *Zeits für Geomorph*, N.F. **5** (3), 211–25.

Mulcahy, M. J. (1964) 'Laterite residuals and sand plains', *Aust. J. Sci.*, **27**, 54–5.

Mulcahy, M. J. (1967) 'Landscapes, laterites and soils in south-western Australia', in *Landform Studies from Australian and New Guinea*, J. N. Jennings and J. A. Mabbutt (eds), A.N.U. Press, Canberra, pp. 211–30.

Mulcahy, M. J. and Hingston, F. J. (1961) 'The development and distribution of the soils of the York-Quairading Area, Western Australia, in relation to landscape evolution', *CSIRO, Soil Publication*, **17**, Canberra.

Nagregall, P. J. C. (1969) 'Microtextures in recent fossil calcretes', *Leiclsche Geol. Mededeelingen*, **42**, 131–42.

Nel, L. T., Truter, F. C. and Willemse, J. (1939) 'The geology of the country around Potchefslroom and Klerksdorp', *Geol. Surv. S. Africa, Expl. Sheet 61*.

Newbold, T. J. (1844) 'Notes, chiefly geological, across the peninsula from Masuitipatan to Goa', *J. Asiatic Soc.* (Benegal), **13**, 984–1004.

Newbold, T. J. (1846) 'Summary of the geology of Southern India: Laterite', *J. Roy. Asiatic Soc.*, **8**, 227–40.

Nicholson, A. and Clebsch, A. (1961) 'Geology and groundwater conditions in southern Lee county, New Mexico', *New Mex. State Bur. Mines and Miner. Res.*, Groundwater Rept. 6.

Ofomata, G. E. K. (1967) 'Landforms on the Nsukka Plateau of Eastern Nigeria', *Nig. Geog. J.*, **10** (1), 1–9.

Oldham, R. D. (1893) *Manual of Geology of India*, Calcutta.

Ollier, C. D. (1975) *Weathering*, Longman, London.

Ollier, C. D. and Tuddenham, W. G. (1961) 'Inselbergs of central Australia', *Zeits für Geomorph.*, N.F. **5**, 257–76.

Ologe, K. O. and Shepherd, I. (1972) 'Landform examples from Nigeria', *Nig. Geog. J.*, **15** (2), 83–6.

Pallister, J. W. (1956) 'Slope development in Buganda', *Geog. J.*, **122**, 80–7.

Patz, M. J. (1965) 'Hilltop hollows – further investigations', *Uganda J.*, **29** (2), 225–8.

Playford, P. E. (1954) 'Observations on laterite in Western Australia', *Aust. J. Sci.*, **17** (1), 11–14.

Prider, R. T. (1948) 'The geology of the Darling scarp-ridge Hill', *J. Roy. Soc. W. Aust.*, **31**, 105–29.

Prider, R. T. (1966) 'The lateritized surface of Western Australia', *Aust. J. Sci.*, **28**, 443–51.

Ruellan, A. K. (1967) 'Individualisation et accumulation décalcaire', *Maroc, Calmer O.R.S. T.O.M. Sér. Pédologique*, **5**, 421–62.

Sanchez, J. A. and Artes, F. (1967) 'Sols sur croûtes calcaires dans les zones côtie res du sud-est de l'Espagne', *Trans. Conf. Medt. Soils*, Madrid, 331–40.

Sanders, C. C. (1968) 'Hydrogeological reconnaissance of calcrete areas in the East Murchison, and Mt Margaret Goldfields', *Department of Mines, Western Australia, Annual Report*, 54–7.

Santos-Ynigo, L. (1952) 'The geology of iron ore deposits of the Philippines', *Symposium gisements de terre du monde*, Alger, tome I.

Schofield, A. N. (1957) 'Laterites of Nyasaland and their indications on aerial photographs', *Road Res. Bull.*, **5**.

Sherman, G. D., Kanehiro, Y. and Matsusaka, T. (1953) 'The role of dehydration in the development of laterite', *Pacific Sci.*, **4** (4), 438–46.

Sillans, R. (1958) *Les gavanes de l'Afrique centrale*, Paris.

Simpson, E. S. (1912) 'Laterite in Western Australia', *Geol. Mag.* **9**, 399–406.

Sivarojasingham, S., Alexander, L. T., Cady, J. G. and Cline, M. G. (1962) 'Laterite', *Adv. Agron.*, **14**, 1–60.

Tennison-Woods, J. E. (1887) 'The desert sandstone', *J. Proc. Roy. Soc. N.S.W.*, **22**, 290–335.

Thomas, M. F. (1966) 'Some geomorphological implications of deep weathering patterns in crystalline rocks in Nigeria', *Trans. Inst. Br. Geog.*, **40**, 173–93.

Thomas, M. F. (1968) 'Some outstanding problems in the interpretation of the geomorphology of tropical sheidls', *Br. Geom. Res. Gr. Publn.*, **5**, 41–9.

Thomas, M. F. (1974) *Tropical Geomorphology: a study of weathering and landform development in warm climates*, Macmillan, London.

Trendall, A. F. (1959) 'The topography under the northern part of the Kadam Volcanics and its bearing on the correlation of the peneplains of south-east Uganda and adjacent parts of Kenya', *Rec. Geol. Surv. Uganda, 1955–56*, 1–8.

Trendall, A. F. (1961) 'Explanation of geology of sheet 45 (Kodam)', *Rept. Geol. Surv. Uganda, No. 6*.

Trendall, A. F. (1962) 'The formation of "apparent peneplains" by a process of combined laterisation and surface wash', *Zeits. für Geomorph.*, Supply Bd. **6** Heft 2, 183–97.

Vann, J. H. (1963) 'Development processes in laterite terrain in Amapa', *Geog. Rev.*, **53** (3), 406–17.

Visser, J. N. J. (1964) *Analyses of Rocks, Minerals and Ores, Rep. S. Africa Geol. Surv. Handbook, No. 5*.

Von Backstrom, J. M. (1953) 'The geology of the area around Lichtenburg', *Union of S. Africa Geol. Surv. Expl., sheet 54*.

Von Sommer, F. (1848) 'A sketch of the geological formation and physical structure of Western Australia', *Quart, J. Geol. Soc., Lond.*, **5**, 51–3.

Williams, M. A. J. (1969) 'Lands of the Adelaide-Alligator area, North Territory', *CSIRO, Land Res. Ser.*, No. 25.

Wilkinson, C. S. (1877) 'Notes of the geology of New South Wales', *Geol. Surv. N.S.W. Dept. Mines, Rept*.

Wolfenden, E. B. (1961) 'Bauxite in Sarawak', *Econ. Geol.*, **56**, 972–81.

Woolnough, W. G. (1927) 'The duricrust of Australia', *J. Proc. Roy. Soc. N.S.W.*, **61**, 24–53.

Wright, R. L. (1963) 'Deep weathering and erosion surfaces in the Daly River Basin, Northern Territory', *J. Geol. Soc. Aust.*, **10**, 151–63.

9 Slopes on non-duricrusted surfaces

The study of slope form, its importance and evolution, has been an important aspect of geomorphology, and has generated most of the methodological disputes in the discipline (Chorley, 1964). The importance of slope studies derives from the fact that slopes, whether erosional or depositional, are ubiquitous on the earth's surface and, as expressed by Young (1964), 'over the large parts of the world, a high proportion of the physical landscape consists of valley slopes'. Apart from being an important component of the landscape, slopes impart the geometrical qualities which determine the characteristics of landform complexes or regional landscapes.

Although slopes constitute such important elements of the landscape, so far there is no consensus as to their descriptive terminology. The need for acceptable terminology in slope study cannot be overemphasized for various reasons. First, it facilitates the measurement of slope profiles in the field, and, secondly, and more importantly, it facilitates valid quantitative comparisons between the measurements of different researchers. It is for these reasons that a comprehensive glossary of slope forms is provided in this book.

Slope studies in the humid tropics

The literature on slopes in the temperate environment date back to the days of Hutton, Playfair and Lyell in the eighteenth and nineteenth centuries. However, this is not the case in the humid tropics, where most observations on slopes are very recent and often made in relation to 'typical' landform complexes like the savanna plains of Cotton (1942).

Chorley (1964) adduced several reasons for the paucity of slope studies in the humid tropics. Firstly, as heavy vegetal cover makes the cost of cutting traverses for slope studies rather prohibitive, it is relatively easier to study slopes in semi-arid regions, the results of which are often uncritically projected into the humid tropics. Secondly, most slope studies in the humid tropics are based on simple visual estimation of slope forms and angles, and it is well known that human eyes are deficient in this regard. Where measurements have been carried out, the methods followed in slope measurements in the field are often questionable, also the equipment used, ranging from the Abney level to the optical reading clinometer and theodolites, have varying levels of accuracy. All these make it difficult to compare the findings of the few researchers engaged in studying slopes in the humid tropics.

Other problems related to slope studies in the tropics include: (1) the belief that slope processes are either too slow or too rapid, hence difficult to measure; (2) lack of knowledge relating to the magnitude and frequency of geomorphic processes; (3) the complexities of present-day processes. Thus our knowledge of process-form responses on slopes is too limited.

There is also the circular nature of the problem related to slope studies. It is possible to explain slope forms by a detailed study of processes operating on them; however, there is also the fact that the magnitude, and often the character of the processes affecting a given slope, can be conditioned by the existing slope geometry.

Except for the measurements carried out by Wigwe (1966), Young (1970, 1972), Swan (1970), Faniran and Ajaegbu (1971), Ofomata (1973), Jeje (1976) and a few others, slope measurements primarily aimed at determining slope forms in the humid tropics are too sparse. Consequently, slope forms in this environment, having been inadequately studied, are not well known.

Slope forms in the humid tropics

The form of any slope is determined by the relationship between the rate of waste generation from the underlying rocks, and the rate at which the debris is removed from the surface. Waste preparation involves both physical and chemical weathering, and these are governed by several factors including the composition and structure of the rocks, and the thickness and state of reduction of the regolith. Structure includes the attitude of the beds, the degree of jointing, the type of deformation, and the presence of features like foliation and lineation.

The rate of regolith removal depends on several factors, among which are the fineness of the regolith, its thickness and the degree of vegetal cover. Regolith removal in the humid tropical environment, as shown in Chapter 4, is effected mainly by running water and mass-movement complexes. These may operate on the slope itself, at the base, or on the crest of the slope in moulding its form. For instance, maximum slope angle on a valley or a hill-slope is governed by the rate of regolith removal at the base, so that a stream channel at the base of the slope may be a dominant control of the slope form.

Several geometric forms are produced by the interaction of the various geomorphic processes, both endogenous and exogenous, upon different lithologies and structures; thus, in the tropics as elsewhere, it is possible to identify slopes on resistant rocks, and valley slopes associated with weathered surfaces and non-resistant rocks in non-duricrusted surfaces.

Slopes on resistant rocks

These comprise major escarpments and slopes on hills and ridges developed on unweathered and partially weathered rocks. The various categories of slopes are examined below:

Escarpments

As mentioned in Chapter 6, major escarpments often separate extensive plains on the cratons of the humid tropics. The coastal zones of Central and south-east Africa provide several examples of such escarpments. In south-east Africa, the coastal escarpment has been severely dissected by rivers which have cut gorges 300–1 000 m deep, marked by near-vertical walls, the most impressive of which are the eastern slopes of the Thousand Hills behind Durban as well as the Tugela valley about 1 000 m deep in the same area. Other examples include the fault scarps in East Africa which rise up to 1 600 m in height and are still remarkably fresh, having suffered little or no erosion since the Tertiary period, due to the resistance of the local rocks. Fresh fault scarps about 300 m high also occur in the Republic of Malagasy, where Lake Alaofra occurs on the scarp foot.

In South America, the more than 400-m-high scarp walls of the Paraiba rift in Brazil are fresh and rectilinear. The northern flank of the Paraiba graben is comprised of the face of the Serra da Mantiquera, nearly 3 000 m high. In Cabo Frio, Brazil, steep monoclinal warping, coupled with the resistant nature of the underlying gneisses, have given rise to the more than 1 000-m-high scarp along the coast of Serra do Mar. A similar scarp stands between Santos and São Paulo.

Apart from tectonically related escarpments, a few scarps in the interior of the continents are erosional. Good examples of these include the Voltaian scarp in Ghana, the Arochukwu–Okigwe–Nsukka scarp in eastern Nigeria and the Bandiagara scarp in Mali.

The Voltaian escarpment developed mainly on shales, sandstones, limestones, conglomerates and sandy-pebbly beds, comprises the Gambaga scarp to the north-east of Ghana, and the Kwahu scarp in the forest to the south. The Ganbaga scarp (140–270 m high) has a free face 120 m high developed on quartz-sandstone; the Kwahu scarp, starting near Koforidua, north-east of Kumasi, is 77 km long and has a free face 120 m high developed on granodiorites (Hunter and Hayward, 1971).

The Arochukwu–Okigwe–Nsukka scarp is formed on a series of shales and sandstones including the Nkporo shales, the Lower Coal measures and the false-bedded sandstones. The Lower Coal measures form the steep free face, while the less resistant false-bedded sandstone forms the crest and the upper slopes. The scarp, varying in height between 150 and 300 m, and locally rising to 450 m, represents the western limb of a breached anticline that trended in a NE–SE direction.

The east-facing Bandiagara escarpment overlooking the Gondo plain in Mali is developed on an extremely resistant, gently dipping sandstone which forms part of the structural basin underlying much of the republics of Mali and Mauritania. The near perpendicular scarp slope, 250 m high, has a low debris slope at the base (Buckle, 1978).

Several other escarpments occur on the crystalline basement rocks in the interior of the continents in the tropics. Examples of such scarps include the more than 500-m-high western and south-western side of the

Jos plateau, the 600-m-high western escarpment of the Mandara mountains, the 300-m-high western scarp of Idanre massif and, on a lesser scale, the 100-m escarpment around Oke–Iho in Nigeria. Other examples include the 600–700-m-high western and northern side of the Adamawa plateau in central Cameroon, the 800-m-high escarpment of the Andringitra granite massif in south-east Malagasy, and the 200-m-high escarpment on the eastern side of the Boborema massif in Pernambuco, north-eastern Brazil.

The scarps are thought to be erosional, as they often separate erosion surfaces of different elevations and possibly of different ages. For instance, King (1962, p. 157) believes such continental scarps in Africa to be retreating at the rate of about 30 cm in 100 years. However, as shown by Thomas (1974), except for minor retreat of the Jos escarpment near Toro, there is nothing to suggest that escarpments associated with extremely resistant rocks are undergoing any erosional bevelling. Thus the scarps are entirely lithological in origin. For instance, the Jos plateau owes its existence to the extremely resistant Younger Granites intruded into the basement complex rocks of the area in the Jurassic period. The rocks represented by a suite of biotite and riebeckite granites are found mainly on the western, south-western and north-western sides where the plateau terminates in steep escarpments. These rocks form the massive hills which rise more than 600 m above the plateau surface. On smaller scales, the plains around the granitic massifs in Idanre and Ibarapa in western Nigeria are developed on undifferentiated schist complex with no erosional bevelling whatsoever into the granites. Also, the plains around the Oke–Iho tableland are developed on amphibolite-schist, while the tableland is underlain by granite syenite (Jeje, 1972, 1974). Thus, as the weaker rocks around these resistant rocks are weathered and eroded, the surrounding plains are gradually lowered while local relief increases along these escarpments.

These escarpments, whether tectonic, erosional or structural in origin, have complex morphology. Some, like the Gambaga and the Kwahu scarps in Ghana, the Arochukwu–Okigwe–Nsukka scarp and the Bandiagara scarp, exhibit the classical four hill-slope elements of King. On the Idanre escarpment and the Bandiagara scarp, free faces constitute up to 80 per cent of the slope profiles. The Bandiagara sandstone impregnated with silica is especially resistant to weathering, with the result that the rate of waste production from the scarp face is by far less than the rate of destruction. This is also the case on the slopes of the porphyritic granites of Idanre. However, some escarpments in crystalline rocks are dominated by massive half-domes undergoing sheeting, as is the case on the Ndomme and Jos plateau escarpments (Thomas, 1974).

Hill-slopes on bare and partially weathered rocks

Hill-slopes on hard and resistant rocks like gneisses and granites have been studied in West Africa by Savigear (1960), Thomas (1966), Kadomura (1977); in Malawi by Young (1972); and in Natal by Fair (1947, 1948), among others. As is evident from the study of Thomas (1967) on the morphology of a single inselberg in the Oyo plain of western Nigeria, the entire slope profile may be bare, or covered by shallow, weathered debris. On granitic hills, it is not unknown for regolith of up to 30 m deep to persist on hill-slopes. However, at about 40°, hill-slopes on massive rocks rarely carry any debris. Thus the nature of the hill-slope is a reflection of the rock structure and the balance between waste preparation and removal.

Several slope forms have been observed on unweathered and partially weathered basement complex rocks in the humid tropics (Buxton and Berry, 1957, 1961; Savigear, 1960; Thomas, 1965, 1966, 1967; Young, 1972; Jeje, 1973). These include the following:

1 convex-rectilinear or dominantly concave profiles often associated with regolith and forest-covered epidiorite and amphibolites, as those measured on these rocks in the Opa basin near Ile-Ife by Jeje (1976).

2 Smoothly convex-concave profiles with a continuous regolith cover, which appears to be the most extensive form on regolith-covered, granitic hills. The upper convex element may be in bare rock, especially where an inselberg surmounts the crest, as those observed in Balos, Sudan, by Ruxton (1958).

3 Convex summit with perpendicular and even overhanging face with no debris slope at the base, as is characteristic of granitic slopes in the Akure–Ado, Ekiti and Igbajo areas in western Nigeria (Jeje, 1973); see also Savigear (1960).

4 Dominantly convex, bare rock slope in which the lower concave elements form a minor proportion of the whole slope profiles, characteristic of some bornhardts in Nigeria. The surface may be paralleled by exfoliating convex rock blocks, as those measured on an inselberg on the Oyo plain by Thomas (1967).

5 Slopes with irregular stepped micro-relief, concave or approximately rectilinear in macro-form, characteristic of granites or gneisses in which horizontal or curvilinear joints form dominant structural elements, as in the case on the dome-on-dome inselbergs recognized in parts of Nigeria by Pugh (1966).

6 Slopes in which the upper parts are cliffed and below which are concave debris slopes formed on massive boulders, with the debris slopes featuring highly irregular micro-relief. Ruxton (1958) recognized nine distinct slope units on this type of profile on granite in Balos, Sudan.

7 Steep and irregular bare rock slopes tending towards rectangular forms conditioned by jointing, especially following the outward tensional collapse of an old inselberg, as those observed by Thomas (1965) in parts of northern Nigeria, and by Jeje (1973) near Eruwa, Oyo State, Nigeria.

8 Irregular bouldery slope with extremely complex micro-relief, and rather complex in macro-form, characteristic of slopes strewn with rounded blocks following the disruption an old tor.

Where dolerite sills form caprocks, slope profiles are dominantly concave. Also, slope form on horizontally bedded rocks are dominantly concave, especially where a resistant stratum forms the upper layer. In the coastal area of south-east Africa in an area underlain by highly resistant Table Mountain sandstone, overlain by weak Dwyka tillite and Ecca shales, the slope profiles exhibit the typical four hill-slope elements of King. The sandstone forms a highly conspicuous scarp about 150 m, above which the weaker tillites and shales form the convex slope elements (Fair, 1948).

Hill-slopes formed entirely on massive quartzites often display the four hill-slope elements of King, with the slope profile dominated by the micro-faceted free face. This is especially evident on the western side of the quartzite ridge tending between Ipetu–Jesa and Oke–Messi in south-western Nigeria.

On the low hills referred to as demi-oranges – *meias laranjas*, developed on weathered granite and gneisses in Rio, Salvador and Recife areas of Brazil – hill-slopes are dominantly convex, the convexity being emphasized as a result of the deposition of colluvial material in the surrounding plains, so that there is an abrupt junction between the hill-slopes and the surrounding plains.

Measurements on weathered quartz-schists in the upper Owena basin of western Nigeria show hill-slopes to be dominantly convexo-concave, with the convex element occupying more than 60 per cent of the measured profiles (Jeje, 1978).

In the sub-humid tropical environment in the Fitzroy basin of Australia, hill-slopes on partially weathered basement rocks have characteristically steep gradients. Although most profiles are diversified by rectilinear micro-slope units, on the macro-level, concave profiles are dominant. In the Daly basin, micro-irregularities on slope profiles are of less importance and slope profiles are dominantly concave (Wright, 1973).

Slope profiles on deeply weathered surfaces

Opinions are divided as to the nature of valley-slope profiles on deeply weathered plains in the humid tropics. Tricart (1972), Birot (1968), Young (1970, 1972), Carson and Kirkby (1972), Journaux (1975), Zonnefeld (1975), So (1976) and Kodamura (1977) are of the opinion that valley slopes in the rainforest areas are characterized mainly by convex forms. However, Mabbutt and Scott (1966), Swan (1970, 1972), Morgan (1973) and Jeje (1976) among others have shown that valley slopes in this particular environment are generally convexo-concave or convex-rectilinear-concave, with the percentage covered by each slope element or segment depending on the intensity of the local geomorphological processes, on the geomorphological history, on the local relief and on the nature of the regolith, especially its texture and content of free silicon (Swan, 1972). Areas underlain by the same rock types can have different slope profiles. Thus in the University of Malaya and the Puchong area, both in south-east Malaysia, underlain by alternating sandstones and shales, slope forms are quite different. In the former area with a local relief of only 61 m, but with the streams subject to recent rejuvenation, valley profiles are dominantly convex-rectilinear, whereas in the latter area with a local relief of 105 m, valley-slope profiles are dominantly convexo-concave (Morgan, 1973).

Based on eighty-two measurements, Young (1970) recognized five

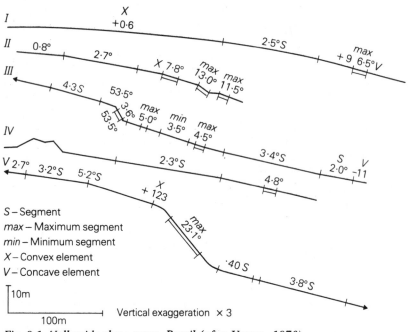

Fig. 9.1 Valleyside slope types, Brazil (after Young, 1970)

different types of valley-side slope profiles and one hill-slope type in the Xavantina and Cachimbo areas of Brazil (Figure 9.1). The characteristics of the valley slope profiles are summarized in what follows.

Slope class 1 Smooth-crested, unincised valley-side slopes, 100–150 m long, consisting of a single slope sequence usually associated with minor valleys. Of the slope profiles 95 per cent comprise a very gentle convexity, steepening from 0–4°. Concavities are generally less than 10 per cent. These profiles form 44 per cent of the total samples.

Slope class II Smooth-crested, incised valleys consisting of two to three sequences but with long upper convexity as in class I. The lowest 50–100 m of the profiles are broken into alternation of steps and benches. These comprise 9.7 per cent of the total samples.

Slope class III Smooth-crested, benched, valley-side slopes, consisting of up to five sequences found along larger rivers. The upper parts are characterized by long convexities terminated by steep maximum segments. The lowest 130–400 m are in the form of steps and benches. The risers are inclined at 10–45°, while the benches are inclined at 1–3°. These constitute 17 per cent of the total measured profiles.

Slope class IV Unincised valley-side slopes with crest formed on laterite comprise only 3.6 per cent of the measured profiles. These are entirely concave.

Slope class V These comprise laterite scarp and ridge slopes and constitute 13.4 per cent of the measured profiles.

Slope class VI These are valley slopes in the Suia–Missu Catchment. The slopes, comprising single profile sequences, are developed on sandstones. These profiles are 1.0–2.5 km long. Convexities occupy 95–99 per cent of the profiles, but these include long segments inclined at 0.3–0.5°.

Field measurements in the Opa basin in the Ife area of western Nigeria on an undulating plain underlain by weathered gneisses, pegmatites, granites, amphibolites and schists, show that valley-slope profiles in the basin can be categorized as follows (Figure 9.2) (see Jeje, 1976):

1 Simple slope profile dominated entirely by convexity from the crest to the stream channel with no concave element. These profiles, 7 per cent of all the samples, have a mean length of 116.0 m with a standard deviation of 71 m. They are mainly associated with first-order streams in which local relief is less than 15 m. The crest section has angle of curvature of 0.5°/100 m which, near the stream, has increased to 5.3°/100 m.

2 Simple convex-rectilinear concave profiles in which the convex and concave elements are of identical length, but in which the rectilinear segments are dominant. These are 150–240 m long, and constitute only 3.5 per cent of all the samples. They are also associated with first-order streams.

3 Simple, smooth-crested profiles with long upper convexity constituting about 80 per cent of the profiles. These are found along first- to third-order streams. Mean length is 248 m with standard deviation of 104 m. Local relief is about 30 m. These profiles constitute 25 per cent of all the samples.

4 Compound slope profiles with two distinct sequences separated by steep segments inclined at 10–22° and less than 30 m long. Each sequence is convexo-concave. The slope profiles are 300–405 m long with a mean of 385 m. These constitute only 5 per cent of the samples.

5 Multi-concave profiles in which a short, upper convexity is followed downslope by concave elements separated by rectilinear segments. These profiles are associated with weathered epidioritic schists in which rock rafters outcrop to form the steep rectilinear segments. The slope profiles, 360–390 m long, are associated with first- and second-order stream segments. The profiles constitute only 3.5 per cent of the samples.

6 Low-angled, complex-slope profiles with three or more convexo-concave sequences in which the convex elements are dominant, often constituting more than 60 per cent of the entire slope profiles. These profiles, with a mean length of 307 m and a standard deviation of 67 m, are found along third- and fourth-order stream segments where cross-sectional valley local relief is between 15 and 30 m. Both the convex and the concave elements are smoothly curved, with angle of curvature about 2.5°/100 m.

The characteristic valley slope profile in the Opa basin is the complex

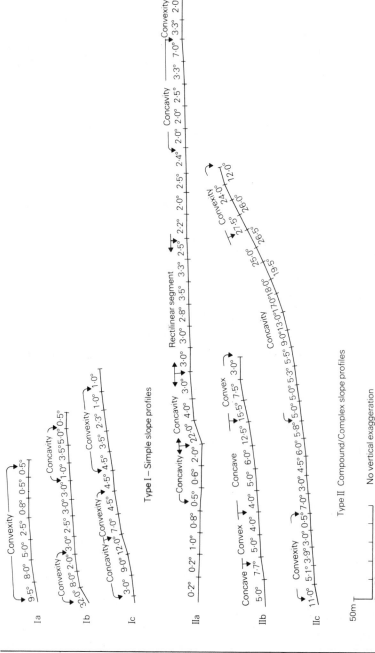

Fig. 9.2 Slope profiles, Ile-Ife area, Nigeria (after Jeje, 1976)

type. Following the methods of determining overall profile form described by Doornkamp and King (1971), it was found that about two-thirds of the fifty-seven measured slope profiles show overall convex forms, while the remaining one-third have overall concave forms. The average slope-profile length is distributed in the ratio of 0.455, 0.305 and 0.240 among convex, concave and rectilinear forms respectively.

There appears to be a relationship between slope-profile form and the textural characteristics of the underlying regolith. In the Opa basin, regolith derived from gneisses, granites and pegmatites is dominated to depths of more than 30 cm by coarse and fine sand, which total more than 60 per cent, while regolith derived from epidioritic schists to depths of 90 cm contains about 57 per cent silt and clay fractions. Slope profiles on the coarser-textured regolith have overall convex form, whereas those on the finer-textured regolith are dominantly concave. This reinforces the observations made by Swan (1972) in Johore, Malaysia, that convexity characterizes coarse-textured regolith because of its greater porosity and lower detachability; whereas clayey regolith gives rise to concave slopes.

In the savanna areas, valley-slope profiles have been observed to be convex-rectilinear-concave (Kessel, 1977) or multi-concave and faceted. The concave elements invariably occupy the greater proportion of the slope profiles, as is evident from the work of Fölster (1969) in the savanna area of Nigeria, or Kadomura (1977) in the savanna area of Cameroon. However, Young (1972) observed that in Malawi the proportions of convexity, maximum segment and concavity are about equal on most slope profiles.

Characteristic slope angles on erosional plains

Characteristic angles are those which most frequently occur either on all slopes under particular conditions of rock and climate, or in a local area (Young, 1961). They appear as modes on graphs of frequency distributions, the classes with the maximum frequencies being the primary characteristic angle. However, the idea of establishing characteristic slope angles in any area should be approached with caution, as the result depends not only on the measurement intervals but also on the slope class selected for the frequency distribution analysis. Thus Gerrard and Robinson (1971) have shown that, by using different measurement intervals on the same slope profiles, both the values of the modal slopes and the percentage frequency in each class differ for different recording intervals used. Also, given the complex nature of the geology and the nature of the terrains, and the size of the drainage basins where slope measurements have been carried out, it is very difficult to compare characteristic slope angles at different locations in the humid tropics.

The characteristic slope angles on an erosional plain in parts of western Malaysia were determined by Morgan (1973), and shown in the following table:

TABLE 9.1   **Characteristic slope angles in parts of Malaysia**

Location	Geology	Relief (m)	Characteristic slope angle
A basin 0.09 km² in the University of Malaya campus, Selangor	Alternating sandstones and shales	61	4°–8°; 10°–13°; 15°–17°; 2°–22°; and 24°–30°
A basin 3.02 km² 2 km east of Tampin, Negri Sembilan	Granite	180	1°–3°; 5°–8°; 11°12°; 18°–25°; and 28°–30°
A basin 0.31 km² 15 km south of Petaling Java in Selangor	Steeply dipping alternating sandstones and shales	105	2°; 7°–14°; 14°–19°; and 21°

As observed by Young (1970) in the Araguaia (Rivero das Mortes) catchment in Brazil, two-thirds of all slopes are less than 3°, and nearly 90 per cent less than 5°, with a characteristic slope angle between 1.8° and 2.9°. In the Xingu (Suia Missu) catchment, over 97 per cent of slope angles measured are less than 3°, with 95 per cent less than 2°, with the characteristic angle less than 1°.

Similarly, in the Opa basin of south-western Nigeria developed on basement complex rocks with local relief mainly between 15 and 45 m, the basin is dominated by gentle slopes varying between 0.5° and 3.5° with a mode of 2° (Figure 9.3).

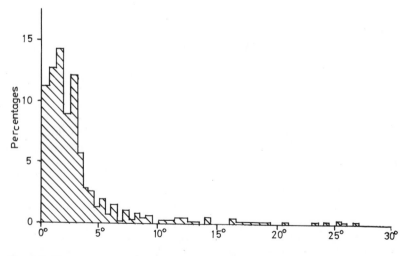

Fig. 9.3 Topographical characteristics, Opa basin, Nigeria (after Jeje, 1978)

Slope evolution

Evolution in this sense concerns the change of slope form with time. Vital to the study of slopes generally is the issue of whether a slope declines, becomes steeper, or remains the same after it has been operated upon by exogenous geomorphological processes for a period of time. Thus the issue is whether slope form is time-bound or time-independent. Accordingly, three deductive theories of slope evolution are widely accepted, namely – slope decline, slope replacement and parallel retreat. These theories advanced to explain slope evolution in different climatic environments apply equally to the humid tropics.

Slope decline

This theory, which essentially views slope evolution as a time-bound phenomenon, was first propounded by W. M. Davis (1899), whose views on this subject can be gleaned from his general discussion of the geographical cycle, which has been outlined in Chapter 6.

According to Davis, following the rapid uplift of a region of any structure, streams can develop consequent on the initial depressions. Aided by the coarse-weathered products, dissection by these rivers is very rapid so that deep valleys with near perpendicular slopes are formed. Rapid downcutting ceases when the valley floors are approaching the base level the elevation of which depends on distance from the river mouth.

Valley widening is controlled by weathering and waste removal under the vegetal cover; weathered material is removed by soil creep and surface wash. At the initial stages when waste removal is faster than waste preparation, valley sides are in the form of bare cliffs and ledges. Later, as more waste is generated and removed, the waste sheet accumulating at the base builds up to cover the ledges and cliffs.

Weathering and removal by sheet wash and soil creep continue on the upper slope surfaces, while waste sheets not only protect the lower valley slope from weathering but increasingly accumulate till the slope becomes convex-rectilinear. The coarse waste sheets on the lower slopes are subject to weathering, so that the comminution of its constitutent material coupled with erosion by surface wash and creep leads to a decline in slope angle (Figure 9.4). As the upper slope profile is subject to continuous weathering and soil creep, more material is removed there than from lower down, so that the upper part becomes markedly convex, whereas on the waste sheet lower down the profile a concavity develops.

Young (1972) observed that the Davisian theory has great flexibility and internal logical consistency. However, some of the assumptions inherent in the theory appear unrealistic. For instance, the assertions that soil creep produces convexities and that more material is removed from the upper than the lower part of the slope are rather difficult to prove.

On deeply weathered surfaces with no cover of duricrust, especially under the rainforest, erosional processes are dominantly sub-surface

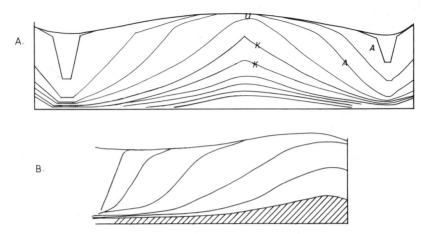

Fig. 9.4 Slope evolution by decline (A after Davis, 1912, B after Davis, 1932)

through solution loss, chemical eluviation and mechanical eluviation (Chapter 4). As theorised by Young (1972), if these processes operate at equal rates all over the slopes and there is no basal erosion, slopes would evolve by parallel retreat. However, observations in areas of low relief underlain by deeply weathered rocks in the humid tropics have shown slope evolution by decline to be the rule, rather than the exception (Swan, 1970, 1972). Also as observed by King (1962), slope decline readily occurs wherever the rocks are weak and highly susceptible to chemical weathering, where local relief is low and texture of dissection is relatively fine.

Slope replacement

This theory was propounded by Penck (1924), and like that of Davis, it is purely deductive. Penck proceeds with several basic assumptions, among which are the following:

1 that the form of an erosional slope is not dependent on the erosional processes affecting it, as these merely serve in removing waste material and thus expose the bedrock to further weathering;
2 that the rate of slope retreat is a factor of its angle of inclination;
3 that the initial slope units are initiated by stream incision under different rates of incision;
4 that weathering and mass movement are the most important processes acting on steep slopes.

According to Penck, the initial slope, probably a steep valley slope unit, is one on which neither erosion nor deposition is active.

The slope unit is subject to weathering, followed by a quasi-instantaneous removal of the weathered material, but the weathered material at the slope base cannot be removed. With constant weathering and removal, the steep slope retreats parallel to itself, but the length of the original unit is progressively reduced as the length of the new basal replacement slope increases (Figure 9.5). Both the upper steep unit and the lower gentle unit will continue to undergo parallel retreat, while a third slope unit of a still lower angle is simultaneously developing at the base of the slope profile. Much later, an almost infinite number of new units, each of gentler gradient than the unit directly above it, will emerge, and the entire profile becomes concave. With the disappearance of the upper steeper segments, the slope will gradually decline.

This theory has been criticized on several grounds. Firstly, the assumptions that physical weathering can operate with equal intensity on a given slope unit, and that the weathered material can be removed instantaneously are difficult to sustain. The latter can occur only if there is vigorous erosion at the base of the slope unit and the weathered debris is not beyond the competence of the eroding river to transport. In such an eventuality, a replacement slope will not form. Secondly, as observed by Young (1972), there is the assumption that the replacement slope is equally exposed to weathering, which is not the case. The period of exposure will normally decrease towards the point of intersection with the steeper slope unit, so that the replacement slope is not subject to equal intensity of weathering all over its surface. Thirdly, it is evident that as slopes become flatter, the intensity of mass movement on such slopes decline so that there can be no instantaneous removal of weathered debris from the low-angled replacement slopes. Based on these observations, the theory may be of limited application except where rock

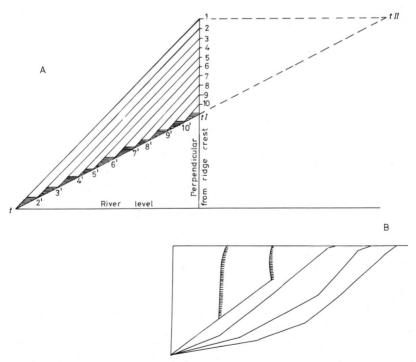

Fig. 9.5 Slopes: A, evolution (Penck, 1924); B, replacement (Young, 1972)

287

fall is an important removal process, and even then, the replacement slope may remain unaffected. However, notwithstanding all these observations, parallel retreat and slope replacement have been noticed on some inselbergs. Ollier and Tuddenham (1961) noticed these processes of slope evolution in association with unloading and sheeting on Mount Olga in central Australia. On the domes developed in conglomerate, unloading is producing thick, curved slabs of rock which on removal further expose dome-shaped rocks beneath. The detached sheets removed rather quickly are subject to rapid disintegration so that there is no accumulation of debris at the base of the slope. As shown clearly on figure 98 of Ollier (1969), the retreating slope is followed below by another slope unit at a lower angle and is separated from the steeper slope above by a distinct break of slope. Ollier and Tuddenham (1961) believe that this lower slope unit will equally undergo unloading and will retreat to form another slope unit below it. However, unlike the dominantly concave profile envisaged by the Penckian theory, slope profiles on Mount Olga are dominantly convex.

As observed all over the humid tropics, parallel retreat and slope replacement normally occur where duricrusts form an important element of the weathering profiles (Pallister, 1956; Savigear, 1960; Moss, 1965; Townshend, 1970; Jeje, 1972; Swan, 1972; Ofomata, 1973; Morgan, 1973; and Thomas, 1974). Wherever duricrusted surfaces are deeply dissected, the typical four-slope elements of King readily develop. Evidence of slope evolution by replacement, as shown by Morgan's (1973) analysis of some measured slope profiles in Malaysia, include the faceted and segmented nature of the slope profiles, and more important, the fact that slope frequency histograms are characterized by peaks at certain low and high angles, which will not be the case if slopes evolve by decline. The latter mode of evolution requires the presence of slopes at all angles in the landscape. Further, the slope units are separated by angular junctions, in contrast to what would be expected if slopes were evolving by declining in angle.

Where duricrusts constitute the free face, the latter is perpetuated by the difference between its rate of disruption and waste removal from the lower slope units. Where the duricrust readily disintegrates and the debris is quickly removed by soil-creep and slope-wash processes, the slope elements persist until the duricrust disappears, after which slope decline supervenes (Jeje, 1970).

Slope retreat

The central point of this theory, which stems from the work of Wood (1942), Fair (1947, 1948) and King (1951, 1953, 1957, 1962), is that slope form is not necessarily time-bound, as the form can be preserved as long as the controlling factors remain unchanged.

Wood (1942) in discussing this theory recognized that a slope profile can be resolved into crest or upper convex element, the free face, the constant or debris slope and the basal concavity. King (1962) not only recognizes these hill-slope elements but defines them as follows:

The crest   This is the summit area of a hill or a scarp, under the dominant influence of weathering and soil creep. It is usually convex, and longest where depth of surfacial detritus is greatest.

The free face   This consists of the bare outcrop of the bedrock on the upper part of the hill-slope, and is the most active element in the backwearing of the slope as a whole.

The debris slope   This is formed by debris falling from the scarp above, especially where debris production exceeds the rate of removal. However, it may be absent if the rate of production equals or is less than the rate of removal.

The pediment   This is the broad concave slope unit extending from the base of the debris slope to the adjacent stream bank or alluvial plain. The profile, approximating an hydraulic curve, is developed as a result of sheet erosion. Where cut across fresh rocks, they may be veneered by a thin layer of weathered material with the bare rocks often exposed towards the base of the hill-slope. The break between the hill-slope and the pediment is rather sharp in hard rocks, but smoothened out into an open curve on weak rocks.

King is of the opinion that the above four slope elements can develop on any part of the earth's surface, provided that the following conditions are satisfied:
1 that there is sufficient relief and fluvial erosion as the most important agent of denudation – with low relief, the slope units will not form;
2 that there are widely spaced streams;
3 that strong bedrocks, like granite, gneiss, quartzites and so on, are present, or that there is a horizontal structure in sedimentary rocks which can afford the development of clear scarp faces.

King believes that these slope elements can evolve on an initial

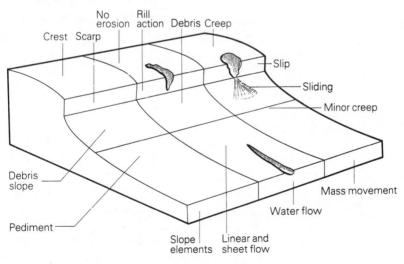

Fig. 9.6 Standard hill-slope elements (after King, 1962)

slope unit 'under the normal flow of meteoric waters across a landscape' (King, 1962, p. 139), or under the action of mass movement, or with both acting in concert. The four slope elements can evolve under the influence of fluvial erosion as follows (Figure 9.6).

The upper slope convexity is mainly under the influence of weathering and soil creep, as little or no erosion takes place because runoff is feeble and its energy insufficient to overcome soil cohesion. Thus the more permeable the bedrock or regolith, the longer the summit convexity. However, as runoff increases down-slope, it becomes more erosive. If the relief is sufficiently high, erosion may be in the form of rills and gullies which steepen the slope profile to make it concave or to expose the bedrock which forms the free face, and which can be subject to weathering. At the slope base, erosion is checked, so that runoff expends its energy in transporting waste to the local streams. The flat surface over which transportation takes place is the pediment. Where the rate of waste production from the exposed bedrock is sparse, there will be a sharp angle between the hill-slope and the pediment, but where it is higher than the removal rate, a debris slope will form.

Following the formation of the four hill-slope elements, scarp retreat is the most important element of parallel slope retreat. Scarp retreat is controlled by weathering and removal of weathered material by surface runoff or by mass movement. However, the debris slope must not grow upwards to cover and eliminate the scarp. The initial angle of the debris slope depends on the angle of repose of the coarse material, and as

Plate 20 Gullying and rilling on exposed surfaces near Jos, Nigeria (M. F. Thomas)

these are weathered to finer grades, removal occurs and the slope retreats in conformity with the scarp face above. Following the retreat of the upper elements, the pediment extends in length and is constantly regarded. Each slope element maintains its form till changes occur in the controlling factors either in terms of structure, lithology or of geomorphic processes.

The claim to world-wide applicability of this theory has been questioned by several geomorphologists who observed that slope forms and their associated processes are so complex in character that it would be foolhardy to hold rigid views as to any unique manner of slope development in any area. The observed modes of slope evolution in the humid tropics further attest to this observation.

In the humid tropics, as elsewhere, parallel retreat has been observed to occur under several situations, most especially where slope evolution is under strong structural and lithological control especially on hard and resistant basement complex rocks. As observed by Ruxton (1958) in Sudan, Savigear (1960) and Thomas (1965) in Nigeria, and Twidale (1962) in the Eyre Peninsula, Australia, slope steepening often occurs in massive, poorly jointed crystalline rocks. Where the joint planes coincide with the slope base, conditions often favour deep chemical weathering through the availability of abundant moisture from up-slope, because of the presence of a vadose zone, and frequent variations of the water table. At such locations, chemical weathering involves granular disintegration followed by the complete decay of susceptible minerals. Chemical decomposition is followed by sub-surface erosion through solution and chemical eluviation of the soluble material and the mechanical eluviation of the finest aggregates, and the removal of the coarse, insoluble quartz waste by sheet flow emanating from the upper slope units, or by rills and gullies at the slope base. Waste blocks resulting from physical disintegration on the upper slope units are removed through free fall, and deposited at the slope base where they are destroyed by chemical weathering processes. If these continue unchecked, the slope may overhang without the rock collapsing as a result of its high internal cohesion. However, where the slope is very high, failure can occur. Such failures can be associated with the gradual opening of the closed vertical joints by chemical weathering so that when the internal cohesion of the rock is overcome by tensional splitting along such vertical joints, outward collapse of such rock masses will occur. Thomas (1962) observed this in some granitic inselbergs in the savanna area of Nigeria. A repetition of basal sapping and rock collapse leads to slope evolution by parallel retreat.

Parallel retreat has also been observed to relate to lithology in horizontally inclined rocks. In south-east Africa in an area underlain by horizontally inclined, weak Dwyka tillites and Ecca shales, in turn underlain by very resistant Table Mountain sandstone into which dolerite sills are intruded, the four hill-slope elements of King are present. The free face is developed on the more resistant dolerite and sandstone, and where the rate of waste preparation is faster than the rate of removal by soil and talus creep, the free face is buried under the debris. Elsewhere, lateral planation by streams and gullying maintain the steep free faces and effect parallel retreat of the slope profiles.

Parallel retreat in deeply weathered rocks occurs especially on artificially steepened slopes like road cuttings, or on steep channel banks developed in cohesive material. In both cases, basal erosion is an important element of parallel retreat. Basal erosion in river channels is most pronounced at concave banks associated with meander bends. Basal erosion is followed by slumping or rotational slipping, which is usually accompanied by steep, rather concave failure planes. Parallel retreat can continue as long as basal undermining lasts; however, once this stops, surface wash supervenes, leading to a decline in the slope angle.

Parallel retreat is common at gully heads all over the humid tropics and is especially noticeable where the gullies have developed on regolith dominated by clay aggregates, as in parts of the savanna area of Nigeria – for example, at Ejiba in Kwara State or the Zaria area of Kaduna State. Parallel retreat is effected at the gully heads by undermining and basal sapping followed by slumping of the undermined regolith.

The observed modes of slope evolution in the humid tropics show clearly that within a single climatic regime, some slopes may decline, whereas others may retreat parallel in the same bedrock material. Also as shown by contemporary studies, slope forms and their associated processes are so multivariate in character that it is no longer possible to hold rigid views as to any unique manner of slope development even in a given region of uniform structure, lithology and climate (Chorley, 1964).

References and further reading

Adejuwon, J. O. (1979) 'Potential primary biological productivity: its implications for irrigation in Nigeria', *mimeo*, Dept. of Geog, Univ. of Ife, Ile-Ife.

Ahnert, F. (1970) 'An approach towards a descriptive classification of slopes', *Slopes Commission Report*, **6**, 71–84.

Baillie, I. C. (1975) 'Piping as an erosion process in the uplands of Sarawak', *J. of Tropical Geog.*, **41**, 9–15.

Birot, P. (1968) *The Cycle of Erosion in Different Climates*, Batsford, London.

Buckle, C. (1978) *Landforms in Africa*, Longman, London.

Carson, M. A. and Kirkby, M. J. (1972) *Hillslope Form and Process*, Cambridge University Press.

Chorley, R. J. (1964) 'The model position and anomalous character of slope studies in geomorphological research', *Geog. J.*, **130**, 503–6.

Cotton, C. A. (1942) *Climatic Accidents in Landscape Making*, Hafner, New York.

Darwin, W. R. B. (1967) in *Water Resources Branch (1967) Annual Report, 1967–68, Northern Territory Administration*, Darwin, N. T.

Davis, W. M. (1899) 'The geographical cycle', *Geog. J.*, **14**, 481–504.

Doornkamp, J. C. and King, C. A. M. (1971) *Numerical Analysis in Geomorphology: an introduction*, Edward Arnold, London.

Douglas, I. (1968) 'Erosion of granite terrains under tropical rain forest in Australia, Malaysia and Singapore', in *Symposium on River Morphology*, General Assembly, Bern, 1967.

Eyles, R. J. and Ho, R. (1970) 'Soil creep on a humid tropical slope', *J. of Tropical Geog.*, **31**, 40–2.

Fair, T. J. D. (1947) 'Slope form and development in the interior of Natal', *Trans. Geol. Soc. of S. Africa*, 50, 105–20.

Fair, T. J. D. (1948) 'Slope form and development in the coastal hinterland of Natal', *Trans. Geol. Soc. of S. Africa*, **51**, 37–53.

Faniran, A. and Ajaegbu, H. (1971) 'Landslopes and farm sizes in a part of Jos plateau', *Nig. Geog. J.*, **14** (2), 199–207.

Fölster, A. (1969) 'Slope development in south-western Nigeria during late Pleistocene and Holocene', *Göttinger Bodenkundliche Berichte*, **10**, 3–56.

Fournier, F. (1967) 'Research on soil erosion and soil conservation in Africa', *African Soils*, **12**, 53–96.

Friese, F. (1936) 'Das Binnenklima von Urwaldern in subtropischen Brasilien', *Petermans*, **82**, 301–7.

Gerrard, A. J. W. and Robinson, D. A. (1971) 'Variability in slope measurements', *Trans. Inst. Br. Geog.*, **54**, 45–54.

Hunter, J. M. and Hayward, D. F. (1971) 'Towards a model of scarp retreat and drainage evolution: Ghana, West Africa', *Geog. J.*, **137** (1), 51–68.

Jeje, L. K. (1970) 'Some aspects of the geomorphology of south-western Nigeria', Ph.D thesis, Edinburgh University.

Jeje, L. K. (1972*a*) 'Landform examples in Nigeria, No. 4: Pediments, *Nig. Geog. J.*, **15** (1), 79–82.

Jeje, L. K. (1972*b*) 'Landform development at the boundary of sedimentary and crystalline rocks in south-western Nigeria', *J. Trop. Geog.*, **34**, 25–33.

Jeje, L. K. (1973) 'Inselberg evolution in a humid tropical environment: the example of south-western Nigeria', *Zeits für Geomorph.*, N.F. 17 (2), 194–225.

Jeje, L. K. (1974) 'Effects of rock composition and structure on landform development: the example of Idanre hills of Western Nigeria', *J. Trop. Geog.*, **39**, 43–53.

Jeje, L. K. (1976) 'Slope profile from a humid tropical environment, the example of Opa basin in Ife area', *Bull. of the Ghana Geog. Assn.*, **18**, 52–67.

Jeje, L. K. (1978) 'A preliminary investigation into soil erosion on cultivated hillsides in parts of Western Nigeria', *Geo-Eco-Trop.*, **2**, 175–87.

Journaux, M. A. (1975) 'Géomorphologie des bordunes de l'Amazonie brézilienne: le modèle des versantes d'évolution paleoclimatine', *Bull. Ass. Geog. Franc.*, **433/423**, 5–19.

Kadomura, H. (1977) *Geomorphological Studies in the Forest and Savanna Areas of Cameroon*, an interim report of the Tropical African Geomorphology Research Project 1975/76, Sapporo, Japan.

Kessel, R. H. (1977) 'Slope runoff and denudation in the Rupununi Savanna, Guyana', *J. Tropical Geog.*, **44**, 33–42.

King, L. C. (1951) *South African Scenery*, 2nd edn., Oliver & Boyd, Edinburgh.

King, L. C. (1953) 'Canons of landscape evolution', *Bull. Geol. Soc. Amer.*, **64**, 721–51.

King, L. C. (1957) 'The uniformitarian nature of hillslopes', *Trans. Edinburgh Geol. Soc.*, **17**, 81–102.

King, L. C. (1962) *The Morphology of the Earth*, Oliver & Boyd, Edinburgh.

Kirkby, M. J. (1969) 'Erosion by water on hillslopes', in *Water, Earth and Man*, R. J. Chorley (ed.), Methuen, London.

Mabbutt, J. A. and Scott, R. M. (1966) 'Periodicity of morphogenesis and soil formation in a savanna landscape near Port Moresby, Papua', *Zeits. für Geomorph.*, N.F., **10**, 69–89.

Mohr, E. C. J. and Van Baren, F. A. (1954) *Tropical Soils*, Interscience, London.

Morgan, R. P. C. (1973) 'Soil-slope relationships in the lowlands of Selangor and Negri Sembilan, West Malaysia', *Zeits. für Geomorph.*, **17** (2), 139–55.

Moss, R. P. (1965) 'Slope development and soil morphology in a part of south west Nigeria', *J. Soil Sci.*, **16**, 192–209.

Ofomata, G. E. K. (1973) 'Aspects of the geomorphology of the Nsukka–Okigwe cuesta, East Central State of Nigeria', *Bull. de l'Inst. fondamental d'Afrique Noire*, **35**, A, 3, 489–501.

Ogunkoya, O. O. (1980) 'Hydrology of third order basins in the upper Owena drainage basin of southwestern Nigeria', Ph.D thesis, Univ. of Ife, Ile-Ife.

Ollier, C. D. (1975) *Weathering*, Longman, London.

Ollier, C. D. and Tuddenham, W. G. (1961) 'Inselbergs of Central Australia', *Seits. für Geomorph.*, **5**, 257–76.

Ong, H. L., Swanson, V. E. and Bisquo, R. E. (1970) 'Natural organic acids as agents of chemical weathering', *US Geological Survey Professional Paper, 700-C.*

Pallister, J. W. (1956) 'Slope development in Buganda', *Geog. J.* **122**, 80–7.

Penck, W. (1924) *Die morphologische Analyse: Ein Kapital der physikalischen Geologie*, Engelhorns, Stuttgart; English Trans. by H. Czech and K. C. Boswell, 1953, Macmillan, London.

Pitty, A. F. (1966) 'Some problems in the location and delimitation of slopes profiles', *Zeits. für Gemorph.*, N.F. **10**, 454–61.

Pugh, J. C. (1966) 'Landforms in low latitudes', in *Essays in Geomorphology*, G. H. Dury (ed.), Heinemann, London.

Rapp, A., Axelson, V., Berry, L. and Murray-Rust, D. H. (1972) 'Soil erosion and sediment transport in the Morogoro river catchment, Tanzania', *Geografiska Annaler*, **54A**, 125–55.

Richards, P. W. (1957) *The Tropical Rainforest*, rev. edn. 1979, Cambridge University Press.

Roose, E. (1967) 'Dix années de mesure de l'érosion et du ruissellement au Sénégal', *Agron. Tropical*, **22**, 123–52.

Rougerie, G. (1956) 'Etudes des modes d'érosion et du façonnement des versants en Côte d'Ivoire equatoriale', *Premier Raport de la commission pour l'étude des Versants*, Union Géographique Internationale (Rio de Janeiro), 136–41.

Rougerie, G. (1960) 'Le façonnement actuel des modèles en Côte d'Ivoire forestière', *Mem. Inst. Fondamental d'Afrique Noire*, Dakar.

Ruxton, B. P. (1958) 'Weathering and subsurface erosion in granite at the Piedmont angle, Balos, Sudan', *Geol. Mag.*, **95**, 353–77.

Ruxton, B. P. and Berry, L. (1957) 'Weathering of granite and associated erosional features in Hong Kong', *Bull. Geol. Soc. Amer.*, **68**, 1263–92.

Ruxton, B. P. and Berry, L. (1961) 'Weathering profiles and geomorphic position on granite in two tropical regions', *Revue Géomorph. Dyn.*, **12**, 16–31.

Savigear, R. A. G. (1960) 'Slopes and hills in West Africa', *Zeits. für Geomorph.*, Supp. 1, 156–71.

Savigear, R. A. G. (1965) 'A technique of morphological mapping', *Ann. Ass. Amer. Geog.*, **55**, 514–38.

Simonett, D. S. (1967) 'Landslide distribution and earthquakes in Bewani and Toricelli mountains, New Guinea', in *Landform Studies from Australia and New Guinea*, J. N. Jennings and J. A. Mabbutt (eds), Cambridge University Press.

So, C. L. (1976) 'Some problems of slopes in Hong Kong', *Geografiska Annaler*, **58A** (3), 149–54.

Sparks, B. W. (1973) *Rocks and Relief*, Longman, London.

Sparrow, G. W. A. (1966) 'Some environmental factors in the formation of slopes', *Geog. J.*, **132**, 390–5.

Swan, S. B. St C. (1970) 'Relationship between regolith, lithology and slope in a humid tropical region; Johor, Malaya', *Trans. Inst. Br. Geog.*, **51**, 189–200.

Swan, S. B. St C. (1972) 'Land surface evolution and related problems with reference to a humid tropical region: Johor, Malaya', *Zeits, für Geomorph.*, **16** (3), 160–81.

Thomas, M. F. (1962) 'On the approach to landform studies in Nigeria', *Nig. Geog. J.*, **5** (2), 87–101.

Thomas, M. F. (1965) 'Some aspects of the geomorphology of tors and domes in Nigeria', *Zeits. für Geomorph.*, **9**, 63–81.

Thomas, M. F. (1966) 'Some geomorphological implication of deep weathering patterns in crystalline rocks in Nigeria', *Trans. Inst. Br. Geog.*, **40**, 173–93.

Thomas, M. F. (1967) 'A bornhardt dome in the plains near Oyo, western Nigeria', *Zeits. für Geomorph.*, **11**, 239–61.

Thomas, M. F. (1974) *Tropical Geomorphology: a study of weathering and landform development in warm climates*, Macmillan, London.

Thornbury, W. D. (1969) *Principles of Geomorphology*, 2nd edn., J. Wiley, London and New York.

Townshend, J. R. G. (1970) 'Geology, slope form and process, and their relation to the occurrence of laterite in the Mato Grosso', *Geog. J.*, **136**, 392–9.

Tricart, J. (1972) *Landforms of Humid Tropics, Forests and Savannas*, Longman, London.

Twidale, C. R. (1962) 'Steepened margins of inselbergs from north western Eyre Peninsula', *Zeits. für Geomorph.*, **6**, 51–69.

Waters, R. G. (1958) 'Morphological mapping', *Geog.*, **43**, 10–17.

Wentworth, C. K. (1943) 'Soil avalanches in Oahu, Hawaii', *Bull. Geol. Soc. Amer.*, **54**, 53–64.

White, S. E. (1949–50) 'Processes of erosion on steep slopes of Oahu, Hawaii', *Amer. J. Sci.*, **247**, 168–86; and **248**, 508–14.

Wigwe, G. A. (1966) 'Drainage composition and valley forms in parts of Northern and Western Nigeria', Ph.D thesis, University of Ibadan.

Williams, M. A. J. (1968) 'Termites and soil development near Brocks Creek, Northern Territory', *Aust. J. Sci.*, **31**, 153–4.

Williams, M. A. J. (1969) 'Rates of slopewash and soil creep in northern and southeastern Australia: a comparative study', Ph.D thesis, Australian National University, Canberra.

Williams, M. A. J. (1973) 'The efficacy of creep and slopewash in tropical and temperate Australia', *Aust. Geog. Studies*, **11**, 62–78.

Wood, A. (1942) 'The development of hillside slopes', *Proc. Geol. Ass.*, **53**, 128–40.

Young, A. (1961) 'Characteristics and limiting slope angles', *Zeits. für Geomorph.*, **5**, 126–31.

Young, A. (1963) 'Soil movement in slopes', *Nature*, London, **200**, 129–30.

Young, A. (1964) In discussion on 'slope profiles: a symposium', *Geog. J.*, **130**, 80–2.

Young, A. (1970) 'Slope form in part of the Mato Grosso, Brasil', *Geog. J.*, **136**, 383–92.

Young, A. (1972) *Slopes*, Oliver & Boyd, Edinburgh.

Zonnefield, J. I. S. (1975) 'Some probelms of tropical geomorphology', *Zeits. für Geomorph*, N. F., **19**, 377–92.

10 Residual landforms

The term *residual* implies, among other things, especially in landform studies, 'remnant of a formerly greater mass of rock or area of land, above a surrounding area which has been generally planated' (*Dictionary of Geology Terms*, 1962). Residual hills in sedimentary environments, including duricrust-capped hills (see Chapter 8), are not included in the present discusion. Accordingly, the term has been used to describe rock hills and mountains such as tors, inselbergs, bornhardts and like features (cf. King, 1948), among other landforms. Although most common and best typified in granitic rock areas, they are also found in some sedimentary environments, especially sedimentary quartzites. Ayers Rock in Australia is an example of a bornhardt in rocks of sedimentary origin. Also, although they have been described more often from warm, humid environments, where they are associated with the process of humid tropical weathering among others (Chapters 3 and 8), they are also widely scattered in extra-tropical areas (cf. the Stone Mountain in Georgia, USA). In fact, a survey by Kessel (1973) found that the examples of inselbergs he quoted came from virtually all climates: 40 per cent from savanna climates; 32 per cent from arid and semi-arid areas; 12 per cent from humid continental and sub-Arctic climates; and 6 per cent each from the humid tropics, sub-tropical and Mediterranean climates (Thomas, 1978).

It is difficult to draw a conclusion, especially from Kessel's figures given above, since the sampling on which they are based is neither scientific nor adequate. It is also doubtful whether one can accept the inference that Kessel's study validates King's (1953, 1957) thesis that the processes of inselberg formation (pedimentation and pediplanation) are universal so that inselbergs would occur 'in practically all climates' (Kessel, 1973, p. 101). Rather, the experience of the present authors suggests that Cotton's (1961, 1962) viewpoint, as well as the position taken by Linton (1951), Ollier (1961, 1969) and Thomas (1966, 1974), among others, to the effect 'that particularly favourable conditions for inselberg formation occur in the tropical wet-dry climates', may be more tenable.

Thomas (1978) has tried to adduce reasons for the data provided by Kessel (1973), particularly the apparent overemphasis on the savanna/semi-arid/arid area with 72 per cent and underemphasis on humid-tropical areas with about 12 per cent. Among the reasons given are:

1 the historical hangover from the initial studies especially of inselbergs in dry regions of Africa and North Africa;

2 the relative inaccessibility of large parts of the humid tropics, which, coupled with dearth of qualified personnel, has meant less intensive study of not only these forms but also other aspects of humid tropical geomorphology (Chapter 1);
3 the fact that studies from higher latitudes have avoided describing these forms as residual hills, for conceptual reasons – similar conceptual problems may have affected the humid tropical areas as well.

Notwithstanding the figures given by Kessel or the controversies in the literature of residual landforms, the authors are convinced of the widespread occurrence of residual rock hills of various shapes and sizes in the present study area; they are also aware of the mutual, if not causal, relationships between the process and forms associated with the landforms. This chapter is therefore devoted to the various aspects of residual landforms in warm, humid climates, including terminology, nomenclatures and description, distribution in space and (possibly) time; and the question of their origins, including process/form relations.

Terminology and nomenclature

Residual rock hills in warm humid climates take different forms and sizes. Unfortunately, the terminology of the various types has not been sufficiently established to avoid ambiguity. Accordingly, effort is made in what follows to define the various terms used.

One problem in this type of assignment is that of boundary – namely, when does one form transform into another, more so when they are seen to overlap one into another? The overlap is caused, among other things, by the common factors and processes of formation (see below) involving rock jointing pattern, continuous and/or deep chemical weathering and differential removal of the weathered products. In situations such as this, all that can be done is to set up quantitative and qualitative criteria as basis for categorization.

This problem is perhaps best exemplified by the use of the terms *inselberg* and *bornhardt*. Bornhardts are defined as 'domed inselbergs'; yet not only are all inselbergs not bornhardts, not all bornhardts may be regarded as inselbergs. In an attempt to resolve this issue, Willis (1934) distinguished between 'true inselbergs', which he termed 'bornhardt', after the German geologist who first described them from Tanzania (Bornhardt, 1900), and other types of hill. The distinguishing characteristics of bornhardts according to Willis were 'bare surfaces, dome-like summits, precipitous sides becoming steeper towards the base, and absence of talus, alluvial cones or soil, a close adjustment of form to internal structure' (1934, p. 124).

However, the term 'inselberg' has been very widely used in the literature to describe virtually all abrupt rises over plains, such as:
1 buttes found in horizontal strata or where duricrust forms a caprock;
2 conical hills, having rectilinear sides and common in arid-regions;
3 convex-concave hills, entirely regolith-covered and transitional with other hills;

4 rock crest over regolith-covered slope, common in savanna regions;
5 rock dome, frequently with near-vertical sides, but merging with low domes and rock pavements ('ruwares'); and
6 kopje or tor formed, at least in part, of large boulders often with a bedrock core (Thomas, 1978, p. 5).

This definition is obviously too wide and cannot apply in the present work.

One other definitional problem here relates to the question of isolation, or individual measurable identity. All these have led to the use of operational definitions, many of which are subjective.

Perhaps it is necessary to add here that fresh rock exposure, particularly of the granitic type, is an important criterion, so that the inselberg problem resolves to that of granite landforms. This is also the view of Thomas (1974, 1978):

> Notwithstanding the attention paid to Ayers Rock, the 'inselberg problem' has become in large measure the granite inselberg problem. A very high proportion of papers on inselberg landscape concern the landforms of granitoid rocks [Thomas, 1978, p. 4].

Nevertheless, experience shows that many operational definitions cannot be extended outside of the specific study area, since landforms, inselbergs inclusive, vary almost infinitely in space. The best that can be done is to avoid a too restrictive set of criteria, and to take care to define terms carefully, as attempted in what follows for residual rock hills or inselbergs.

The terms commonly used in the literature in connection with residual rock hills include *tors* or *boulder inselbergs*, *bornhardts* or *domed inselbergs*, *stacks*, *rock towers* or *castle koppies* (*kopjes*) and *sugar loaves* or *whalebacks* (Faniran, 1971). Each of them is described briefly in what follows.

Tors

Characteristically, tors are groups of spheroidally weathered rock boulders rooted in bedrock, which may be exposed as a basal rock surface or concealed by a waste mantle. Although sometimes confused with kopjes or stacks, the rounded shapes and the way in which the blocks are 'piled up or perched in rather improbable positions on the massive unweathered rock base' (Faniran, 1971a, p. 105) distinguish tors from these other landforms. The forms that the boulders take give them such descriptive terms as *woolsacks*, *niggerheads*, *boulderfields*, *boulder-streams* and *blockfields* (Ollier, 1969). Thomas (1965, 1976) recognized three broad groups:
1 tower-like forms defined by a blocky, rectilinear jointing;
2 tabular forms induced by strong lateral fractures; and
3 hemispherical forms with domical profiles.

The constituent blocks vary widely in size, depending on the jointing pattern of the original rock and the type and degree of subsequent chemical weathering. The most common sizes range between 1 and 3 m in diameter, although block diameters may reach 10 m. In fact, boulders of between 5 and 30-m diameter are known, which, if stacked one on top of another, still qualify to be called tors. If, however, individual blocks reach 10 m in diameter, they are bornhardts or domed inselbergs. Twidale (1968) described the relationship between jointing pattern on the one hand and tors and related landforms on the other as follows:

> Clearly, the shape and size of tors is closely related to the geometry of the joint system. Where joints are widely spaced and tight, castellated tors occur. Where there are massive monolithic blocks subdivided into arched sheets, domed *inselbergs* or *bornhardts* are formed. Where the joint blocks are elongated or tabular, the tors weathered from them are like-wise elongated or tabular.... Where joint blocks are closely spaced, tors are very small. And where the rock is so readily attacked by weathering agencies . . . corestones (and hence tors) are not formed [p. 489].

Domed inselbergs or bornhardts

The word 'bornhardt' has been bestowed on geomorphology by the German geologist who in 1900 first described the landforms which later came to be so called. The landforms were originally called *inselberge*, island mounts, apparently because of their prominence in the plain or shield topography of the host country. Thomas (1976, p. 429) described

Plate 21 Granite dome in the savanna near Kusheriki, Kaduna State, Nigeria (M. F. Thomas)

Plate 22 A castle kopje on the plain near Ike-Iho, Oyo State, Nigeria (L. K. Jeje)

bornhardts as 'larger spheroids that rise as monolithic forms to heights varying between a few metres to 500 m or more'. Although bornhardts take different forms, the type examples are the regular domed-shaped and in the form of inverted bowls. The Aseke hill in the savanna country of Oyo State has features which qualify it for a bornhardt. It has recently been described as 'having a typical domed shaped, slightly oval in plan' (Faniran and High, 1969, p. 141). Other forms taken by bornhardts are the elongate *whalebacks*, the *cupola-like* bornhardts and the *sugar loaves*. Generally speaking, the regular dome-shaped types are common in savanna country, while the elongated types – whalebacks, sugar loaves and so on – occur widely in the forested parts, especially in Nigeria. However, geology is a very important factor here, especially the extent and nature of secondary structures such as foliation, jointing and sheeting. Thomas (1976, p. 429) puts it thus:

> Many of these forms [elongate whalebacks and cupola-like bornhardts] are found in foliated rocks such as gneisses and migmatites. . . . Most domical forms are associated with sheeting.

Castle kopjes

These are irregular, often castellated forms, which exhibit few if any signs of spheroidal modification of constituent blocks. The difference between tors and castle kopjes is essentially in terms of the shape of the rock blocks or stacks. Whereas in tors the blocks are rounded to sub-rounded, in kopjes the blocks are in the form of elongate slabs and are aligned at variable angles, depending on the dip of the parent rock. Ollier (1969, pp. 201–2) puts the situation as follows:

> Where the weathering took place on moderately or steeply dipping schist a landscape described as 'fretted' is exposed . . . but where the dip of the schist was less than about 20° a tor landscape is found. In some places the boundary between the tor and the fretted landscape is sharp, but in other places is not clearly defined.

Ackermann (1962) has described features similar to kopjes from Zimbabwe, calling them *penitents* or *monk rock* or *Bussersteine*. Illustrations of different types of residual rock hills are shown in Figures 10.1*a* and *b*, which are henceforth referred to as inselbergs.

Distribution in space

The spatial distribution of natural phenomena is studied for several reasons, the most important of which relate to possible causal factors of origin. Thus, in the study of residual rock hills, emphasis has been on geology, topography (relief), and climate, among other environmental variables. These are the same variables which have featured prominently in the controversies surrounding especially the origins of the landforms. Faniran (1973, 1974) has, for instance, studied the spatial pattern of these forms in relation to both themselves (inter-inselberg distance) and other landscape features (regional watershed), while Burke and Durotoye (1971) related the rock hills to local stream network – all in an attempt to elucidate their origins. We shall discuss each of these variables briefly.

Fig. 10.1a Inselberg landscape at Rio Nuia, Angola

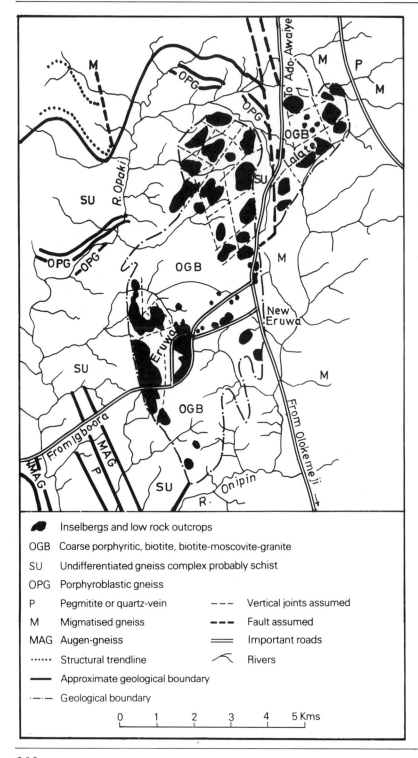

Inselbergs and low rock outcrops
OGB   Coarse porphyritic, biotite, biotite-moscovite-granite
SU   Undifferentiated gneiss complex probably schist
OPG   Porphyroblastic gneiss
P   Pegmitite or quartz-vein   --- Vertical joints assumed
M   Migmatised gneiss   ▬▬▬ Fault assumed
MAG   Augen-gneiss   ═══ Important roads
•••••• Structural trendline   ⌒ Rivers
▬▬ Approximate geological boundary
·—·— Geological boundary

0   1   2   3   4   5 Kms

Climate

As in the case in duricrust–topography relations, the literature on the topic of climate–inselberg relationships can be grouped into about three categories (Faniran, 1974a, p. 152). First, there are those who attribute inselbergs to specific climatic environments. Among these are Passarge (1928) and Davis (1930), who ascribed them to deserts; Davis (1933) described them as *desert domes*, and believed them to be indicative of the 'penultimate state of desert erosion' (Thornbury, 1954, p. 292; see also his Figure 11.9, p. 293). King (1948, 1953, 1966) described them as the result of the semi-arid denudational process of pedimentation and pediplanation; Bornhardt (1900) ascribed them to savanna processes; while Ollier (1960), Cotton (1961, 1962), Thomas (1966) and Faniran (1974a, b) ascribed them to the twin processes of deep weathering and exhumation. To many of these people, inselbergs found outside of the specified climatic zone are relict and indicative of climatic shift. This argument is similar to that which regards duricrust as relict in extra-tropical regions.

Secondly, there are those who appear to deny any significant role for climate. This viewpoint is perhaps best illustrated by the work of Kessel (1973), quoted earlier, who claimed that 'inselbergs, including bornhardts, are found in practically all climates' (p. 101). One possible implication of this viewpoint seems to be that different processes are capable of producing identical features in the form of the inselberg, thus exemplifying, *par excellence*, the 'principle of "convergence"'... whereby similar forms have been produced from separate origins via the operation of different processes... recognised as the principle of "equifinality" (Bertallanffy 1950)'. (Thomas, 1974, p. 166).

Finally, there are those who distinguish between the inselbergs of different climates. This can be inferred from the emphasis given to current processes in inselberg regions (cf. Jeje, 1973; Thomas, 1976, 1978). It is claimed by these workers that these processes are capable not only of producing 'different' types of inselbergs but also of modifying 'inherited' forms. Thus, it is claimed that desert processes tend to produce faceted forms owing perhaps to intense physical weathering and wind erosional activities, while the inselbergs in humid regions are smoother with convex to convex-concave slopes. This position does not appear tenable in view of the variety of shapes and sizes assumed by the inselbergs in different climatic regions. Other factors, including the nature of the rocks, especially their jointing pattern and the prevalent mode of slope evolution, are capable of explaining most of the shapes and sizes assumed by known inselbergs all over the globe.

The present authors feel, also, that the views expressed in the literature on climate–inselberg relations are not necessarily exclusive. In fact, almost every landform, except the very few ephemeral ones, shows evidence of both relict and contemporary processes. Very many examples of both incipient and exhumed basal surfaces turned inselbergs have been described (Faniran, 1971b, 1974b; Thomas, 1978) to show that the

← *Fig. 10.1b Residual rock hills*

process of deep weathering acting on variably jointed rocks is capable of producing irregular basal surfaces, the rises or domed parts of which can be exhumed following periods, whether continuous or periodic, of waste removal. Moreover, the established magnitude of climatic shifts since the Tertiary period, especially in the last two million years, appears sufficient to explain the latitudinal range not only of inselbergs but also of several other natural phenomena – for example, duricrusts and pediments to name only a few. Finally, and as hinted earlier, the climatic influence is itself conditioned by other factors, including rock type, diastrophism or crustal stability and time.

Geology

If the definition of inselberg as rock hill is accepted rather than Young's (1972) all-inclusive connotation, the role of geology in inselberg formation becomes obvious. This is because, with the exception of a few cases, such as the Ayers Rock, virtually all inselbergs occur in crystalline rocks, particularly granitic rocks – granites, gneisses, migmatites.

Nevertheless, not all granites, gneisses and migmatites have inselberg landscape, which is known to develop in other rocks besides these three types; it thus seems obvious that certain characteristics of crystalline rocks, rather than crystallinity *per se*, are decisive. Among those that have been specifically studied are mineralogy, texture, microfeatures, joints, faults and porosity.

Unfortunately, many of these studies are not detailed enough to be conclusive. For example, whereas Thorp (1967*a*, *b*, *c*, 1975), among others, has used gross mineralogical differences between granites to explain the distribution of granitic inselbergs, King (1948, 1962) and Kessel (1973) have reported no change of rock composition across the piedmont angle. Similarly, despite the frequent reference in the literature to joint control on inselberg formation, including directional accordance (Twidale, 1962, 1976; Thomas, 1966, 1974; Thorp, 1967*a*, *b*, *c*; Jeje, 1973; Faniran, 1974*a*, *b*), these have not been based on detailed field measurements because of the attending problems. Thomas (1978, p. 10) describes the problem thus:

> the measure of [joint] frequency is more difficult and has seldom been attempted; both the practical difficulty of identifying joint plains beneath the piedmont, except where there is bare rock, and the theoretical problem of boundary conditions affecting joint spacing on the hillforms inhibit research into this question.

Finally, King (1975), among others, studied the rock characteristics of porosity, and claimed that 'bornhardts arise in rocks with a low porosity, consequent upon metasomatic infusion of a host rock' (cf. Thomas, 1978, p. 12, who also reported that Birot, 1958, found that low porosity corresponded with lack of fissuration and actual jointing in the domes around Rio de Janeiro). Similar observations have been made in Nigeria, where inselbergs have been associated with the migmatites in the metamorphic basement complex, as well as with the granitic intrusions. In Sierra Leone, massive inselbergs occur in the porphyroblastic granites, in which homogeneous resistant masses have been emplaced by doming (Marmo, 1956).

In other words, several rock characteristics have influenced the evolution of inselbergs wherever they occur. Among these are the mineralogical, including chemical, composition; textural, including grain-size, characteristics; and structural, including jointing, foliation, faulting and attitudinal characteristics. They severally and collectively influence the course(s) of denudational processes of rock-weathering removal and deposition, as they do in the evolution of other landforms.

Topography

This is a factor which is often relegated by geomorphologists, obviously because it is the element that is being studied, with a view to establishing its origin. It did not, until very recently, occur to geomorphologists, as it does to other environmental scientists, that the elements themselves are parts of the total environment which influences all operations within it.

Among the first to realise the significance of the topography in geomorphological studies are fluvial geomorphologists, who recognize the importance of basin shape, among other variables, in drainage basin analysis. This is a wide departure from the Davisian school of landform being a function of structure, process and stage, or the rearranged scheme of process, stage and structure by King (1953). There is no doubt that the initial form, as well as the other geographical aspects of landforms, including their arrangement in space, can exert significant influence on their ultimate forms.

However, the study of topography in relation to inselbergs has taken slightly different forms. Thomas's (1978) recent review shows that this has taken the form of relating inselbergs to other landscape features, namely:
1 escarpments (Pugh, 1956, 1966; King, 1948, 1966; Kessel, 1974);
2 zones of deep dissection (Jeje, 1974; Thomas, 1974);
3 stream courses (Burke and Durotoye, 1971);
4 watersheds (Faniran, 1973); and
5 other inselbergs (Faniran, 1975a).
Although it is difficult yet to draw conclusions from these studies, which are still in their initial stages, they have succeeded in highlighting the complexity of the problem. It is this complex situation that has been resolved into the concept of randomness, which is implicit in the work of Faniran (1973, 1974a, b) among others.

The study of inselbergs has benefited from the controversies surrounding their origin. It has moved from the simplistic position of the Davisian-cum-Kingian era of single-process explanation to that in which explanations are being sought from all possible sources of information – geology, climate, topography and so forth. In all these, one thing stands out: that the humid tropical environment is conducive to the evolution of inselbergs. In other words, the residual rock hills found therein, and possibly others where evidence for climatic shift can be established, have evolved from the operation of humid tropical processes under the influence of local geological and topographical elements.

Origin and evolution

Although it is still being argued to what extent one can generalize over the entire globe, the role of rock weathering cannot be overemphasized in the discussion of the evolution of inselbergs in the humid tropics. In fact, to whichever one or more of the described models of inselberg evolution one subscribes (cf. Faniran, 1974a; Thomas, 1978), one comes up with the necessity for pre-weathering of fresh granitic rocks in order to evolve into discrete landforms. What is yet to be resolved is the necessity for deep weathering for inselberg formation. Thomas (1978, p. 18) puts it thus:

> Few authors would deny the influence of physico-chemical weathering processes on the evolution of inselbergs and other hill forms in crystalline rocks. The crucial issue remains, however, the role of deep weathering as a precursor to inselberg formation. The literature still reflects a marked divergence of viewpoint, between those who consider the weathered mantle to be superficial and insignificant ... and those who require deep penetration of weathering ... before the inselberg is excavated by surface erosion.

Fortunately, however, the humid tropical process is capable of accommodating virtually all the possible versions and interpretations of weathering–landform relations. Perhaps the most exhaustive discussion of this viewpoint is that by Thomas (1974a, b, 1976, 1978), who concluded that inselberg landscapes in crystalline, granitoid or gneissic rocks are affected by rock weathering in the following ways, among others:

1 Deep pre-weathering of the more jointed or fissured rocks, leading to the formation of the proverbial 'dome-and-basin' type of basal surface, with the domes and cores forming in the more resistant or massive rocks and the basins in the weaker rocks.
2 Deep penetration of weathering below a jointed or permeable cover rock, or accompanying the removal of such covers by slope retreat.
3 Continuous decomposition of less resistant and mantled rock during ground-surface lowering around excavated dome summits.
4 Decay of joint blocks within the interior of boulder inselbergs, as a result of water penetration by way of tension joints.
5 Formation of shallow waste mantles on more massive rock surfaces subject to sheeting in a humid climate.
6 Lateral encroachment of the weathering front in the piedmont angle zone, leading to the steepening of the hill-slope and hill-slope retreat.
7 Exploitation (direct) of joints by weathering, including wind action, to

form surface features such as weathering pits and gullies on gentle slopes and taffoni on steep slopes.

Whereas it is possible to isolate each or a combination of these processes in different climatic regions, observations in humid tropical regions show that they all operate or are likely to have operated in the past to produce the associated forms. Also, although Thomas (1974, 1978) suggests that the different inselberg types have different origins, this need not necessarily be so, especially if they are inherited from pre-weathering events followed by excavation. The various forms and sizes assumed by inselbergs in the humid tropical environment have in fact been attibuted largely to inheritance – that is, they are thought to have been developed during the process of deep weathering, when joints and other zones of weakness could have been exploited (cf. Faniran, 1972). In other words, the evolution of inselbergs in the humid tropics can be described within the ambit of the deep weathering removal process in typical granitoid rocks.

Although few people would today seriously deny the influence of physico-chemical weathering processes on the evolution of residual rock hills in crystalline environments, arguments still rage in the literature, between those who consider the weathered mantle to be superficial and insignificant in relation to the known heights of many inselbergs (for example, the Idanre hills measure more than 600 m above the plain) and those who believe that inselbergs are fashioned during the operation of differential weathering of variably jointed rocks, before their excavation. Of course, King (1948, 1957, 1975) continues to deny this process, especially in respect of the South African inselbergs which he studied. We shall not be bogged down by the counter-arguments of these opponents of the significance of rock weathering in inselberg formation, since several authors have done justice to them (cf. Thomas, 1965, 1966, 1974a, b, 1978; cf. also Faniran, 1972, 1974a, b). The points being made by King and his followers are also not very strong in our study area, while Mabbutt (1961), among others, has weakened these points by his description of typcial etchplains in the Australian desert regions.

The importance of deep weathering in the formation of inselbergs has been stressed since the time when these landforms were first described, at the close of the nineteenth century. For instance, Branner (1896) identified this close relationship in the case of the inselbergs he studied in Brazil. Also, and perhaps more significantly, Falconer (1911), who worked in Nigeria, gave perhaps the first and most explicit account of the exhumation (excavation) process in the formation and development of inselbergs. He wrote as follows:

A plane surface of granite and gneiss subjected to long continued weathering at base level would be decomposed to unequal depths, mainly according to the composition and texture of the various rocks. When elevation and erosion ensued the weathered crust would be removed, and an irregular surface would be produced from which the more resistant rocks would project . . . a repetition of these conditions of formation would give rise to the accentuation of earlier domes and kopjes and the formation of others at lower level [p. 246].

Several workers have given similar accounts from other parts of Africa – namely, Holmes and Wray (1913) from Mozambique; Willis (1936) from East Africa; Ollier (1959, 1960) from Uganda; and Thomas (1974, 1976, 1978) from Sierra Leone, among other places. Even the greatest antagonist of the exhumation hypothesis, L. C. King, indirectly recognizes the possible role of weathering generally in the evolution of some inselbergs (King, 1966) and particularly in the Valley of a Thousand Hills in Natal (King, 1975), where he explained the local inselbergs in terms of active dissection within deeply incised valleys, rather than as a result of pedimentation and pediplanation. Indeed, Willis (1936, p. 120) considered rock decay a required condition for inselbergs. The other conditions listed include:

1 crystalline rocks, especially gneiss or schists intruded by granite and traversed by veins of aplite and quartz;

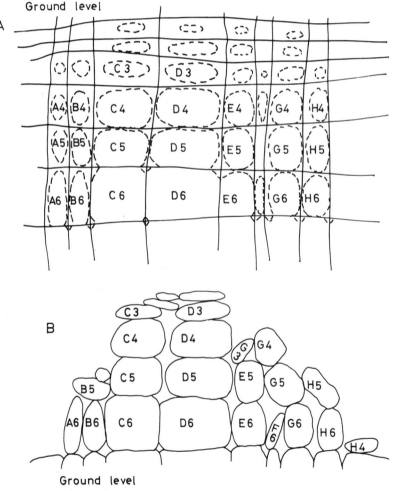

Fig. 10.2 Linton's model of inselberg formation

308

2 vertical or steeply dipping schistosity and jointing, which in general facilitates the decomposition of the rock and the steep-to-vertical slopes of the rock hills;

3 warm and humid climate favouring abundant vegetation growth and deep weathering; and

4 geological instability or uplift.

Outside Africa, workers who have advanced similar views on the significance of deep weathering include Cotton (1917) from New Zealand; Linton (1955) from Britain; Büdel (1957) from Surinam; Birot (1957) from Brazil; and Mabbutt (1961) and Twidale (1964) from Australia, to name only a few. Consequently, Ollier (1961, 1969), following on Linton's (1955) work proposed a two-stage model of inselberg formation (see Figure 10.2), a model which has been widely applied albeit with modifications.

Thomas (1974, 1978) has introduced at least two forms of modification. Firstly, he tried to distinguish between tors and domed inselbergs; secondly, he submits that the evolution of these features can be continuous rather than staged. In other words, the processes of weathering and removal can occur *pari passu*, or concurrently at least during part of the evolutionary history (Figure 10.3). The present authors do not see any necessity for separating the evolution of tors from that of bornhardts, but definitely subscribe to the possibility that at least in certain, if not in all, cases, and especially in humid tropical environments, weathering and removal can proceed at the same time, but not necessarily at the same rate. Depending on the environment, the rate of weathering may exceed that of removal, resulting in the accumulation of weathered debris. This is what has happened over large areas when regoliths of the

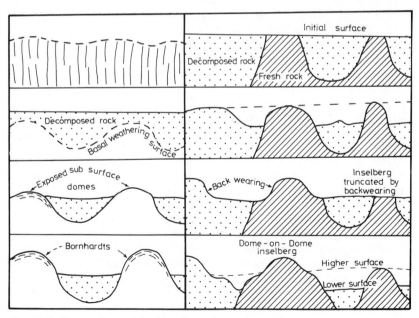

Fig. 10.3 Thomas's model of inselberg formation

order of 10 m and more occur (cf. Falconer, 1911; Faniran, 1972; Thomas, 1965, 1966, 1974, 1978). Among the factors favourable to deep chemical weathering are easily weatherable rocks either on account of the constituent minerals or on account of secondary events such as jointing, faulting and so on; low relief, which minimizes the rate of removal; and abundant rainfall, warmth and vegetation (De Swardt, 1964, Faniran, 1973, 1974b; Thomas, 1974). Variations in the operation of these environmental influences explain not only regional variations in weathering depths but also local variations, including the irregular basal surface (the dome-and-basin topography of the weathering front). As observed earlier, depending on the jointing pattern among other things, the basal surface may take the form of tors, kopje or domed inselbergs, the form they take on exposure. On the contrary, the rate of removal may exceed that of weathering, especially of the domed parts or rises in the weathering front, resulting in the excavation (exhumation) of parts of this front in the form of ruwares or flat-rock surfaces, tors, kopjes and bornhardts of varying shapes and sizes.

The following are a few of the evidences provided in the literature for a sub-surface origin of inselbergs or residual rock surfaces and hills; they are based essentially on the works of Thomas (1974), supplemented by the experience of the present authors.

1 Deeply weathered profiles are widespread over wide areas of the humid tropical shields.
2 There is wide variation in the extent, depth and degree of alteration of the country rock from place to place. This is confirmed by, among other things, the large coefficients of variations obtained for available data on weathering depth (Chapter 3).
3 Information available on the weathering front or basal surface shows that in broad terms it consists in places of discrete basins and rises (Enslin, 1961; Thomas, 1966), and in detail a series of irregularities, depending on the rock type and most especially the jointing pattern. The number of examples of incipient as well as emerging corestones described from road and railway cuttings as from other artificial excavation continues to rise (cf. Faniran, 1971a, for Nigerian examples; and Boye and Fritsch, 1973; and Boye and Seurin, 1973; for examples from Cameroon).
4 The transition between fresh rock and regolith in many profiles is sharp, reminiscent of the smooth surface of many inselbergs. Examples include those described from south-western Nigeria by one of the present co-authors (Faniran, 1971a); and by Rougerie (1955), Birot (1958) and Thomas (1965, 1966), among others. The observed sharpness facilitates the stripping of the rock surface.
5 The fact that some stream channels develop on thick regoliths and others on fresh rock outcrop indicates the possibility of the superimposition of drainage upon an irregular basal surface (Figure 10.4). Furthermore, in some areas, weathering depth increases from the regional base level, represented by the channel of the regional stream, to the interfluve. A typical example is the middle Oshun valley, south-western Nigeria, where the weathering depth data have been subjected to a polynomial trend surface analysis. The result

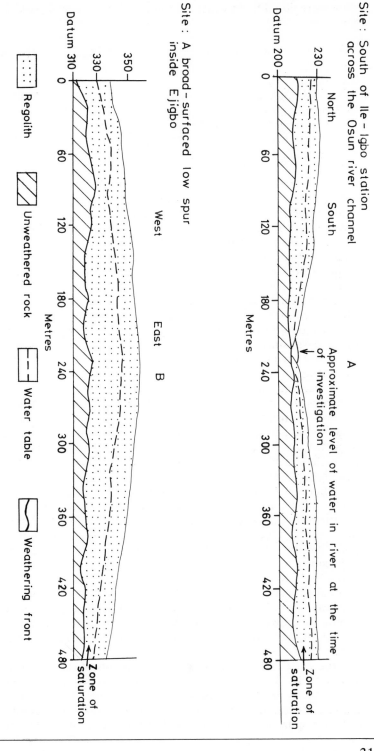

Site : South of Ile – Igbo station across the Osun river channel

Site : A broad – surfaced low spur inside Ejigbo

Fig. 10.4 Basal surface forms

shows a regional trend which is exampled by a quadratic equation of the form:

$$Y = 2.1532 - 0.3624X_1 + 0.4531X_2 + 0.0837X^2_1 - 0.1029X^2_2 + 0.0367X_2,$$

Y = depth of weathering
X_1X_2 = perpendicular geographical coordinates (X_1 = E–W, X_2 = N–S coordinate)
Goodness of fit R^2 = 0.7744
Correlation coefficient (R) = 0.8800
F – Statistic = 67.9822, significant at the 0.01 level

In other words, the model explains 77 per cent of the variations in the data used (Omorinbola, 1979; Faniran and Omorinbola, 1980). By contrast, on the Jos plateau a similar analysis but for a much smaller area gives a cubic surface (Faniran, 1974b; Faniran and Omorinbola, 1980). It is pertinent to note that the grid used in the second set of data (Jos plateau) is much finer, and so recognizes local variations within a broad regional trend. In both cases, however, analysis of residuals shows clearly the influence of both the irregular basal surface and post-deep-weathering dissection by streams.

6 The occurrence of weathering basins or troughs contiguous with domal outcrops (tors, kopjes, bornhardts); the steep plunge in places of the basal surface around many inselbergs; and the occurrence of truncated (stripped) deep-weathering profiles within widely stripped landsurfaces (etchplains) suggests spatial association between inselbergs and deep weathering.

7 The presence of regolith, and in some cases duricrust blocks and duricrust profiles, on some inselbergs (cf. Falconer, 1911, for Nigerian examples, and Eden, 1971, for Guyanese examples) indicates the former existence of a deeply-weathered landsurface above the level of at least some present hills, and from which they have been exhumed.

8 The shape of many inselbergs appears to be such that can only be reasonably explained by reference to intimate structural control and the operation of several processes. Examples are vertical slopes or faces, high degree of asymmetry, and morphological variations at the plain–hillslope junction, all of which argue against a straightforward application of any of the theories of direct erosional attack, particularly King's theory of parallel slope retreat.

9 A number of sections that have been studied indicate the presence of corestones within the regolith around individual domes, or resting against the side of some domes; these are reminiscent of the tors and bornhardts observed at the surface following exposure (cf. Thomas, 1965, 1974).

10 Inselbergs occur as discrete, isolated elements of the landscape (cf. Faniran, 1974a; cf. Figure 10.5a and b). This phenomenon is referable to granitic intrusions and other rock structures, where closely spaced domes occur. Indeed, the close correspondence between granitic rocks and inselbergs casts great doubts on any

explanation involving the systematic modification of larger rock masses – for example, by pedimentation and pediplanation.

11 Although no dates have been suggested for inselbergs, they appear to be very old, in fact relict, features wherever they occur today. Many of them have not undergone any noticeable degree of modification under the present processes, thus suggesting that they were formed as corestones within the weathered mantle and later excavated. The occurrence of inselbergs at virtually all positions in the landscape (Faniran, 1973; Thomas, 1974), and with distributional patterns conforming to the random distribution (Faniran, 1974b) provide convincing evidence for structural control on sub-surface weathering, and negate any claim to the universality of the combined processes of pedimentation and pediplanation, the only other seriously canvassed model of processes of inselberg formation.

12 Finally, inselbergs are known to have been exhumed from beneath sedimentary covers along basement complex–sedimentary rock junction. The best examples are from south-western Nigeria (Falconer, 1911; Savigear, 1960; Jeje, 1972). In such places, although deep weathering is known to penetrate the underlying basement complex, this does not prevent the occurrence of inselbergs in the stripped locations.

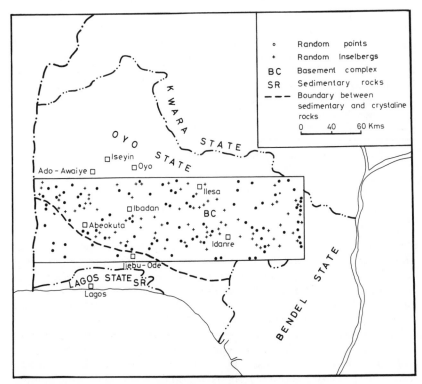

Fig. 10.5a Distributional patterns of inselbergs

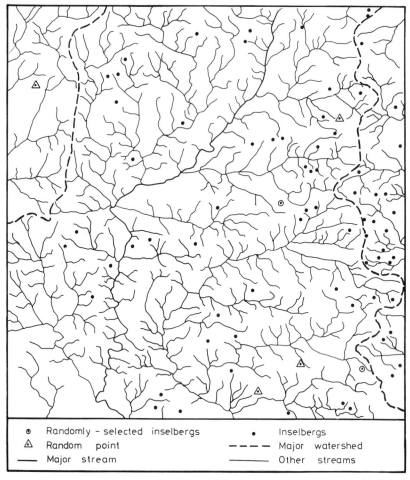

⊚	Randomly – selected inselbergs	• Inselbergs
△	Random point	– – – – Major watershed
——	Major stream	—— Other streams

Fig. 10.5b Distributional patterns of inselbergs

Conclusion

In this chapter, we have used the concept of deep weathering discussed in Chapter 3 to elucidate the formation of perhaps the most interesting landforms of the humid tropical shields. Although no claim is made for a universal application of the deep-weathering model of inselberg formation, the relevance of the model in the humid-tropical shields is no more in great doubt, given the flood of both empirical and theoretical work done so far. Apart from the several field descriptions of both buried and emerging corestones, the application of the nearest-neighbour technique shows that the distribution of the inselbergs in the crystalline region of south-western Nigeria is random, when only the nearest neighbours (NND I) are considered. With other orders (NND 2–NND 6) the distributional pattern approaches regularity or uniformity (Figure 10.5a, b). No clustering was detected. A similar exercise for the sedimentary rock areas shows that although the residual hills within a locality may be randomly to regularly distributed, a tendency to cluster occurs as the area is extended to include entire drainage basins where the hills cluster around the watersheds. They have been completely destroyed elsewhere.

Nevertheless, there is still need for further work in several directions, especially on the affective factors of geology, climate, relief and time. Of these, geology, because of its apparent overwhelming importance, ranks highest. There is need to study further the stress conditions and specific rock properties that influence rock weathering and decay. The significance of the climatic parameters is no more in serious doubt, but there is the question about boundary conditions in relation to the relative rates of rock weathering and removal processes. On relief, it needs to be established what range of relief is necessary to produce different inselbergs, as well as to study the nature of the processes operating in different relief regions – for example, in the piedmont zone of the inselberg landscape. Also, more detailed morphometric analysis of surface and sub-surface forms promises to throw further light on the inselberg controversy. Finally, the time factor in deep weathering and removal has not been seriously considered. Although many geological formations have been dated, the rate at which they are affected by deep weathering, the time and nature of the changes that lead to accelerated removal and exposure of corestones and the nature and rate of the post-exhumation processes are yet to be clarified. Fortunately, all these can be easily undertaken in the humid tropics where all the possible factors and processes of inselberg formation are widely exemplified.

References and further reading

Ackermann, E. (1962) 'Butserstein-Zeugen vorzeitlecher Grundswasserschwankungen', *Zeits für Geomorph.*, **6**, 148–82.

Bertalanffy, J. von (1950) 'An outline of general systems theory', *Br. J. Phil. Sci.*, **1**, 134–65.

Birot, P. (1957) 'Les domes crystallines', *Centre Nat. Rech. Sci.,(C.N.R.S.) Mem. Documents*, **6**, 8–35.

Birot, P. (1968) *The Cycle of Erosion in Different Climates*, trans. C. I. Jackson and K. M. Clayton, Batsford, London.

Bornhardt, W. (1900) *Zur oberflachengestellung und Geologie Deutsch Ostafrikes*, Berlin.

Boye, M. and Fritsch, P. (1973) 'Dégagement artificiel d'un dome crystallin au sud Cameroun', *Travaux Documents Geogr. Tropicale* (Bordeaux), **8**, 31–62.

Boye, M. and Seurin, M. (1973) 'Les modalités de l'altération à la carrière d'Ebaka', *Travaux Documents Geogr. Tropicale* (Bordeaux), **8**, 31–62, 65–94.

Branner, J. C. (1896) 'Decomposition of rocks in Brazil', *Bull. Geol. Soc. Amer.*, **7**, 255–314.

Büdel, J. (1957) 'Die *Doppelten Einebnungsflächen* in den feuchten Tropen', *Zeits. für Geomorph.*, N.F., **1**, 201–88.

Burke, K. and Durotype, B. (1971) 'Geomorphology and superficial deposits related to late Quaternary climatic variation in south-western Nigeria', *Zeits. für Geomorph.*, **15**, 340–44.

Cotton, C. A. (1917) 'Block mountains in New Zealand', *Amer. J. Sci.*, **194**, 249–93.

Cotton, C. A. (1961) 'Theory of savanna planation', *Geog.*, **46**, 89–96.

Cotton, C. A. (1962) 'Plains and inselbergs in the humid tropics', *Roy. Soc. N.Z. (Geol.)*, **1** (18), 269–77.

Davis, W. M. (1930) 'Rock floors in arid and in humid climates', *J. Geol.*, **38**, 1–27, 136–58.

Davis, W. M. (1933) 'Granite domes of the Mojave Desert, California', *Trans. San Diego Soc. Nat. Hist.*, **7**, 211–58.

De Swardt, A. M. J. (1964) 'Lateritisation and landscape development in equatorial Africa', *Zeits. für Geomorph.*, N.F., **8**, 313–33.

Dictionary of Geological Terms, (1962) Amer. Geol. Inst.

Eden, M. J. (1971) 'Some aspects of weathering and landforms in Guyana (formerly Br. Guiana)', *Zeits. für Geomorph.*, N.F., **15** (2), 181–98.

Enslin, J. F. (1961) 'Secondary aquifers in South Africa and scientific selection of boring sites in them', *Intern. Afr. Hydrol. Conf., Nairobi, CCTA Publn. 66, Sect. 4.*

Falconer, J. D. (1911) *The Geology and Geography of Northern Nigeria*, London.

Faniran, A. (1971*a*) 'Landform examples from Nigeria no. 3: differential weathering 1: Corestones and stonelines', *Nig. Geog. J.*, **14** (1), 105–9.

Faniran, A. (1971*b*) 'Implications of deep weathering on the location of natural resources', *Nig. Geog. J.*, **14** (1), 59–69.

Faniran, A. (1972) 'Depth and pattern of deep weathering in the Nigerian basement complex area', in *African Geology*, T. F. J. Dessauvagie and A. J. Whiteman (eds), Ibadan Univ. Press.

Faniran, A. (1973) 'Quantitative analysis of topographical maps', *W. Afr. J. Educ.*, **17** (2), 253–61.

Faniran, A. (1974*a*) 'Nearest-neighbour analysis on inter-inselberg distance: a case study of the inselbergs of south-western Nigeria', *Zeits. für Geomorph.*, Suppl. Bc., **20**, 150–67.

Faniran, A. (1974*b*) 'The extent, profile and significance of deep weathering in Nigeria', *J. Tropical Geog.*, **38**, 19–30.

Faniran, A. and Hugh, C. (1969) 'Landform examples from Nigeria No. 1., an inselberg', *Nig. Geog. J.*, **12**, 141–44.

Faniran, A. and Omorinbola, E. O. (1980) 'Trend-surface analysis and practical implications of weathering depths in basement complex rocks of Nigeria', *Nig. Geog. J.*, **23**, in press.

Holmes, A. and Wray, A. (1913) 'Mozambique: a geographical study', *Geog. J.*, **62**, 143–52.

Jeje, L. K. (1972) 'Landform development at the boundary of sedimentary and crystalline rocks in south-western Nigeria', *J. Tropical Geog.*, **34**, 25–33.

Jeje, L. K. (1973) 'Inselberg evolution in a humid tropical environment: the example of south-western Nigeria', *Zeits. für Geomorph.*, N.F., **17**, 194–225.

Jeje, L. K. (1974) 'Relief and drainage development in Idanre hills of western Nigeria', *Nig. Geog. J.*, **17**, 83–92.

Kessel, R. H. (1973) 'Inselberg landform elements; definition and synthesis', *Rev. Geom. Dyn.*, **22**, 97–108.

King, L. C. (1948) 'A theory of bornhardts', *Geog. J.*, **112**, 83–7.

King, L. C. (1953) 'Canons of landscape evolution', *Bull. Geol. Soc. Amer.*, **64**, 721–52.

King, L. C. (1957) 'The uniformitarian nature of hillslopes', *Trans. Edinburgh Geol. Soc.*, **17**, 81–102.

King, L. C. (1962) *The Morphology of the Earth*, Oliver & Boyd, Edinburgh.

King, L. C. (1966) 'The origins of bornhardts', *Zeits. für Geomorph.*, N.F., **10**, 97–8.

King, L. C. (1975) 'Bornhardt landforms and what they teach', *Zeits. für Geomorph.*, N.F., **19**, 299–318.

Linton, D. L. (1955) 'The problem of tors', *Geog. J.*, **121**, 470–80.

Mabbutt, J. A. (1961) 'A stripped landsurface in Western Australia', *Trans. Inst. Br. Geog.*, **29**, 101–14.

Marmo, V. (1956) 'On the porphyroblastic granite of central Sierra Leone', *Acta Geog* (Helsinki), **15**, 1–26.

Ollier, C. D. (1959) 'A two-cycle theory of tropical pedology', *J. Soil Sci.*, **10**, 137–48.

Ollier, C. D. (1960) 'The inselbergs of Uganda', *Zeits. für Geomorph.*, **4**, 43–52.

Ollier, C. D. (1975) *Weathering*, Longman, London.

Ollier, C. D. and Tuddenham, W. G. (1961) 'Inselbergs of Central Australia', *Zeits. für Geomorph.*, N.F., **5**, 257–76.

Omorinbola, E. D. (1979) 'A quantitative evaluation of groundwater storage in basement complex regoliths in south-western Nigeria', Ph.D thesis, University of Ibadan.

Pallister, J. W. (1956) 'Slope development in Buganda', *Geog. J.*, **122**, 80–7,

Passarge, S. (1928) *Panoramen Afrikanischer: Inselberg-landschaften*, Reimer, Berlin.

Pugh, J. C. (1956) 'Fringing pediments and marginal depressions in the inselberg landscape of Nigeria', *Trans. Inst. Br. Geog.*, **22**, 25–31.

Pugh, J. C. (1966) 'Landforms in low latitudes', in *Essays in Geomorphology*, G. H. Dury (ed.), Heinemann, London.

Rougerie, G. (1955) 'Une mode de dégagement probable de certain domes granitiques', *Compt. Rend. Acad. Sci.*, **246**, 327–9.

Savigear, R. A. G. (1960) 'Slopes and hills in West Africa', *Zeits. für Geomorph*, N.F., Suppl. Bd., **1**, 156–71.

Thomas, M. F. (1965) 'Some aspects of the geomorphology of domes and tors in Nigeria', *Zeits. für Geomorph.*, N.F., **9**, 63–81.

Thomas, M. F. (1966) 'Some geomorphological implications of deep weathering patterns in crystalline rocks in Nigeria', *Trans. Inst. Br. Geog.*, **40**, 173–93.

Thomas, M. F. (1974a) *Tropical Geomorphology: a study of weathering and landform development in warm climates*, Macmillan, London.

Thomas, M. F. (1974b) 'Granite landforms: a review of some recurrent problems of interpretation', in *Progress in Geomorphology*, E. H. Brown and R. B. Waters (eds), Inst. Br. Geog., **7**, 13–37.

Thomas, M. F. (1976) 'Criteria for the recognition of climatically induced variations in granite landforms', in *Climatic Geomorphology*, E. Derbyshire (ed.), J. Wiley, London.

Thomas, M. F. (1978) 'The study of inselbergs', *Zeits. für Geomorph.*, N.F., **31**, 1–41.

Thornbury, W. D. (1954) *Principles of Geomorphology*, J. Wiley, New York.

Thorp, M. B. (1967a) 'Closed basins in younger granite massifs, northern Nigeria', *Zeits. für Geomorph.*, N.F., **11**, 459–80.

Thorp, M. B. (1967b) 'Joint patterns and the evolution of landforms in the Jarawa granite massif, northern Nigeria', in *Essays in Geography*, R. Lawton and R. W. Steel (eds), Univ. of Liverpool.

Thorp, M. B. (1967c) 'The geomorphology of the Younger Granite Kudaru Hills', *Nig. Geog. J.*, **10**, 77–90.

Thorp, M. B. (1975) 'Geomorphic evolution in the Liruen Younger Granite Hills, Nigeria', *Savanna*, **4**, 139–54

Twidale, C. R. (1964) 'A contribution to the general theory of domed inselbergs', *Trans. Inst. Br. Geog.*, **34**, 91–113.

Twidale, C. R. (1968) 'Inselberg', in *Encyclopedia of Geomorphology*, R. W. Fairbridge (ed.), Reinhold, New York, pp. 556–9.

Willis, B. (1934) 'Inselbergs', *Ann. Assn. Amer. Geog.*, **24**, 123–9.

Willis, B. (1936) *Studies in Comparative Seismology: East African plateaux and rift valleys*, Carnegie Inst., Washington, D.C.

Young, A. (1972) *Slopes*, Oliver & Boyd, Edinburgh.

11 Karst landforms in the humid tropics

The term *karst* is derived from the Slav word *'krs'* or *'kras'*, meaning rock (Sweeting, 1968). It was originally used as a name for a region of massive limestone country bordering the Adriatic Sea in Yugoslavia. The district is characterized by bare limestones, sinkholes and underground streams. The term is now used to describe terrains underlain by soluble rocks and on which solutional processes are dominant. Such terrains must also be dominated by unique landform assemblages resulting from the solutional processes and exhibit absence of surface drainage and the predominance of vertical and underground drainage. Apart from limestone and dolomite, these features can occur in other soluble rocks like gypsum and rock salts, and in insoluble types like quartz-diorite (Feininger, 1969).

Sparks (1971) objects to the use of the word 'karst' as synonymous with limestone relief, firstly, because this usage tends to perpetuate the concept that there is one typical kind of limestone relief and, secondly, because it is a term derived from one particular area, and there appears to be no point in having a type area of limestone relief. He therefore proposed that the term 'karst' be reserved for areas similar in relief to the type area of northern Yugoslavia, while the term 'limestone relief' be employed in describing limestone terrains. However, as the term 'karst' already has a worldwide acceptance, it will be retained in this chapter as a collective term for landform assemblages on limestones and dolomite.·

Location of karst relief in the humid tropics

The literature on temperate karsts date back to the nineteenth century. Interest in limestone relief in the temperate environment stemmed first from the aesthetic values and the challenge posed to explorers; later, interest became more scientific, with emphasis on the description of the landforms and the explanation of the formational processes. Such is the interest generated in the subject that the International Geographical Union set up a special commission to study it. Probably, this interest in the subject and advances in geological mapping have facilitated the identification of most karst terrains in the temperate zones (Thornbury, 1969, p. 304).

Interest in tropical karst is very recent, presumably owing to the poor advances in geological mapping in most tropical countries. This notwithstanding, karst topography has been identified in Jamaica, Cuba, the Bahamas and Puerto Rico in the West Indies; northern Yucatan,

Tabasco in Mexico, central Florida, Honduras and Belize in Central America; Java, Sarawak and Sulawezi in Indonesia; the districts of Selangor, Perak, Perlis and Kelantan in Malaysia; central New Guinea and Fitzroy ranges in Australia; the Goba district in Ethiopia; Kasai and Mount Hoyo areas in Zaire; Tanga in Tanzania; Ankarana and the Mahafaly plateaux in Malagasy.

Conditions favouring the development of karst

Conditions favouring the development of karst relief under different morphoclimatic regimes are already well documented (Thornbury, 1969; Sweeting, 1968; Sparks, 1971). These are geological, climatic and topographic.

Geological conditions include the following:

1 the presence over large areas of soluble rocks near or at the surface;
2 these soluble rocks must be of considerable thickness (hundreds of metres);
3 the rocks which should have low primary porosity should be of a high degree of purity, and the joints must be widely spaced to provide mechanical strength and to facilitate movement of water through the rocks.

The rock that possesses these attributes is limestone, especially the massive, hard types; chalk, though soluble, is extensively characterized by primary porosity and permeability. Dolomite possesses some of the attributes, but it is not as readily soluble as limestone, whereas rock salt and gypsum are highly localized in occurrence.

The importance of lithology in the development of the limestone landforms has been emphasized by many authors. A few examples will suffice to illustrate this. For instance, micro-relief like the 'giant grikeland' of Jennings (1967); the linear, narrow grooves of Wilford and Wall (1965) and Cooke (1973); and the linear depressions of McDonald (1976) have been shown to have developed on massive, coarsely jointed beds of limestone aligned along the joint planes. Sweeting (1958) also shows that in Jamaica conekarsts and the associated cockpits have developed on the coarse, crystalline, well-jointed limestone, whereas dolines are associated with marly limestone which she considers as the main causative factor. Wilford and Wall (1965) also associate massive towers with pure limestone; while low hills with convexly rounded and gentler soil-covered slopes are associated with impure limestone in Sarawak. In Puerto Rico, Monroe (1964) shows that true tower karst is developed only on thick, massive, pure homogeneous limestone. The same is true of Belize, where the lower karsts are associated with granular, well-jointed dolomitic limestone.

In Sulawesi, Indonesia, McDonald (1976) shows that low hills, less than 100 m, are developed in thinly bedded impure limestones while plateau-like hills, more than 300 m high, are developed in massive coral limestones. As observed by Sweeting (1968) and Versey (1972) among others, complex cave systems and poljes tend to be most developed in folded and faulted regions.

The dip and strike of the rocks also affect landform development. In Sarawak, Wilford and Wall (1965) observed that where the limestones dip moderately to steeply, hills are elongated and parallel to the strike. Jennings (1971) also noted that uvalas are usually elongated along the strike in steeply dipping rocks, whereas in horizontal beds they are more lobate. Also, as will be shown later, lithology and structure affect drainage in karst terrain.

The climatic requirements include the availability of moderate to heavy rainfall coupled with high temperature, so that solution can readily take place. In association with the rainfall, the existence of dense vegetation assists solutional weathering and therefore karst development. Vegetation provides organic matter which forms humus, and which, together with plant-root respiration, can raise the level of carbon-dioxide concentration in the soil to about 30 per cent. The diffusion of this carbon dioxide into water passing through the soil leads to a high intensity of solution, especially in the humid tropics.

Given the differences in temperature, rainfall and vegetal cover over the earth's surface, it is to be expected that different types of karsts will occur in different limestone areas. It has thus been observed that while features such as sinks, dolines and poljes characterize karst terrains in the humid middle latitudes, the most conspicuous features in the humid tropics are cockpit–cone karsts and tower karsts (Thornbury, 1969; Panos and Stelcl, 1968; Corbel and Muxart, 1970). Jennings and Bik (1962), studying karst terrain in New Guinea at different altitudes, also found the low-level areas, less than 200 m a.s.l., dominated by cockpits, conekarsts and 'giant grikeland'; the intermediate areas at 1 580 m a.s.l. dominated by dolines, closed depressions and residuals; and the high levels above 3 300 m dominated by dolines, pinnacle and arete karst. However, as indicated by Jennings and Bik (1962), at each altitudinal level the picture is complicated by the influence of structure, lithology and evolutionary history to the extent that no clear-cut morphoclimatic influences could be discerned.

This is also the position taken by Jennings (1972), who observed that tropical karsts basically contain the same features as those of temperate karsts. Those he observed in the north-east Langkari Islands, western Malaysia, are apart from the usual conekarst, characterized by the classicial Yugoslavian holokarst, especially dolines, poljes, uvalas and so on. Also, according to Sweeting (1976), data on limestone solution worldwide fail to support the idea that solution varies with climate; rather, waters in ordinary non-dolomitic limestones throughout the world tend towards a fairly uniform value of between 250 and 300 m/litre of calcium content, so that limestone dissolution is mainly a question of speed and length of contact of acidulated water with the rock. She further stresses that the essential differences between temperate and tropical karst may be one of speed of landform development, karstic evolution being most rapid in hot, wet climates where water and biogene carbon dioxide are at a maximum. Also Ford (1975), who studied under different climatic regimes in North America, argues that the majority of the features found in sub-Arctic and alpine environments 'would not be out of place in the tropics or temperate areas'. Thus the supposed correlations between climate and landforms may not hold with regard to karst landforms.

Other important conditions for the development of karst terrain include relief and drainage. Available relief must be considerable to permit free circulation of underground water and the full development of karst landforms. According to Thornbury, there must be a system of entrenched valleys below uplands underlain by the soluble, well-jointed rocks to permit free drainage of water through the limestone, as moving water apart from transporting the dissolved carbonate also carries on the solutional process.

Processes of karst formation: solution

The most important geomorphological process operating in a limestone area is that of solution, and the most important catalysts of this process are rainwater and carbon dioxide. The latter is present in the atmosphere where the pressure is about 0.3 per cent, or in the soil where the pressure could be more than 100 times that of the atmosphere. Carbon dioxide (CO_2) dissolves in water to form carbonic acid (H_2CO_3), which reacts with calcium carbonate ($CaCO_3$) to form calcium bicarbonate, which is an aqueous solution. The equation is given as:

$$CaCO_3 + CO_2 + H_2O \rightleftharpoons Ca(HCO_3)2.$$

Ions of carbonate are dissociated and substituted with H^+ ions. The dissociated ions can then be transported away by water.

The chemical equilibria of calcium-carbonate solution, as discussed by Bogli (1964), Garrels and Christ (1965), Kern and Weisbrod (1967) and Sparks (1971), is a little more complex. According to Sparks (1971), the process of solution and the formation of carbonic acid involve about four phases, as follows:

1 In the first place, $CaCO_3$ can dissociate in water in which CO_2 is absent:

$$CaCO_3 \rightleftharpoons Ca^{++} + CO_3^{--}$$

2 the CO_2 that is dissolved in water to form carbonic acid will dissociate in two stages:

$$H_2CO_3 \rightleftharpoons H^+ + HCO_3^-$$
$$\text{and } H\,CO_3^- \rightleftharpoons H^+ + CO_3^{--}$$

3 Water itself dissociates to a certain extent:

$$H_2O \rightleftharpoons H^+ OH^-$$

4 Finally a certain amount of CO_2 dissolved in the water combines to form carbonic acid:

$$CO_2 + H_2O \rightleftharpoons H_2CO_3$$

Bogli (1964) had earlier identified four phases in the process of limestone solution, which Sparks (1971) examined extensively.

The rate of reaction and the amount of calcium carbonate dissolved per unit of time are high for all the first three phases and rise with increases in temperature. A rise of about 10°C doubles the rate of solution. Thus the rate of solution caused by these phases is about four times as high in the humid tropics as in the cold climates. Phase 4 takes a long time, and keeps in operation the chain of reactions of the first three phases until equilibrium is reached. The time taken by this phase varies, depending on the thickness of the water layer, the nature of flow, whether laminar or turbulent, or whether flow is in gravity or pressure channels, and the available surface area of the rocks. Generally, the rate of solution is low and decreases with time.

The overall rate of limestone denudation has been determined from the amount of dissolved load in rivers. Owing to the faster rate at which carbon dioxide is dissolved in cold water, it will be expected that solution will be greater in a cool climate than in warmer climates. In fact Corbel (1959) believes that limestone solution is more rapid in the colder temperate zones than in the humid tropics. He puts the rate of ground lowering resulting from solution at 200 mm/1 000 years in periglacial regions, 100 mm/1 000 years in temperate regions, and 30 mm/1 000 years in hot, wet climates. However, detailed measurements by Groom and Williams (1965) show that the rate of carbonate solution increases with temperature, being higher in summer than in winter. Table 11.1 further shows that net denudation rate in limestone actually increases with increasing temperature and availability of water.

TABLE 11.1 Rate of chemical denudation in limestone

Area	Koppen climatic type	Mean annual precipitation (mm)	Net rate of denudation $m^3/km^3/yr$ or mm/1000 yrs	Source
Kissimes river, Florida, USA	Cfa	1 200	5	Corbel (1959)
Yucatan, Mexico	Aw	1 000–1 500	12–44	Corbel (1959)
Indonesia	Afl	200–3 000	83	Balazs (1968)

Source: Jennings, 1971, p. 181

As is well known, the most important factor in the solution of limestone is the availability of carbon dioxide and water. Under the tropical rainforest, there is a great concentration of organic carbon dioxide in the soil. This, together with the high temperatures and high rainfalls marked by high intensities, can effect rapid solution of limestone, solution which, according to Sweeting (1958, 1972), may be about 400 per cent greater than the rate in the cooler climates. Due to rapid plant growth and decay, and intense microbial activity generating high carbon-dioxide pressure in the soil, generally soil water is very aggressive in the humid tropics. As soil water passes slowly through decaying organic matter, carbon dioxide is diffused into it, and the acid so formed can cause intense solution in the underlying rocks. Thus, soil-covered limestones are subject to continuous chemical action.

Tropical karst terrains
Drainage and associated landforms

Rapid infiltration of rainwater is characteristic of karst rock outcrops; however, intensively heavy rains often generate long overland flow. Due to its secondary permeability, water flows over a limestone surface until it encounters joint planes into which it gradually disappears, so that the frequency and openness of these joint planes and fissures influence the rate of infiltration into the rock. Where joints are few and tightly closed, infiltration is hindered and surface runoff generated, but where open joints are common, drainage is mainly sub-surface. Even where the rocks are covered by soil, high rates of infiltration are still characteristic except where the soil is the acid, dense, clayey types. Surface covers like colluvium and alluvium also promote a high rate of infiltration, as does forest cover.

With the characteristic high rate of infiltration in limestone terrain, surface streams are very few. However, it has been observed that in the humid tropics, surface drainage can develop under certain circumstances. Rivers may develop where the rocks are impure; even then the drainage density is invariably less than on other rocks (Jennings, 1971). Such rivers are often allochtonous, especially where their head-streams are located on impervious rocks. This is the case in the karst terrain in Sulawesi, where the rivers traversing the limestone have their sources on volcanic rocks immediately east of the limestone outcrop (McDonald, 1976). Such allochtonous rivers often lack tributaries throughout their length in the karst region. To cross the limestone, the rivers are floored by alluvium, also of allochtonous origin, which seals off the permeable rocks. Some tropical karsts consist largely of broad alluvial plains with perennial rivers meadering over them, as is the case in the Kinta valley in Malaysia. In any case, rivers entering karst are liable to lose all or part of their drainage underground. The streams disappear underground either through cave entrances, lateral or vertical opening on the river channel (Figure 11.1), or through narrow fissures in the bedrock, often referred to as sinkholes or ponors.

The Ankarana plateau, underlain by massive limestone in the Republic of Malagasy, exemplifies karst drainage in the humid tropics (Buckle, 1978). The plateau shows no trace of surface drainage; even

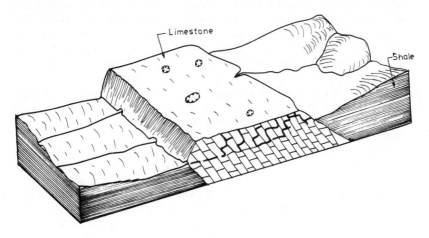

Fig. 11.1 Karst topography — sinkholes

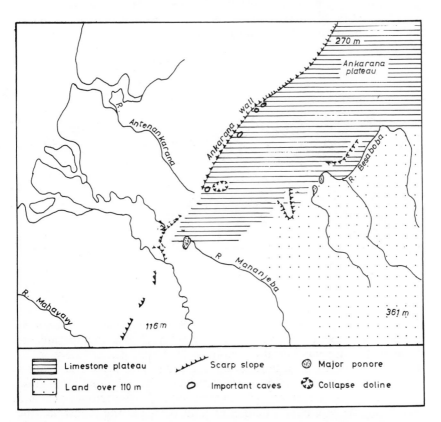

Fig. 11.2 Rivers on the Ankarana plateau, Malagasy (after Buckle, 1978)

sinkholes and dolines are very few on it while grykes are common. However, at the boundary to the east with non-carbonate rocks are several ponors and sinkholes which relate to an extensive network of subterranean drainage and caves. Several rivers taking source on non-carbonate rocks east of the plateau disappear into the limestone through the ponors and sinkholes. The Mananjeba follows a subterranean course for 1.5 km before resurfacing at the base of the high vertical scarp bounding the limestone to the west. The Basaboba river also flows underground for about 5 km before reappearing as the Antenankarana to the west (Figure 11.2).

Another example is provided by the white limestone in Jamaica, where stream-flow is mainly subterranean with the rivers upsurging as springs near the north coast (Sweeting, 1958). The water issues in the form of springs where the White Limestone is in contact with less fissured and less permeable beds. Also, springs occur where coastal limestone (Pliocene) formations abut against the white limestone.

Blind valleys

A blind valley is one closed at the lower end by a rock wall at whose base there is a sink through which the river disappears underground (Figure 11.3). Such a valley usually results from the lowering of the river bed by solution. They are very common around limestone hills in the Kinta valley of Perak in Malaysia (Paton, 1964).

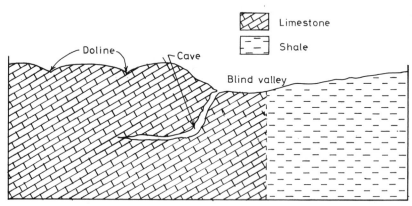

Fig. 11.3 A blind valley

Dry valleys

These are like ordinary river valleys except that the channels contain no water. Dry valleys are formed where a river on crossing an impermeable formation disappears into the limestone. However, where a network of such valleys develop on limestone, they could have originated from a previous cover of impervious rocks (such as shales), so that after these were removed the rivers got incised into the limestone beneath. Following the solutional opening of joint planes in the limestone, the drainage can disappear underground. Examples abound on the limestone

Ankarana plateau in the Malagasy Republic, which is dissected by several steep-sided, narrow, dry valleys (Buckle, 1978). Dry valleys are also common in the white limestone country of Jamaica. However, among the best examples is the Sadeng dry valley in south-central Java, which winds through the Gunung Sewu conekarst to the south coast. According to Lehman (1936), this valley is the former course of the Solo river now flowing out, north of the karst area in the Batoeretno basin. The valley is about 250–300 m wide and 150 m deep. Its long profile is interrupted by shallow lakes separated by gentle rises up to 30 m high.

Cockpit–conekarst terrains are characteristically drained subterraneanly. The storm runoff is drained through shafts or sinks located on the floor of the cockpits. These sinks have been ordered and analyzed in much the same way as ordinary surface streams in the Strahler tradition. Ordering is based on the hierarchical status of the highest-order channel draining into the sinks (Figure 11.4). As discovered by Williams (1966, 1973) in the karst terrain of New Guinea, higher order depressions are associated with highland regions, while lower orders are associated with low surfaces, which is an indication of the role of slopes in drainage development even on limestone. Williams (1972) also found that the frequency of second-order sinks was higher than the first- or third-order sinks. Generally, the frequency of stream sinks varies inversely and geometrically with sink order, whereas the mean area of sink catchment varies directly and geometrically with sink order. Williams (1972) also

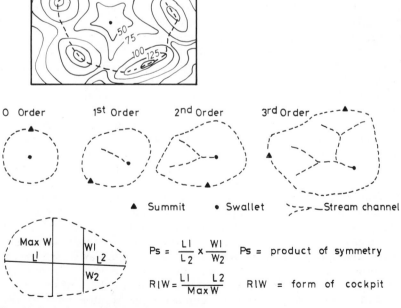

$$Ps = \frac{L1}{L2} \times \frac{W1}{W2} \quad Ps = \text{product of symmetry}$$

$$RlW = \frac{L1}{MaxW}\frac{L2}{} \quad RlW = \text{form of cockpit}$$

Fig. 11.4 Ordering of karst drainage system (after P. W. Williams, 1972)

found that the mean distance apart of sinks of the same order increases geometrically with sink order. All these observations show the measure of organization in stream sink systems.

Limestone gorges

Occasionally, a river entering a karst terrain may have sufficient volume to maintain its flow and thus cut deep gorges. These are deep, steep-sided valleys which are deepened as a result of channel erosion through solution and the removal of the dissolved material by the rivers. The steep sides are maintained as a result of the low rate of valley widening by slope processes. Examples of such gorges are common in the karst terrain of north-east Tanzania, where they are cut and maintained by allogenic streams (Cooke, 1973). Other examples include the Manambolo gorge in the western part of the Malagasy Republic where the Manambolo river has cut through the limestone massif of the Causse de l'Antsingy (Buckle, 1978). The steep-sided gorge, several hundreds of metres high, is over 20 km in length. Its walls rise almost perpendicularly from the river's edge, and along it are several caves – for example, the Anjohikinakina cave. Gorges are also common in the Strickland mountains in New Guinea.

Although minor gorges could have originated from the collapse of cave roofs and exposure of cave passages, some, like the Manambolo gorge, appear to have resulted from antecedence.

Caves

The most important sub-surface features in any limestone terrain are caves or caverns which are natural subterranean chambers in the limestone and joined to the land surface by a system of interconnecting shafts and galleries. Although caves can form in other rocks through wave and fluvial erosion, landslides and weathering, those formed in limestone are the largest, most numerous and most complex.

Limestone caves vary from simple rooms, short passages or open shafts to very intricate systems of passages linking rooms, shafts, chambers and halls of all shapes and sizes (Figure 11.5a and b). Some caves are dry, whereas the floors of others are occupied by running streams or lakes. Some have deposits of various kinds in them, while others have none. Some show evidence of corrasion. The entrances may be vertical shafts or lateral openings in slopes and cliffs, although in some cases small entrances can lead into large, intricate cave systems.

Caves are common in humid tropical limestone terrain. Examples include the Salaga caves in Kasai province and more than twenty caves on the fault block Mount Hoyo in Zaire; and the Omar caves near Goba, south-east Ethiopia, where the Web river, an upper tributary of the Juba, disappears underground for a distance of 1.5 km (Buckle, 1978). The Salaga caves, comprising a series of stalactite and stalagmite chambers, extend underground for a distance of more than 2 km. Caves are also very common in the white and yellow limestones in Jamaica. These caves occur at the base of the white limestones, in the top beds of the yellow limestones, and also at the zone of contact of the Montpellier beds with the crystalline white limestones in Jamaica.

Caves can form in different ways, most prominent of which are:
1 through tectonic widening of existing joints;
2 through corrosion by water and hydraulic action along joint planes in the non-phreatic zone within the limestones;
3 through solution in the phreatic zone below the water table by slowly moving groundwater (Moore, 1968; Jennings, 1971).

The most important processes of cave formation are solution and erosion by underground streams. Slow-moving water in the phreatic zone can enlarge joint and bedding planes by solution into networks of small galleries and shafts. If the limestone is uplifted, the water table goes down, but the vadose water percolating down the rock can continue the process of enlarging the cavities. Later, subterranean streams may increase the size of the cavities to form caves. Such joint-controlled caves often appear at different levels in the rock. Besides, some caves are cut irrespective of rock structure.

According to Jennings (1971), most solution in limestone occurs near the top of the permanent phreatic zone and in the intermediate zone through which the water table oscillates. Fast, turbulent, water-table streams can erode big sub-horizontal tunnels irrespective of structure. The roofs and walls of such tunnels are subject to solution so that current markings occur on the ceilings. These pressure passages can assume various forms depending on the extent of structural control. Cylindrical tubes develop where there is little structural influence; vertical or horizontal elliptical passages develop where joint planes occur; and rectangular passages, where beds differ markedly in their response to

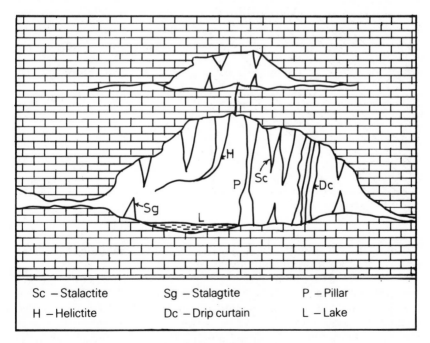

| Sc – Stalactite | Sg – Stalagtite | P – Pillar |
| H – Helictite | Dc – Drip curtain | L – Lake |

Fig. 11.5a Limestone caves — schematic diagram

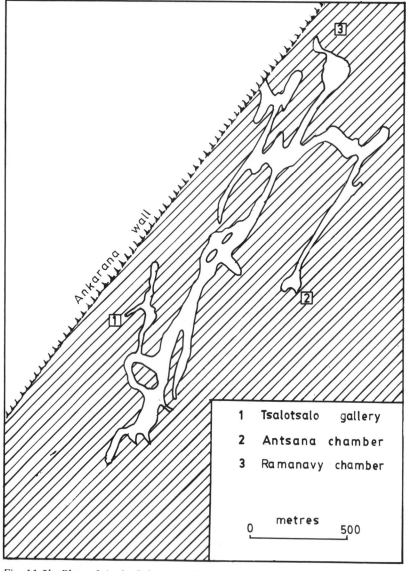

Fig. 11.5b Plan of Andrafiabe cave, Malagasy (after Buckle, 1971)

erosion. Such pressure passages may be subsequently modified by vadose flow or collapse to form well-developed caves.

When caves are finally formed, evaporation and diffusion of carbon dioxide from water percolating through the limestone ceilings load to the precipitation of calcium carbonate on the ceilings, walls and floors. Ceiling pendants are known as stalactites, while the upward-growing masses are known as stalagmites. Columns or pillars are formed when the preceding features grow together. In addition to these are helictites, which start growing from the ceiling and extend in various directions vertically, obliquely or in curves. Also several slender columns, collectively known as drip curtains, may be formed. All these depositional features are collectively known as *speleothems*, and massive deposition of these is characteristic of humid tropical limestone caves – for example, of the Salaga caves in Kasai province, Zaire.

Tropical karst surface landforms

As remarked earlier, several authors are of the opinion that limestones in the humid tropics are characterized mainly by conical or tower karst and the associated cockpit or polygonal karst. However, as observed by Jennings (1972), the distinction of the landforms into just two main types oversimplifies humid tropical karst, which comprises a great variety of both minor and major landforms.

The minor features include lappies on bare limestone surfaces, and features like grikes, clints and zanjones associated with limestone pavements. Major forms include enclosed depressions like dolines and cenotes, uvalas, poljes, and tropical karst residuals which include cockpit–conekarsts, tower karsts, and isolated hills and ridges.

Lapiés

These are solution grooves (rillen karren) separated by knife-edged ridges on bare limestone surfaces. They have been extensively described from the Tanga district of Tanzania, where the limestone surfaces are carved into sharp pinnacles and ridges by the rillen karren (Cooke, 1973); and in Sarawak, where the exposed limestone surfaces are dissected by irregular rills (0.6–1.3 m wide, 1.6–6.2 m deep), most of which are partially filled with limestone debris. Many lapiés are V-shaped, becoming wider with greater amplitude; where several of them occur, they can coalesce to form a dendritic pattern (Wilford and Wall, 1965). When the serrated, knife-edged ridges in between the grooves become rounded, the terrain is known as 'rundkarren'.

The formation of these rills has been linked to the initial phase of limestone solution. According to Bogli (1964) and as described by Sparks (1971), a limited amount of calcium carbonate can dissociate in pure water in the absence of carbon dioxide. Thus rainwater in contact with bare limestone can produce the closely spaced runnels.

Limestone pavements and associated features

Limestone pavements comprise relatively flat, bare, limestone surfaces diversified by different types of depressions. Such pavements are common in the sub-humid tropics where rainfall hardly exceeds 600 mm per annum, or where the rainfall is concentrated to within a few months in the year, as on the Mahafaly and Ankarana plateau in the Republic of Malagasy or in the Fitzroy basin in north-west Australia. However, such pavements have been observed in perennially humid parts of the tropics, as in New Guinea.

The pavements observed by Jennings in the Fitzroy basin are

diversified by large solution grooves (up to 20 m in depth, 3–10 m wide and hundreds of metres long), occurring in networks described as 'giant grikeland'. The Kange plateau and the Kionioni platform in Tanga, Tanzania, are also criss-crossed by shallow, narrow grikes (0.5–1 m wide and about 1.5 m deep), separated by roughly rectangular clints, and dissected by narrow, shallow joint-aligned grooves about 30 cm deep and 15 cm wide. The coastal limestone pavement in La Habana in Cuba has also been described as displaying extensive clints separated by grikes about 9 m deep (Panos and Stelcl, 1968). Finally, in the Darai hills of New Guinea, the chalky, porous limestones display pavements creviced by grikes (about 18 m deep, and 0.6–6.0 m wide) separated by wide, honeycombed clints (Williams, 1972).

Zanjones occur on limestone pavements in Puerto Rico and are somewhat different from grikes. Although both are joint-aligned, zanjones are long, vertical-walled trenches, a few centimetres to several metres wide, and about 1–4 m deep. They occur as parallel trenches oriented in one direction, as opposed to grikes, which intersect following several joint sets (Williams, 1972).

Grikes can form as a result of direct rain action on limestone pavements, especially where rainwater collects along vertical joint planes. However, they develop best under intense solutional action along joint planes under a cover of soil and vegetation, and can be exposed later by erosion following deforestation (Thornbury, 1969).

Dolines and cenotes

Apart from grikes and clints, limestone pavements in the humid tropics are usually diversified by dolines and cenotes (Figure 11.6). Dolines are

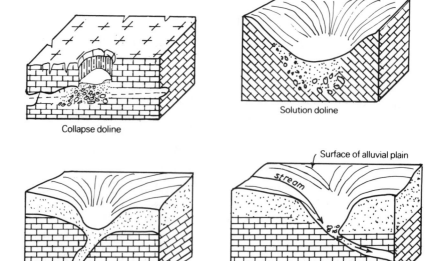

Collapse doline

Solution doline

Subsidence doline

Alluvial doline

Surface of alluvial plain

Fig. 11.6 Different types of dolines

closed depressions, circular to oval in plan, with depths varying with the diameters of the depressions. Other shapes are dish, bowl, conical and cylindrical. The term 'cenote' originates from the Maya language in Yucatan, and is used to describe rock-walled, circular, enclosed hollows with depth diameter ratios of more than one. It can contain water and may be confused with dolines (Marker, 1976).

Dolines are formed in many different ways, and based on their mode of formation. Sweeting (1972) identified four different types: surface rock dolines or solution dolines; collapse dolines; suffosion collapse dolines; and alluvial dolines. Jennings (1971) added a fifth type, the subsidence doline, while Thornbury (1969, p. 310) mentions the karst window dolines.

Solution dolines are formed primarily by pronounced surface solution at joint intersections. The solutes and the insoluble residues are removed through solution-widened joint planes at the base of the depression. Such dolines are characterized by conical forms, but they could have wide, flat, swampy floors if residues accumulate at the bottom at a faster rate than removal down the widened joint planes.

Collapse dolines characterized by very steep, near-vertical or overhanging walls are formed when the roof of a cave formed by underground solution collapses. Depth–width ratios are often greater than in solution types. However, unless further collapse occurs, these dolines tend to change progressively to the conical or bowl-shaped types as the side slopes are worn down and the bottoms filled up.

Suffosion collapse dolines are formed as a result of solution and formation of a cave beneath a coherent caprock, followed by the collapse of the cave roof. Steep-walled dolines are formed initially, becoming conical as the slopes are worn down.

Alluvial dolines are formed by solution beneath incoherent alluvial cover, while subsidence dolines are formed where superficial deposits or thick residual soils overlie karst rocks. Such dolines can develop through irregular subsidence and more continuous piping of the materials into widened joint planes and solution pipes in the bedrock beneath (Jennings, 1971).

Of the above, the solution and the collapse varieties appear to be the most common. Dolines have been reported from many areas in the humid tropics. Jennings (1972) observed that the limestone plateau between Perlis in Malaysia and the Thailand border displays an exaggerated Yugoslavian holokarst with the plateau pockmarked by dolines and poljes of greater dimensions than their classical prototypes. Similar forms have been reported from Sarawak by Wilford and Wall (1965), while Cooke (1973) observed four types of dolines in Tanga, north-east Tanzania. These include small, grass-covered, flat-floored, enclosed depressions on the Kiomoni plateau and at Maweni area; large solution dolines, varying in size from 1 200 m² to 20 hectares and aligned along dominant joint direction; funnel-shaped depressions 10–30 m in diameter on the Kiomoni platform; and collapse dolines on the Kange plateau. By contrast, the dolines observed in New Guinea by Jennings and Bik (1962) are of moderate dimensions and range from conical to dish-shaped types. Dolines also occur in parts of northern and central

Jamaica on marly limestone, especially on the yellow limestone series and Montpelier beds in the drier coastal zones. They vary in size from 15 to 130 m in diameter and about 30 m in depth on the yellow limestone. Those on the Montpelier beds are shallow, open and V-shaped (Sweeting, 1958).

Dolines are particularly prominent on the Mahafaly plateau in the Malagasy Republic. The commonest are the collapse type. They are concentrated in a belt about 15 km wide west of the plateau and north of Itampolo, where more than fifty have been identified. These vary in diameter from 10 m to 100 m and about 50 m deep. One of the largest is the Ranofoty, about 225 m wide and 70 m deep. The floors of these dolines are covered by large rock blocks and occasional pools, the existence of which depends on the depth of the local water table.

Poljes

The coalescence of more than two adjacent dolines leads to the formation of uvalas, which occur in association with dolines on the Mahafaly plateau in the Malagasy Republic. Poljes are thus large, closed depressions with flat floors across which rivers flow, although the river may disappear into a sink at the base of the limestone wall. Poljes, locally called *wangs* in Malaysia and *hojos* in Cuba, are usually elongated along the strike of limestones, but they could be compact or irregular in plan. The steep sides are usually separated from the alluviated floor by a sharp break of slope. Streams rise as springs at the margins and then flow over the alluviated floor to low points where they sink into ponors. Small poljes may be occupied by one stream, whereas large ones can contain many streams. Where the ponors cannot drain the streams efficiently, shallow lakes can accumulate on the polje floor.

Poljes have been observed in Jamaica, where they occur as interior valleys in northern Trelawny. Examples include Duanvale, Hamshire valley, Cambridge valley and Queen of Spains Valley (Sweeting, 1958; Versey, 1972). However, the best examples are those described by McDonald (1976) from south Sulawesi, Indonesia. Two large poljes occur here in the vicinity of Daimanggala and Banto Banto, 2.5 km and 1.0 km wide respectively and several kilometres long. Their floors are alluviated and traversed by streams taking source farther away on non-carbonate rocks. The enclosing limestone walls have foot-slopes inclined at 30° to 50° and rising to 100 m from the polje floors.

Poljes can form in two ways, through erosion and by tectonism. Sweeting (1968) believes that long-continued denudation of limestones by lake waters through solution abrasion may form rock platforms. Such eroded rock floors can then give rise to locally bevelled basins which become alluviated to form poljes. However, several authors, like Thornbury (1969), Jennings (1971) and Versey (1972) among others, are of the opinion that most poljes are primarily structural in origin, as they tend to coincide with structural basins formed by folding or faulting. As observed by Versey (1972), those in Jamaica were initiated by faulting, which caused the underground drainage to be impeded and forced to the surface to cause solution at the foot of the fault blocks. Jennings (1971) observed that the polje on which Lake Tibera in New Guinea lies is located on a tectonic depression formed as a result of high angle reverse faulting in a Plio-Pleistocene orogeny (Figure 11.7).

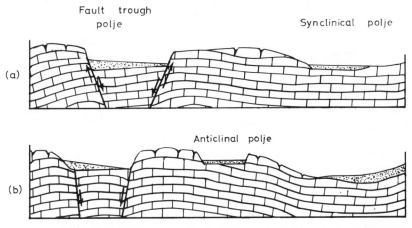

Fig. 11.7 Poljes in New Guinea: A, primary tectonic poljes in syncline and fault; B, enlargement of tectonic poljes and development of secondary poljes by corrosion (after Jennings, 1971)

Residual landforms

As already shown in Table 11.1, karst denudation in the humid tropics appears to proceed at a faster rate than in other environments due to the prevailing high temperatures and rainfall. Most of the features already described result from rapid solution and eventual collapse. Features such as dolines can expand to form larger depressions separated by erosional residuals, while features like uvalas and poljes can expand to form flat, alluviated plains on which erosional residuals occur. These residuals are referred to as cockpit–conekarsts, and tower karsts. In some cases, isolated ridges are formed in carbonate rocks while the surrounding plains are formed in non-carbonate rocks. These features are briefly described in what follows.

Cockpit–conekarsts

These comprise 'a succession of cone-like hills with alternating enclosed depressions or cockpits' (Sweeting, 1958). The depressions, which are star-shaped, are often referred to as 'honeycomb' or 'polygonal' karst (Williams, 1966, 1972). These landforms, which are believed to typify limestone relief all over the humid tropics, have been observed and described from Jamaica, the central highlands of New Guinea, northern Puerto Rico, Java, Malaysia and parts of Zaire.

The cockpits in Jamaica are among the deepest in the world, being between 100 m and 166 m in depth. Those in New Guinea are relatively shallower, at 30 m to 70 m in depth. The depressions are usually conical, but may be bowl-shaped. The sides may be near vertical or overhang. Many are floored with boulders, whereas others have solid floors. Almost all have vertical gaping shafts or sinks on their floors through which storm runoffs are drained underground. In most places, the cockpits are covered by vegetation, under which soils up to 2 m deep may be present.

Although there is little or no surface flow in the depressions, valley-like forms, often stepped in long profiles, usually descend to the central basin from the perimeter. Although these dispose of storm runoff, they do not have well-developed stream courses. Such valleys, focusing on the lowest part of the depressions, impart the characteristic star-shaped contour form to the cockpits (Williams, 1972). The cockpits are often not simple enclosed depressions, in some cases small basins nestle on the slopes of the major depressions to form relatively compound features.

Morphometric data on cockpits are too scanty. However, available data show that they occur at the rate of 30 per 1 km² in Java (Jennings, 1971). In New Guinea, the depressions show a tendency towards uniformity of spacing. The mean area of the 1 228 depressions studied by Williams (1972) is 0.065 km², with a standard deviation of 0.060 km² and a coefficient of variation of 93 per cent. Most of these are simple first- or second-order depressions and, as shown by Williams, area tends to increase steadily with order and with the number of the residual hills on their boundaries.

The hills around these depressions, which are either conical or hemispherical, are remarkably locally uniform in area and height. In Jamaica, they are between 100 and 133 m high; in New Guinea 30–100 m; and on the Kissenga plateau in Zaire 50 m. In Jamaica, the mean diameter of the hills is about 300 m, while the hill-slopes are inclined at 30–40°. In New Guinea, the hill-slopes are inclined at 45–50°, with resistant bands often forming steps and cliffs up to 10 m high. Overall, the hill-slopes, which are veneered with thin organic skeletal soils, appear smooth and convex at the top, straight down the sides and concave at the base, with no sharp break in slope. However, the slopes may be grooved and fretted by solution.

The consensus of opinion is that cockpits and the associated hills evolved through solutional weathering of exposed limestones (Jennings and Bik, 1962; Aub, 1964; Sweeting, 1968; Corbel and Maxart, 1970). Williams (1972) among others postulated that they are initiated at the intersection of tension and shear joints on exposed limestones. Formation is preceded by the development of sinks, which are subsequently expanded by concentrated runoff from the intense tropical rainstorms. Expansion can also occur through the capturing of minor secondary depressions at the peripheries of the major depressions (Figure 11.8). However, some authors relate the mode of expansion to sub-surface processes. Thus Versey (1972) postulates that the great rise in water level during and after heavy rains brings groundwater circulation close to the bottom of the cockpits and thus erodes them, subsequently leading to collapse of the depression floor and widening of the sides.

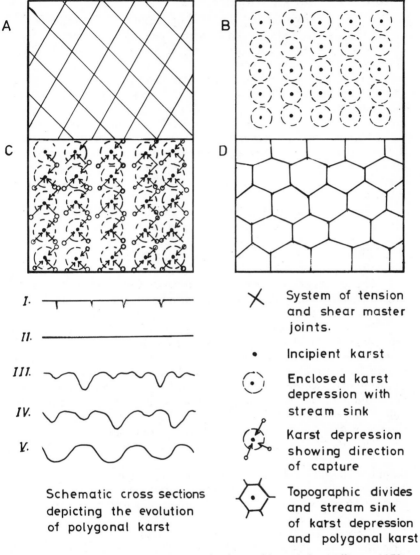

Fig. 11.8 Some stages in the evolution of polygonal karst (after Williams, 1972)

Finally, Lehman (1936) ascribed the formation of the cockpits in Gunung Sewu, Java, to rejuvenation of surface drainage by tectonic updoming of the planation surface on limestone, and the subsequent breaking up of the valley systems into chains of closed depressions. However, Lehman failed to explain the mechanism through which this could have occurred.

Tower karst and karst plains

Tower karst, otherwise called 'mogotes' or 'pepino hills', are isolated hills left standing above flat, alluviated plains. Such hills are common in the karst areas of Puerto Rico, Cuba, Malaysia, Sulawesi, Sarawak, Tabasco in Mexico, West Irian in New Guinea, Jamaica, Belize and Ankarana plateau of Malagasy Republic. In Malaysia, many mogotes stand abruptly although unconformably on steep granite slopes, while in New Guinea they often rise from supporting conical or pyramidal limestones. In Cuba, the plains surrounding the towers are sometimes developed on non-carbonate rocks (Panos and Stelcl, 1968); but generally they occur as isolated hills on well-drained or flooded limestone plains, on alluvial plains, on marine terraces, on cuesta backslopes, as outliers of sharp-edged cuesta faces, and as isolated hills on old surfaces cut across limestones.

The towers vary in size. In Belize, they are about 122 m high and 200–400 m in diameter, while in Jamaica they are 91–152 m high. In Malaysia they range in height from a few metres to over 300 m in the case of the Bukit Batu hill; while the Api mountain in Sarawak is 1 510 m high and covers an area of 38.0 km².

The towers are invariably characterized by steep slopes of 60–90°, which often overhang. The slopes may also be irregular, marked by steps and rises. Some towers in Sarawak have cliffs up to 610 m high. Where they are located on well-drained sites, scree slopes inclined at 25–60°, depending on the size of the constituent material, are common. Such scree slopes, however, are rather insignificant where they rest on peaty swamp due to rapid decay. The cliffs are honeycombed, marked by caves and irregularly winding furrows, while their bases are pitted by swamp slots, solution notches, recesses, caves and spring heads. In some cases in Belize, the towers are surrounded by basal depressions (about 6 m wide and about 4 m deep) through which water drains into sinks.

The poorly drained alluvial plains around the towers tend to be hummocky due to uneven removal of the underlying limestone and the slumping of alluvium into solution hollows. The limestone surface beneath the alluvium is generally smooth, but are sometimes broken by linear depressions; in some cases, sub-circular alluvium-filled depressions occur.

There is no consensus of opinion on the mode of tower karst formation. While Panos and Stelcl (1968) and Corbel and Muxart (1970) among others believe that tower karsts result from differential erosion of isolated bodies of limestones interbedded with rocks like marls, shales and sandstones, others, like Lehman (1953), Sweeting (1958, 1972), Paton (1964), Wilford and Wall (1965), Jennings (1971, 1976) believe that conekarsts are usually transformed into towers either by slope retreat or by recurrent flooding and alluviation of cockpits or both. However, there is no consensus as to the mechanics of slope retreat. Thus, while Paton (1964) thinks that the basal notches on the towers of Kinta valley and Perlis in Malaysia were formed by wave action, Lehman (1953) and Wilford and Wall (1965) believe that such notches may result from corrosion by lateral streams or by slow-moving swamp water, or by soil water in the surrounding alluvium; Jennings (1971, 1976) on his own part believes that gravity collapse of the rock wall above basal caves is responsible for slope steepening, if not basal undercutting; while Sweeting (1958, 1972) ascribed the transformation to spring sapping.

Finally, Wilford and Wall (1965) listed several other possibilities based on their Sarawak experience. Among these are:

1 the removal of less competent strata from above and below steeply dipping limestones;
2 the removal of less competent strata from one side of a fault – that is, differential erosion of weaker strata along faults;
3 the removal of less resistant intruded igneous rocks from within the limestone; and
4 the removal of less resistant beds into which the limestone passes laterally (Figure 11.9).

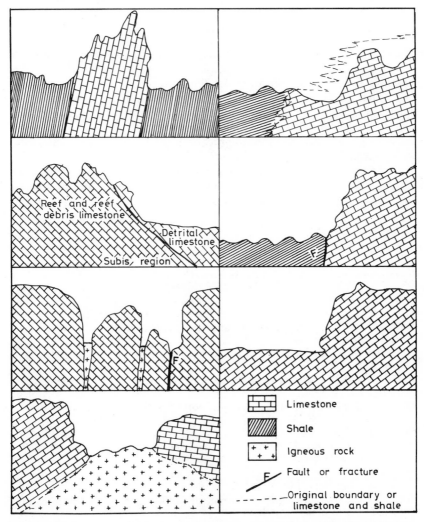

Fig. 11.9 Formation of limestone towers in Borneo (after Wilford and Wall, 1965)

In some places, exposed limestone on the towers are covered by crystalline and well-indurated patina of insoluble calcrete which protects the limestone from further solutional weathering. Monroe (1966) and Panos and Stelcl (1968) stressed the importance of case hardening in preserving the towers after they have formed. For instance, marked asymmetry in the mogotes of Puerto Rico is due to case hardening involving greater superficial induration of weak limestone on the windward side (Monroe 1966). However, Jennings (1971) attributed the asymmetry of the same mogotes to the prevailing winds, greater rainfall and consequent solution on the windward side. Panos and Stelcl (1968) ascribed the asymmetry in the Cuban mogotes more to the dipping of the limestones than to other factors.

Karst ridges

These comprise isolated ridges developed on limestones but surrounded by non-carbonate rocks. Examples include the Cordillera de Guaniguanico and the Sierra de los Organos in Cuba.

The Sierra de los Organos is composed of narrow, parallel, steep-sided limestone ridges separated from one another by gently rolling longitudinal lowlands or deep, narrow transversal defiles. The crest of some ridges are dissected into conical hills, dry valleys, deep cockpits, sinkholes and small elongated depressions.

These ridges represent erosional residuals as the less resistant surrounding non-carbonate rocks have been weathered and eroded so that the steeply dipping limstones form monoclinal or hogback-like ridges.

References and further reading

Aub, C. F. (1964) 'Karst problems in Jamaica with particular reference to the cockpit problem', *20th Intern. Geog. Cong., Karst Symposium.*

Balazs, D. (1968) 'Karst regions in Indonesia', *Karszt-es Barlangkutatas* (Budapest), **5**.

Bogli, A. (1964) 'Kalklosung und Karrenbildung', *Zeits. für Geomorph.*, Suppl. B. **2**, 4–21.

Buckle, C. (1978) *Landforms in Africa: an introduction to geomorphology*, Longman, London.

Cooke, H. J. (1973) 'A tropical karst in north east Tanzania', *Zeits für Geomorph.*, N.F., **17** (4), 443–59.

Corbel, J. (1959) 'Erosion en terrain calcaire', *Annales de Geog.*, **68**, 97–120.

Corbel, J. and Muxart, R. (1970) 'Karsts des zones tropicales humides', *Zeits. für Geomorph.*, **14**, 411–74.

Feininger, T. (1969) 'Pseudo karst on quartz diorite, Colombia', *Zeits. für Geomorph.* N.F., **13** (3), 287–96.

Ford, D. C. (1975) 'Castleguard cave, an Alpine cave in the Canadian Rockies', *Studies in Speleology*, **2**, 299–310.

Garrels, R. M. and Christ, C. L. (1965) *Solutions, Minerals and Equilibria*, Harper & Row, New York.

Groom, G. E. and Williams, V. H. (1965) 'The solution of limestone in South Wales', in *Denudation in Limestone Regions: a symposium, Geog. J.*, **131**, 37–41.

Jennings, J. N. (1967) 'Some karst areas of Australia', in *Landform Studies from Australia and New Guinea*, J. N. Jennings and J. A. Mabbutt (eds), Cambridge Univ. Press, 256–92.

Jennings, J. N. (1971) *Karst*, MIT Press, Cambridge, Mass.

Jennings, J. N. (1972) 'The character of tropical humid karst', *Zeits. für Geomorph.*, N.F., **16** (3), 336–41.

Jennings, J. N. (1976) 'A test of the importance of cliff-foot caves in tower karst development', *Zeits. für Geomorph.*, N.F. Suppl. Bd., **26**, 92–7.

Jennings, J. N. and Bik, M. J. (1962) 'Karst morphology in Australian New Guinea', *Nature*, **194**, 1036–8.

Kern, R. and Weisbrod, A. (1967) *Thermodynamics for Geologists*, Freeman, San Francisco.

Lehman, H. (1936) *Morphologische Studies auf Java*, Stuttgart.

Lehman, H. (1953) 'Der tropische Kegelkarst in Westindien', Verh df. *Geographisce Tage.*, **29**, 126–31.

Marker, M. E. (1976) 'Cenotes: a class of enclosed karst hallows', *Zeits. für Geomorph.*, N.F. Suppl. Bd., **26**, 104–23.

McDonald, R. C. (1976a) 'Hillslope base depression in tower karst topography', *Zeits. für Geomorph.*, N.F. Suppl. Bd., **26**, 98–103.

McDonald, R. C. (1976b) 'Limestone morphology in South Sulawezi, Indonesia', *Zeits für Geomorph.*, N.F. Suppl. Bd., **26**, 79–91.

Monroe, W. H. (1964) 'Lithologic control in the development of a tropical karst topography', *20th Intern. Geog. Conf. Karst Symposium*.

Monroe, W. H. (1966) 'Formation of tropical karst topography by limestone and solution reprecipitation', *Caribbean J. Sci.*, **6**, 1–7.

Moore, G. W. (1968) 'Limestone caves', in *The Encyclopedia of Geomorphology*, R. W. Fairbridge (ed.), Reinhold, New York, pp. 582–7.

Panos, V. and Stelcl, O. (1968) 'Physiographic and geologic control in development of Cuba mogotes', *Zeits. für Geomorph.*, N.F. Bd., 12, 117–73.

Paton, J. R. (1964) 'The origin of the limestone hills of Malaya', *J. Tropical Geog.*, **18**, 134–47.

Sparks, B. W. (1971) *Rocks and Relief*, Longman, London.

Sweeting, M. M. (1958) 'The Karstlands of Jamaica', *Geog. J.*, **124**, 184–99.

Sweeting, M. M. (1968) in *The Encyclopedia of Geomorphology*, R. W. Rairbridge (ed.), Reinhold, New York.

Sweeting, M. M. (1972) *Karst Landforms*, London.

Sweeting, M. M. (1976) 'Problems in karst geomorphology', *Zeits. für Geomorph.*, N.F. Suppl. Bd., **26**, 1–5.

Thornbury, W. D. (1969) *Principles of Geomorphology*, 2nd edn., J. Wiley, London and New York.

Versey, H. R. (1972) 'Karst of Jamaica', in *Karst*, M. Herak and V. T. Springfield, (eds), Elsevier, Amsterdam.

Verstappen, H. Th. (1960) 'Some observations on karst development in the Malay Archipelago', *J. Tropical Geog.*, **8**, 40–9.

Wilford, G. E. and Wall, J. R. D. (1965) 'Karst topography in Sarawak', *J. Tropical Geog.*, **21**, 44–70.

Williams, P. W. (1969) 'The geomorphic effects of groundwater', in *Water, Earth and Man*, R. J. Chorley (ed.), Methuen, London.

Williams, P. W. (1972) 'Morphometric analysis of polygonal karst in New Guinea', *Bull. Geol. Ass. Amer.*, **83**, 761–96.

12 Tropical coasts

Introduction

There is no consensus of opinion as to what constitutes the coastal zone. Bird (1968) defines it as 'a zone of varying width, including the shore and extending to the landward limit of penetration of marine influences: the crest of a cliff, the head of a tidal estuary or the solid ground that lies behind coastal dunes, lagoons and swamps'. Ketchum and Tripp (1972) are less definitive, viewing the coastal zone as 'a band of variable width which borders the continents, inland seas and Great Lakes'. Inman and Brush (1973) delimit the coastal zone to include the coastal plain and the continental shelf. According to these authors, 'the coastal zone is composed of the coastal plain, the continental shelf and the water that covers the shelf; it also includes other major features such as large bays, estuaries and deltas'. They distinguished the shoreline as the sedimentary and solid surface associated directly with the interaction of waves and wave-induced currents on land; while the shore zone includes the beach, the surf zone, and the near shore waters where wave action moves the bottom sediments. Zenkovich (1967) defines it as the 'scene of interaction between the hydrosphere, the lithosphere, the biosphere and the atmosphere'.

Bird's definition appears appropriate for the delimitation of coastal zones in the tropics, because with the exception of a few places where solid bedrocks form rocky promontories, as in eastern Brazil and some parts of West Africa, the coastal zone is in the form of low, swampy terrain characterized by extensive fluvio-marine plains and parallel beach ridges with intervening lagoons and mudflats.

The tropical coast

Wave action and shore drifting operate in all climatic zones where the sea is not frozen over, so that coastal processes tend to be generally polyzonal. However, as observed by Tricart (1972), certain littoral processes are zonal, having resulted from a combination of zonal influences. Sediment yield is known to be at a maximum in zones with highly seasonal rainfall and sparse vegetal cover (Langbein and Schumm, 1958); the savanna areas, which constitute some of the greatest zones of fluvial erosion in the world (Fournier, 1960), can thus be expected to supply readily large amounts of fluvial sediments into the adjacent coastal areas. Also, accumulation of coastal sediments is favoured by the perhumid tropical conditions which favour deep weathering of the exposed continental shelf (Fairbridge, 1961) during periods of low sea levels. Most parts of the coastal areas in the tropics are thus characterized by large accumulations of sediments on account of the large material brought to the coast by rivers passing through semi-arid areas, or dredged up from the continental shelf during periods of rising sea levels. Thus with the possible exception of the periglacial regions, tropical coastal zones are fringed by the most extensive fluvio-marine plains (Figure 12.1).

The high temperatures prevalent in the humid tropics play dominant roles in the evolution of certain features peculiar to the tropics alone. These include coral reefs and algal ridges, which form distinctive coastal features in the humid tropics, and mangroves, which thrive in the

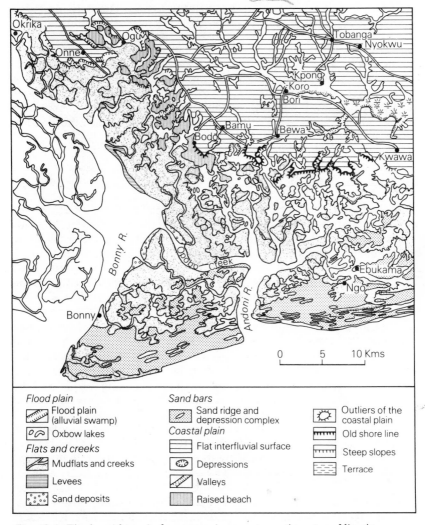

Fig. 12.1 The humid tropical coast environment, south-eastern Nigeria

brackish lagoonal environments and in the tidal mudflats. The extensive mangrove swamps characteristic of the humid tropical coasts are absent in other climatic zones.

Other forms rather peculiar to the humid tropics include spray corrosion features on rocks along the littoral zones in savanna and semi-arid areas characterized by long dry seasons. Spray corrosion pittings have also been observed under tropical forest climates where there is a high enough daily variation of humidity in the littoral zone for corrosion to occur – for instance, at Salvador in Brazil. It is also well known that tropical climates with well-marked dry seasons favour the formation of beach rocks.

Factors of coastal landforms

As in other parts of the world, the characteristics of the coastline in the humid tropics depend on certain factors, among which are the activities of waves, currents and tides, the structure of the local rocks, the past pattern of sea-level movement, the work of man, the climatic conditions favouring the development of certain organisms and other geomorphic agents. The first four are briefly examined below, while the other two are discussed later in this chapter.

Waves and currents

These are the main geomorphic agents along the coastlines. Waves consist of undulations in the sea surface caused by the friction of winds against the sea surface. This friction causes individual molecules of water to oscillate in a series of upright paths that translates into a slight forward movement during each oscillation (Figure 12.2). The water itself experiences a limited displacement, while the main movement is in the shape of the water, so that in the open sea, waves are just shapes. Wave forms advance in ever-widening circles.

Distinctions have been made between sea waves and swells. Sea waves are those initiated in the storm belts of the temperate latitudes and are usually referred to as high-energy waves, while swells are waves which have a long fetch – that is, they have travelled over long distances. These waves are generated by storms in the mid latitudes along the paths of the stormy westerlies, but they travel to the low latitudes where they affect most of the coastlines. For example, much of the West African coastlines are influenced by south-westerly swell generated by storms thousands of kilometres away in the latitudes of the 'Roaring Forties'. Thus most of the tropical coasts are affected by long, low swells of medium energy except where waves are generated by on-shore winds and hurricanes, as happens occasionally in East Africa or Central America. In West Africa, where the swell is south-westerly, waves are highest during the wet season when the south-west winds are prevalent. They are about 3 m high in July to August, hence highly erosive (Usoroh, 1977). Based on Price's (1955)

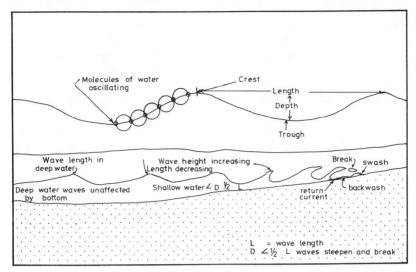

Fig. 12.2 Ocean waves and swells

coastal classification, all coasts in the humid tropics are low-energy coasts.

Longshore drift

Where waves are oblique to the shoreline, the swash moves sand particles upslope normal to the direction of the approaching waves, but the particles are brought down in the direction of the steepest beach gradient by the backwash. The result is that particles move along the beach with a zigzag motion (Figure 12.3). This drift of material along a coast as a result of waves breaking at an angle is known as longshore drift.

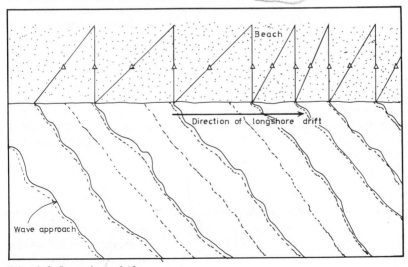

Fig. 12.3 Longshore drift

Tides

Tides consist of the periodic rise and fall of ocean, sea and other coastal waters in response to the gravitational attraction of both the moon and the sun on the rotating earth. The heights of tides are in addition affected by the size and depth of the oceans, the shape of their shorelines and latitudinal location. Seasonal winds and atmospheric pressure patterns also influence the tides.

In many parts of the world, tide rises and falls about twice daily. The tidal range varies very widely around the world's coasts. In some places in the mid latitudes, it is up to 20 m, in other places especially in the low latitudes, it is hardly measurable. For instance, along the West African coast, it is 1 m in Lagos, rising to 3 m in the Cross River estuary, 4–5 m along the Gambian estuary and as much as 7 m on the coast of the Republic of Guinea. It is about 5 m in the Amazon estuary. Apart from raising and lowering the level of wave erosion on the coast, tides can also generate strong currents, especially in estuaries where tides can drag the bottom in a process described as tidal bore. Tidal bore, which is capable of extensive erosion and transportation of sediments, is also of significant importance in the sedimentation of tidal mudflats, which are ubiquitous features of the humid tropical coasts.

Structural control

The effectiveness of erosion on coastal rocks depends on such rock properties as hardness, susceptibility to weathering, frequency of joints, height of cliffs and orientation, relative to the dominant waves. Generally speaking, the weaker the rocks, the easier they are to erode. Thus cliffs composed of weak or severely fissured rocks are liable to basal scouring and slumping. By contrast, where cliffs are relatively high and composed of resistant rocks like massive quartzites, erosion is relatively slow. For instance, the low cliff on the severely fissured and steeply dipping metasediments at Moree near Cape Coast in Ghana is subjected relatively to more rapid recession than the high, forest-covered cliff developed on massive igneous rocks just west of Takoradi.

Where structural trends lie at an angle to the coast, the shore tends to be indented, as is clearly evident on ria-shorelines in Sierra Leone and elsewhere. Where, however, structure and shorelines are parallel, the coasts are straight with few indentations. An example is the west coast of Deccan in India, where fault scarps more than 1 000 m high, plunge into the sea.

Past pattern of sea-level changes

Although eustatic changes occurred in the past geological periods, those of the Quaternary period seem to have left the greatest impact on the coastal areas. While glacio-eustatic changes in sea level have been estimated to occur between +100 m (Termier and Termier, 1963) and

−145 m (Ewing *et al.*, 1960) since the Pleistocene, the actual pattern of eustatic sea-level movements, is, however, still not clearly resolved, as shown in Table 12.1.

TABLE 12.1   Pattern of sea-level movement since the upper Pliocene

Period	Valentin (1952)	Fairbridge (1961)
Upper Pliocene	+ 100 m	+ 200 m, + 130 m
Gunz glacial advance	− 20 m	− 10 m
Gunz – Mindel interglacial	+ 60 m	+ 100 m, + 80 m
Mindel glacial advance	− 90 m	− 40 m
Mindel – Riss interglacial	+ 30 m	+ 55 m, + 30 m
Riss glacial advance	− 110 m	− 55 m
Riss – Wurm interglacial	+ 15 m	+ 18 m; + 8 m, + 3 m
Wurm glacial advance	− 95 m	− 100 m
Post Wurm	−	+ 2 m

Source: Valentin, 1952; Fairbridge, 1961

This pattern of shoreline movement is complicated by coastal instability all over the world, especially since the Pleistocene (Neuman, 1968, p. 152). For instance, Tjia (1970), working around the tectonically unstable Sunda Platform in South-east Asia, recognized twelve submerged shorelines at depths of −82–90 m, −67 m, −60 m, −50–51 m, −45 m, −36 m, −30–33 m, −28 m, −18–22 m, −13 m, −10 m and −7 m; and four elevated beaches at +10–12 m, +16–18 m, +30–33 m and +50 m. While Fairbridge (1961) believes that the sea level rose to its present level some 5 000 to 6 000 years ago, and has only fluctuated within +3 m of its present level, Tjia *et al.* (1977), using C^{14} evidence, establish periods of high sea levels around the tectonically stable peninsula of Malaysia at between 2 500 and 2 900 yr B.P., 4 200 and 5 700 yr B.P. and 8 300 and 9 500 yr B.P.

Whatever the pattern, changes in sea level have several morphological implications. Firstly, they caused waves to operate at different levels and thus led to the formation of features (for example, raised beaches) that can now be observed at different levels along the shorelines. Secondly, they gave rise to submerged and emerged coastlines.

Man's influence

Man has influenced the character of the coastal scenery in several ways, especially by building drainage canals, breakwaters, groyne and artificial harbours, which have often led to changes in the geomorphic processes and in the course of coastal development. A good example is the construction of moles on either side of the entrance to Lagos harbour, which have caused coastal progradation behind the western mole and severe erosion behind the eastern mole. This will be further examined later.

Coastal processes in the humid tropics

The most important exogenetic coastal processes in the humid tropics include sub-aerial processes of chemical and biological weathering, erosion, and deposition by waves, currents, tides and rivers.

Weathering

Sub-aerial processes, especially weathering and erosion in the form of mass wasting and gullying, are operative on certain coasts, and can promote even more changes on the shoreline than waves. Alternating wetting and drying by sea spray promotes chemical weathering and the production of debris from granitic cliffs. The rock minerals are affected by various weathering processes. The alkali components are subject to hydration, while the ferro-magnesian minerals are liable to oxidation. Decomposition of the silicate minerals thus leads to the disintegration of the rocks. As chemical processes proceed relatively rapidly under high temperatures, chemical weathering is very intense on coastal rocks, especially as sea water has been shown to be more powerful than fresh water in dissolving and decomposing silicate and ferro-magnesian elements (Zenkovich, 1967). Also, as shown by Tricart (1972), crystallization of salts in sea water in a tropical climate, above the tidal level, is very active in granular disintegration, especially in granites, primarily as a result of the decomposition and mechanical destruction of the micas they contain.

Wherever coral-reef limestones and aeolian calcarenites occur along the shoreline, they are subject to both biochemical and biophysical erosion. The former involves both solution and biological action. Rainwater with pH less than 7 can interact with limestone at the supra-littoral zone to form minor pits, pans and hollows on the bare cliff. Biochemical erosion is affected in the inter-tidal and the shallow sub-tidal zones by the myriads of organisms growing or living on the reef limestones. As observed by Hodgkin (1964) and Neumann (1968), these organisms act in two ways. They can generate carbon dioxide which, if undissipated, can dissolve in sea water and thus become capable of dissolving limestone. Also as discussed by Fairbridge (1968b) and Neumann (1968), microscopic boring algae can so densely grow together as to form a mat or encrustation over the surface of the limestone. Their organic waste below the mat can react with the limestone surface and thus help to dissolve it. The dissolved material seeps deeper into the rock, and on exposure to insolation it is precipitated between the grains and so 'case hardens' the upper layer of the limestone. Thus the outer layer develops a rind. Such case-hardened limestones are subject to inter-tidal mechanical fragmentation during strong waves. The broken fragments may then be used in abrading the limestone.

Direct biological erosion takes the form of boring, browsing and burrowing by several organisms, among which are crabs, chitons, sponges, echinoids, gastropods and pholad pelecypods. Neumann (1968, pp. 78–79) gives a comprehensive list of these organisms and their functions in effecting biological erosion. The surface of the affected limestone is etched, jagged and hackly.

Solution of limestone occurs in the inter-tidal zone, but the exact mechanism is still not well known. Theoretically, tropical sea water is supersaturated with $CaCO_3$ and hence cannot dissolve more. However, experiments have shown that slow complexing under diurnal heating and cooling takes place, rendering the surface sea water nocturnally undersaturated with respect to $CaCO_3$, so that more $CaCO_3$ can be dissolved (Fairbridge 1968a, pp. 653–57). Thus on tropical coasts where coral limestones are extensive, sea-water solution has led to the formation of various features, especially coastal lapies, as on the coast of Barbados.

The net effect of these weathering and erosion processes is the development of cavities and honeycomb features on the limestones and the formation of notches at the base of limestone cliffs. Such notches are up to 3 m deep on most limestone cliffs, but as observed by Neumann (1966) in the Harrington Sound, a quiet lagoon in Bermuda, the notch at the base of the surrounding cliff is 5 m deep. It may even be up to 10–20 m deep, as observed by Kuenen (1933) on limestone cliffs in the East Indies.

The rate of cliff recession resulting from biological and chemical erosion on limestone appears to be relatively fast. Experiments by Hodgkin (1964) on aeolian calcarenite coasts, and a number of measurements by other workers, show the mean recession rate at about 1 mm/yr. However, Neumann (1968) estimates the rate at about 10 mm/yr along tropical coasts.

Wave erosion

Waves are highly erosive. Wave energy depends on its height, length, velocity and steepness. Waves exert a tremendous amount of energy on exposed shoreline. This is especially the case during storms, when the impact of breaking waves can be as great as 25 tonnes per square metre. According to Twidale (1976), the spray thrown up by breaking storm waves can travel at speeds up to 110 km/hour. Also as shown by King (1972, p. 451), for a storm wave 3.05 m high and 45.8 m long, the observed pressure is 0.593 kg/cm². Such powerful waves, armed with rocks and pebbles, can cause more changes in a few hours than ordinary waves can do in several months.

Factors of wave erosion include the following:
1 the nature of rocks along the shoreline;
2 tidal range;
3 openness of shore to wave attack;
4 depth of water offshore;
5 configuration of the coastline;
6 presence or absence of a protective beach;
7 abundance and size of abrasive tools;
8 stability of sea level.

Apart from the above, wind constitutes an important factor in wave erosion. High-frequency destructive waves are usually wind driven. As shown by Usoroh (1977), onshore winds usually raise the level of the water near the shore, leading to the formation of return currents seawards. Such currents disturb sediments on the sea floor, and such sediments are taken up by waves and used for erosion.

As in fluvial action, waves erode through hydraulic action, abrasion and solution. Hydraulic action takes several forms. It can occur when waves surge into weakly consolidated rocks loosening the binding material and removing the sediment. It can also affect well-jointed rocks by widening the cracks and loosening the rock blocks. It has been noted in coral reefs that under heavy surf action, air can be trapped in undercuts and blowholes and become compressed by successive surges. The resultant hydraulic pressure build-up may be strong enough to break off 10–20-tonne blocks. Such blocks, referred to as negroheads, are common on reef shore platforms (Fairbridge, 1968b). Of more importance is cavitation, which occurs where cliffs are fissured by joints and the sea offshore is deep. At such places, waves reach the shoreline and are reflected without breaking. The reflected waves interfere with oncoming waves to create a pattern of standing waves. Such waves may set up oscillations that may break up the jointed rocks due to the sudden increase and decrease of pressure (Sparks, 1960). Also, where waves break against fractured cliff rocks, extremely high pressures are exerted on the fissured zones. Air is compressed between the water and the rocks within the fractured zone at very high pressures, which may be so high as to break off 10 to 20 tonnes of rocks (Fairbridge, 1968a).

Abrasion occurs when waves use the rock debris, detached through hydraulic action and cavitation or deposited along the shoreline by streams, or derived from the sea floor, to erode the shoreline. However, in the humid tropics, the high rate of chemical decomposition causes most of the materials to be fine and thus of limited use as abrasive tools (Tricart, 1972). Abrasion can occur in three ways:

1 it can occur at the point where erosive waves are crashing on the shoreline as the sandy particles in the breaking waves are made to strike against stationary shore rocks;

2 it can occur in swash and backwash when the rock debris are moved up and down the shoreline by longshore drift; and

3 it can occur where rock blocks are detached from fissured cliff rocks, and the cavities formed subjected to potholing.

Soluble rocks like limestones are susceptible to chemical erosion by solution, as already examined. However, the exact mechanism is still not well known. As evident in the literature, nearly all sandy beaches in the humid tropics are currently experiencing serious erosion. They are being undermined and steepened into cliffs which retreat at relatively fast rates. Examples have been described along the coast of West Africa (Pugh, 1954a; Webb, 1958; and NEDECO, 1954), and along the coasts of Espíritu Santo and Bahia in Brazil (Tricart, 1972). The example of the Nigerian coastline is described below.

Usoroh (1971, 1977) carried out a study of coastal erosion in the Lagos area. He discovered two main factors to be responsible for the severe rate of erosion on the shoreline. Firstly, during the rainy season between March and October when the south-west monsoon winds are onshore, severe storm surges generate powerful destructive waves which erode the beaches so that the shoreline at this period is characterized by a steep, low cliff up to 2 m high. Secondly, he observed a disequilibrium between the supply and removal of sand from the beach as more sand is being removed than supplied by longshore drift. In addition, there is the readily erodible nature of the unconsolidated sands: Table 12.2 shows the rate of erosion at selected stations on the beach around Lagos (Figure 12.4).

TABLE 12.2 Annual rates of beach recession and accretion around Lagos between 1972 and 1975

Station	1972	1973	1974	1975	Mean
Kweme Beach	−2.23	−4.33	−5.24	−4.73	−4.14
Angorin Beach	−1.98	−6.78	−4.39	−5.57	−4.43
Ganyingbo Beach	−5.36	−6.97	−6.89	−5.00	−6.06
Topo Beach	−5.49	−3.62	−9.73	−7.13	−6.49
Iworo Beach	−4.61	−6.24	−5.43	−5.36	−5.41
Igbo-Oja Beach	−4.73	−2.97	−7.63	−7.08	−5.60
Agaja Beach	−5.17	−4.67	−9.32	−7.43	−6.65
Okun-Ibese Beach	−6.20	−3.34	−9.18	−7.26	−6.50
Inogbe Beach	−5.70	+7.24	+5.39	+3.88	+2.73
Light House Beach	+21.78	+21.29	+29.04	+26.54	+24.91
Victoria Beach	−22.29	−26.63	−30.43	−23.33	−23.17
Alagutan Beach	−21.06	−23.18	−23.06	−20.99	−22.07
Ijede Beach	−6.72	−4.28	−9.00	−4.14	−6.04
Oworosoki Beach	−1.30	0	−1.98	0	−0.82
University of Lagos Beach	−7.42	−6.38	−9.82	−6.36	−7.50
Pedro Village Beach	−6.84	−4.92	−11.79	−6.49	−7.05

Source: Usoroh, 1977, p. 101

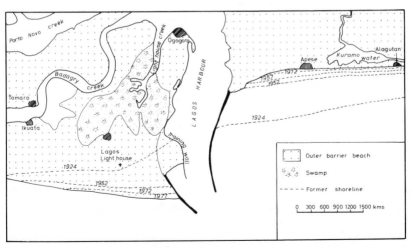

Fig. 12.4 Sand accretion and erosion around the entrance to Lagos harbour, 1924–77 (after Usoroh, 1977)

351

As is obvious from the above table, the Light House Beach experienced a high rate of accretion, while the Victoria Beach equally suffered severe erosion. This has resulted from man's interference with the natural configuration of the shoreline. Severe erosion started on the Victoria Beach following the construction of breakwaters at Lagos Harbour entrance. The moles were built between 1907 and 1922. The western mole arrested the eastward drift of sand along the coast, leading to accretion behind the breakwaters on the Light House Beach. Between 1924 and 1958, the beach advanced by about 400 m (Webb, 1958), and by 1975 it had advanced by another 600 m (Usorch, 1977). Behind the eastern mole, erosion was so severe that between 1912 and 1975 Victoria Beach retreated by more than 1.5 km. However, the intensity of erosion decreases eastwards because, at a point 1.5 km away, the beach has receded by only 760 m.

Various attempts have been made to halt the erosion on the Victoria Beach; among these are (1) the inland extension of the east mole at the western end of the Victoria Beach, and (2) the regular pumping of about 450 000 m³ of sand annually from Lagos Harbour on to the beach. This, however, is not sufficient to replace the annual loss, which is estimated at about 574 000 m³. In any case, pumping of sand on to the beach has been discontinued so that erosion is ravaging the beach unchecked.

Eroded sediments are transported along the shoreline. The mode of transportation depends on the nature of the load. Silt-size particles are removed in suspension in longshore currents and breaking waves in the surf zone. Sand-size particles are moved by longshore drift, while gravels, cobbles and even boulders are moved by sliding, rolling and saltation within and seaward of the surf zone.

As shown by Tricart (1972), longshore drift can reach down to a depth of 20–30 m, and can account for an enormous volume of littoral drifting which, at Vridi in the Ivory Coast, is estimated at 800 000 m³ annually since the Vridi Canal was opened in 1950.

Longshore drift favours the formation of offshore bars and spits. In fact, much of the debris transported by beach drifting is deposited to form beaches. The morphology of the beaches themselves change rapidly from day to day or even hour to hour, as they are built during calm weather when constructural waves are operating and eroded during storms under the influence of destructive waves.

Deposition

Waves are important agents of deposition as well as erosion. However, on some coasts, wind is also an important depositional agent (see Chapter 5). The most important factors relating to deposition are types of waves, wave energy, supply of sediments and nature of the shoreline.

Deposition takes place where swash is more powerful than backwash. In such relatively low-energy waves, swash moves particles up the beach while the weak backwash is unable to return the material carried forward by the swash.

Shallowness of sea depth near the shoreline encourages deposition. Waves tend to break when the ratio of its height to its length, which increases as the waves enter shallow water, reaches 1:7. At such points breaking waves can deposit sediments scoured up from the sea floor to form submerged offshore bars which may later emerge with further deposition.

Most of the depositional landforms along coasts in the tropics are only partly built with materials eroded locally from the coast itself. The rivers supply most of the sediments used to build the various coastal features. Also as shown by Tricart (1972), onshore movement of sediments from the continental shelf is especially important in the recent geological past.

The most ubiquitous coastal deposits in the tropics are sand and mud, both of which are highly mobile. Shingles are extremely rare. Most of the sand is fashioned into beach deposits like barrier bars, beaches and spits, while the mud is deposited in vast tidal marshes and swamps.

An important aspect of coastal deposition in the humid tropics concerns the role of mangroves. Mangroves, especially *Rhizophora sp.* and *Avicennia*, grow in the brackish waters in old lagoons and on estuarine banks. Their root network not only reduces stream currents and induces estuarine deposition, it also protects muddy tidal flats from erosion.

All the processes examined above have combined to give rise to several coastal landforms, some of which are discussed in that follows.

Coastal landforms in the humid tropics

Most of the early classifications of coasts were based upon apparent relative movement of land and sea. Thus Davis (1902) divided coasts into those of emergence and submergence. Johnson (1919) followed suit, except that he added two other categories of neutral and compound shorelines. Neutral shorelines are those with no clear evidence of emergence or submergence – for example, deltaic alluvial plain, volcanic plain and coral reef coasts – whereas compound coasts are those whose features reflect the effects of both emergence and submergence.

Shepherd (1968) classified coasts into primary and secondary types. The former comprise those where the sea rests against a surface that owes its topography to some terrestrial agency – for example, rias, fiords, deltas, alluvial plain coasts, volcanic coasts and fold coasts. The secondary types are those that owe their form to marine processes, such as wave-erosion coasts, barrier coasts, mangrove and coral coasts.

Several other criteria have since been used in classifying coasts, among which are structure, motion, agency, material, wave energy, geometric pattern, coastal equilibrium, transverse profile, erosion/deposition, stage, climate, ecology and time (Tanner, 1960). However, as observed by King (1972), no completely satisfactory classification of coasts has yet been proposed, as some are purely descriptive whereas others are genetic. King (1972) is of the opinion that three main variables should be taken into account in classifying coasts; (1) the relative movement of sea level; (2) the effect of marine processes, and (3) the form of the land–sea contact zone.

Based on these criteria, the following major coastal landforms can be recognized in the humid tropics:

1 barrier coasts;
2 mangrove coasts;
3 delta coasts;
4 submerged coasts;
5 emerged coasts;
6 rocky coasts;
7 coral coasts.

Each of these main types and their associated features are described below.

Barrier coasts

These are characterized by lagoons blocked by barrier beaches both of which are parallel to the coast. As observed by Galliers (1968), some of the barriers are stable forms supporting settlements and roads, and almost all have mature palms and well-developed woody plants growing on them. As observed along the coast of Ghana, the lagoons are rather elliptical bodies with the main axis parallel to the shore or narrow, elongated, sometimes sinuous bodies often running transversely to the coast, such as Ologe Lagoon near Lagos. Some of the small lagoons, as those observed by Galliers (1968) along the coast of Ghana, have no outlet to the sea except during the rainy season when the barriers are breached. Such lagoons vary in size during the year. In the dry season, they are reduced to intricate systems of small channels, but during the wet season they are inundated with water and overflow the barrier beaches. Larger lagoons, like the Lagos Lagoons, are directly linked with the open sea. Other examples along the Guinea coast include those in Ivory Coast and between south-eastern Ghana and Nigeria. Outside West Africa, the coasts of the Congo Republic and Gabon are traversed by several bars and lagoons, especially the Ndogo Lagoon. On the east coast of the Republic of Malagasy, south of Tamatave, the Nosive and the Ampitable lagoons and their associated barriers are very prominent. Barriers also occur on the Yucatan peninsula and on the east coast of Brazil from Rio Grande do Sul to Paraiba; they are also present in the Caratasca area in Belize, and on the eastern coasts of India, Java, Sri Lanka and Borneo.

A major feature of the coast of south-western Nigeria is the series of sandy barrier bars and lagoons especially between the Niger Delta and the Republic of Benin boundary (Pugh, 1953; Webb, 1958; and Usoroh, 1977). West of Lagos, the landward side of the lagoon-barrier beach complex is marked by an old degraded cliff about 15 m high. Between this cliff and the sea four morphological units are distinct (Figures 12.5 and 12.7), namely:

1 the broad in-filled alluvial plain (an old lagoon);
2 the inner parallel sand ridges;

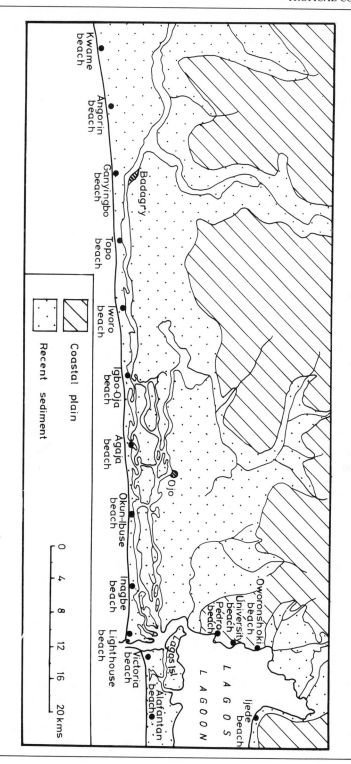

Fig. 12.5 Some stations on the Lagos coast at which beach recession and accretion were observed (after Usoroh, 1977)

3 a zone of narrow lagoons and creeks;

4 the active barrier beach.

All these units occur more or less in a sub-parallel form, with the sandy deposits alternating with the lagoons.

The active barrier beach is about 3 m high and is composed of two sections, the unvegetated surf-beaten, eroding beach, behind which is a sand bar varying in width from 500 m to 1.5 km. This sand bar comprises several narrow sand ridges, each representing a former strandline. These strandlines are separated by muddy depressions. Individual ridges vary in width from 50 m to 150 m, while some are about 7 km long. Near the village of Kweme, several ridges merge into a wide sand complex (Areola, 1977; Usoroh, 1977).

Behind the barrier bar and parallel with it are two parallel systems of creeks, the more prominent of which is the Badagri creek which in the west continues as far as Porto Novo, and in the east merges with the Lagos Lagoon. However, between Ologe Lagoon and Apapa, the creeks form a network of meandering and anastomosing channels separated by an extensive mudflat. A number of these channels have silted up.

Behind the creeks are a succession of parallel sand ridges separated by mud-filled depressions trending E–W. The settlements of Badagry, Ojo and a large part of Lagos are located on these inner sand ridges, as also are the roads between Lagos and Badagry (Figure 12.7). Between Ojo village and Lagos, four inner sand ridges can be identified at heights of 9.0, 7.0, 5.5 and 4.0 m above sea level respectively from the innermost to the outermost, so that their heights decrease steadily from the interior towards the shoreline. Between the inner sand ridges and the ancient cliff lies an extensive swampy depression now partially filled with fluvio-lagoonal deposits.

Analysis by Usoroh (1977) shows that the sands of the modern barrier beach are well sorted, and generally coarse with this particular fraction about 80 per cent. The sand of the interior bars are medium grained, while those farther inland are fine grained. Sands from the outer barrier beach are predominantly angular, while 90 per cent of those from the inner barrier bars are sub-rounded to rounded. The angularity of the outer barrier bar is attributed to (1) derivation from terrigenous material of the arenaceous sands of the interior, and (2) the possibility of their being derived from exposed weathered sea bed during past periods of pronounced regression. Opinions are divided about the origin of barrier beaches. Two main theories of barrier beach formation will be considered here:

1 development and shoreward migration of emergent offshore bars from material supplied by the landward movement of sediments on the gently sloping continental shelf;

2 growth of spits parallel to the coastline from material supplied by longshore drift.

Elie de Beaumont (1845), the proponent of the first idea, theorized that waves approaching the shore stir up the sea floor at the breaking point where waves lose much of their energy. The sediments accumulate to form a bar which later builds upward to become a barrier beach enclosing a lagoon. By contrast, Pugh (1953) attributed the formation of the barrier beaches along the coast of south-west Nigeria to regression during Quaternary glacial periods. He suggested that the barrier bars were formed under conditions of low sea level as offshore bars which later drifted and merged with the existing coastline. Galliers (1968), who studied the Ghanaian coastline, observing the sands in the barrier bar to be poorly sorted, sub-rounded to rounded, is not convinced that the material could have been derived from the continental shelf offshore. However, Buchanan (1958) and Bruckner and Morgan (1964) are of the opinion that as sand could be found as far as the edge of the continental shelf, most of the sediments for the barrier bars in the area studied by Galliers were derived from the continental shelf. Guilcher (1959) before then even provided evidence for an offshore source for the barrier bar sands of Benin Republic.

Tricart (1972) believes that most of the sand was derived from the weathering of exposed continental shelf. According to him, during the Wurm glacial advance when the sea level was -100 m than now, a large part of the continental shelf in the humid tropics was exposed and subjected to intense chemical weathering 1–2 m deep. During the subsequent transgression, the ocean progressively invaded these plains. The rising sea removed the weathered material, dispersed the clay and piled up the sand into large offshore bars which are well developed, especially where the rocks are rich in quartz (granites, gneisses, mica schists and so on), as along the Guinea coast or in eastern Brazil.

Proponents of the growth of spits to form barrier bars believe that these spits are formed by longshore drift which move sand parallel to the coast. The movement of sand follows the line of breakers instead of the water's margin, so that the sediment is built into a continuous outlying ridge at some distance from the water's edge. Webb (1958) is of the opinion that the barrier beaches along the Nigerian coast resulted from longshore drift of sand from west to east, more or less at a fixed angle of 45° to the dominant south-west wind, so that they developed as spits attached to the basement rocks far away in south-western Ghana. According to Bird (1968), the recurved part of the bars are indicative of stages in the growth of the spits. Usoroh (1977) used the evidence of the recurved sand ridges and spits at Okoun-Seme, Monebego, Houtakpevi and Glogbo in Benin Republic and at Ganyingbo Beach near Badagry to suggest that the barrier beaches along the coast of south-western Nigeria originated as spits, and because spits grow relatively rapidly, series of them can develop over a short geological period.

Many authors are convinced that the speed at which spits grow may explain why sandy beaches and barrier bars are common along inter-tropical coastal plains. Examples of such fast-growing spits include the Langue de Barbaries, an enormous sand spit parallel to the coast at the mouth of the Senegal river deflecting the river southwards. It is about 0.5 km wide and grows at a mean rate of 100 m annually. It has increased by 12 km in the last 70 years. The Lobito spit is another example. It advances at the mean annual rate of 20 m as a result of the annual northward drift of more than 25 000 m^3 of sand (Buckle, 1978, p. 212). The spit at Surinam advanced at a mean annual rate of 62 m, advancing for 1 900 m in 30 years. Other examples include the Ras Luale, a spit

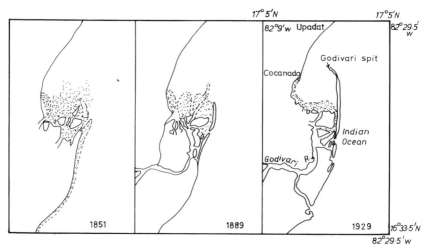

Fig. 12.6 The development of the Godivari spit, east coast of India (after King, 1972)

50 km north of Dar-es-Salaam on the Tanzania coast; the Banjul spit in Gambia; and the Godivari spit on the eastern coast of India, about 0.6 km wide, and growing at a mean rate of 160 m per annum (Figure 12.6).

Spits invariably shut off small bays, as at Copa Cabana, Ipanema and Niteroi in Brazil.

Where the spits link the coast with a former offshore island, it is described as a *tombolo*. Examples include the Ras Hafun tombolo in Somalia, linking the mainland near the town of Hordio to the island of Ras Hafun; and the Lumley Beach tombolo in Sierra Leone, which joins the mainland at Lumley to Aberdeen Island.

Swift (1968) however, is of the opinion that the two main theories of barrier beach formation are not mutually exclusive, as both longshore movement and shoreward movement of sediment can play a part in barrier formation. He believes that a barrier bar can form wherever there is more sediment arriving in an area than leaving it, whether the sea level is rising, stationary or falling, provided the conditions are favourable.

Whatever their origin(s), beaches on the sand bars often display features of geomorphic interest, the most important of which are beach cusps and beach rocks.

Beach cusps consist of a series of low mounds separated by crescentic-shaped troughs spaced at more or less regular intervals along the beach face. Such beach cusps have been observed on the sand bar adjacent to the Fosu Lagoon at Cape Coast in Ghana, where the cusps occur for 10 km along the beach. They have also been observed south of Cape Lopez near Port Gentil, Gabon.

Beach rocks, however, are composed of quartz sand, pebbles and sometimes rock debris and conglomerate cemented together by $CaCO_3$, most especially aragonite (Driscoll and Hopley, 1968). They are found in tropical and sub-tropical environments, especially in areas with alternating wet and dry seasons. They are very common on the eastern coast of Brazil, the West Indies especially in Barbados, the Bahamas and Antigua, in Florida, and the Townsville area, Queensland. They have also been described near Hong Kong (Tschang Hsi Lin, 1962).

They vary in thickness, between 0.3–0.5 m where the tidal range is small, but about 3 m where the tidal range is large, as at Puerto Rico (Tricart, 1972). In some places, as in Townsville area, they can occur at several levels. They have been recognized at heights of 7.4 m, 4.3 m and 2.9 m above Townsville Tide Datum (Driscoll and Hopley, 1968). Where exposed, they are subject to chemical weathering and erosion so that they are perforated by groves and sinuous cavities. They can also be eroded into ridge-and-furrow features, forming micro-cuestas facing inland.

The origin of beach rocks is not well known. The main argument over their formation has concerned the origin of the $CaCO_3$ cement. Whereas authors like Gardiner (1903, 1930), Kuenen (1933), Daly (1924) and Ginsberg (1953) believe that the $CaCO_3$ is precipitated between tide levels from sea water supersaturated with $CaCO_3$, others, like Field (1919) and Russell (1959), consider that the cementing matrix is precipitated from groundwater at about the level of the water table following solution of the $CaCO_3$ from the calcareous upper beach sands. Russell (1962) states that 'the seepage of fresh groundwater into calcareous materials of the beaches would cause some dissolution [of $CaCO_3$] and precipitation'. Guilcher (1961) proposes physico-chemical explanations involving a combination of evaporation of water in the beach at low tide, and the decomposition of $CaCO_3$ associated with organic compounds in shells which would dissolve some calcium that will precipitate under favourable conditions.

Tricart (1972), however, is of the opinion that most beach rocks are relict, having formed during the dry climatic period corresponding with the Dunkirkian regression.

Whatever may be their formational processes, beach rocks are important on tropical beaches, especially in effectively breaking the surf and protecting the upper beach from erosion except at high tides during onshore storms. As observed by Tricart (1972), the rocks effectively prevent wave erosion along the coast of Bahia, but where absent, the coastline has retreated 100–300 m to form concentric baylets.

Mangrove coasts

Extensive mangrove flats can develop along tropical coasts where the following conditions are satisfied: (1) high tidal range; (2) a large supply of detrital sediments especially from the hinterland; (3) low coastal relief with extensive continental shelf; and (4) low coastal wave energy to prevent a rapid dispersal of the sediments.

Mangrove coasts are usually characterized by minor sandy deposits near the sea which may form a chain of islands behind which are myriads of anastomosing creeks and wide expanses of tidal flats colonized by different types of mangroves. Examples of such coasts abound all over the humid tropics, especially along the northern and southern coast of Borneo; eastern Sumatra; the south-western coast of New Guinea; the eastern and western coasts of Johore, Malaysia; the Quan Long coastal area of south Vietnam; western Burma; the Philippines; the Kyankpyr islands; between Chinde and Beira in Mozambique; the eastern coast of Yucatan and Belize; the delta of the Orinoco in Venezuela; the Pacific coast of Colombia in the Guayaquil estuary of Ecuador; and on either side of the Niger Delta, especially between Bonny river and the mouth of the Imo. In fact, most tropical deltaic coasts are vast swamps colonized by mangroves between which the distributaries meander .to reach the sea (Figure 12.1). The example of south-eastern Nigeria is described below.

As in the Lagos area, close to the sea are series of barrier bars which occur in the form of low ridges trending W–E and WSW–ENE, and divided into separate bodies by the Andoni river (Figure 12.7). The sand ridges are separated by series of narrow linear depressions, most of which are colonized by *Rhizophora racemosa*. The sand ridges are so narrow that as many as eighteen could be recognized on 1-km transect drawn on 1:40 000 aerial photograph south of Finima village near Bonny (Jeje, 1978). At an elevation of 3 m a.s.l., these sand bars are higher than the extensive mudflats behind them. They are, however, cut off from the mudflat by the Opobo Creek to form a chain of islands fronting the mudflats.

The extensive mudflats are separated from the coastal plains by a low scarp about 5 m high. The mudflats, which are virtually at sea level, are subject to daily tidal inundations. At low tides, the flats stand out about 1 m above the level of the creeks, while at high tides they become a mass of water hardly distinguishable from the creeks. The flats are colonized by stunted *Conocarpus*, *Rhizophora racemosa*, less than 5 m high and *Avicennia*, which can survive inter-tidal salt-water immersion. These trees are specially adapted to the mudflat. The adaptations involve specialized root structure, especially prop roots and aerial stilt roots that grow downwards from the upper branches, and buttressed trunks to support the trees in soft mud, pneumatophores to permit oxygen to reach the roots buried in the anoxic mud, and viviparous seedlings that remain attached to the parent tree until they have developed leaves and a proto root (Bloom, 1978, p. 464). These roots and the spike-like pneumatophores aid in entrapping sediment and thus in the aggradation of the mangrove swamps.

The swamps are underlain by dark brown peat to a very great depth. The peat is usually composed of sandy to clayey, slimy mud, mixed with decayed and partially decayed organic debris. The local communities cut, dry, and use the peat to build houses. The peats, which constantly ooze water, are poorly consolidated, so that at low tides they are subject to basal erosion by the creeks.

Associated with the mudflats and swamps are the numerous creeks through which the tides invade the mudflats. The creeks often exhibit a rather intricate pattern of orientation, meandering, reticulation and anabranching. They vary in size from minor ones a few metres wide to Bonny and Andoni creeks, each 3–5 km wide. Some, like the Oroberekiri, Bonny and Ogu, have levees along their channels.

Fig. 12.7 Morphology of the coastal zone around Badagry, Nigeria ➝

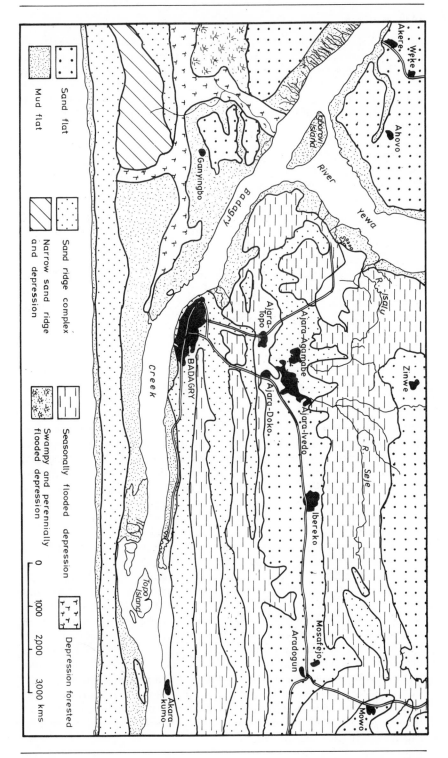

Delta coasts

These consist of alluvial deposits at the mouths of rivers. As the different types are already described and their mode of formation explained in Chapter 5, it is unnecessary to describe them further here. It will be sufficient to recall that in Chapter 5, four types of deltas were recognized: namely, the arcuate, the bird's-foot, the cuspate and the estuarine types. Most of these are characterized by extensive tidal mudflats and colonized by mangroves.

Examples of the arcuate type include the Niger Delta, the deltas of the Manhada and Godavari in eastern India, the delta of the Mangoky in the western part of the Republic of Malagasy and the delta of the Orinoco in Venezuela. The best-known example of the bird's-foot type in the humid tropics is the Krishna Delta in eastern India. Examples of the cuspate type include the Manberamo Delta in New Guinea, the Coco Delta between Honduras and Nicaragua, and the delta of the Río Magdalena in northern Colombia. Estuarine types appear to be very common all over the humid tropics. Some of them include the deltas of the Ord, Victoria and Fitzroy in north-western Australia; the Aly Delta in New Guinea; the Kampar in Sumatra; the Mentarang in Borneo; the Irrawaddy in Burma; the Mekong in south Vietnam; the Congo Delta in Zaire; the Betsiboka in the Republic of Malagasy; the Amazon and the Rio do Para in Brazil.

Submerged coasts

Major changes in sea level since the Pleistocene period have been examined earlier, and as demonstrated by Fairbridge (1961) the present period is marked by eustatic rise in sea level in most places. Shepherd (1973) is also of the opinion that most coasts are now submerged due to the paucity of ice cover all over the earth's surface. Evidence for submergence include cliffs and wave-cut platforms preserved at depths 8–13 m below present sea level (Zenkovitch, 1967, p. 537), submerged coastal fortresses and forests (Monkhouse, 1972).

Distinctions are usually made between submerged upland and submerged lowlands; the former are characterized by rias and longitudinal types of coast, whereas the latter are typified by broad, shallow estuaries, marshes and mudflats uncovered at low tide and several anastomosing creeks. Where ridges and their intervening valleys run at an angle to the coast, submergence leads to the formation of ria coasts. The valleys form rias while the ridges form headlands, so that the coastline is highly indented. In the same general area erosional residuals form islands. Examples of ria coasts in the humid tropics include those between King Sound and Cape Londonderry, north-western Australia, the western coast of Sierra Leone and that of Guinea Bissau, the coast of East Africa between the harbours of Mombasa and Dar-es-Salaam, and between Lurio and Maaia in Mozambique, north-west Malagasy between Cap d'Ambre and Cap St André.

The coastline of Sierra Leone, especially the lower valleys of the Sherbro and the Rokel rivers, constitutes among the best examples along the West African coast. The Sherbro trench at the mouth of the ria is about 30 m deep, decreasing to 10 m at the head of the ria. Indeed, Sierra Leone perhaps has the best natural harbour along the West African coast.

In some places the trend of the hills and ridges runs roughly parallel to the coast, in which case submergence will lead to the flooding of the valleys while the hills and ridges form long, narrow islands. The best example of this type of coastline in the humid tropics is at Yampi Sound in north-western Australia, where the general trend of the shore runs parallel to the structural grain in NW–SE orientation (Twidale, 1976).

Finally, the submergence of lowlands leads to the rising of the base level, which encourages deposition by rivers. Submerged lowlands are thus typified by broad, shallow estuaries, marshes, mudflats with several anastomosing creeks, the types of terrain described between Bonny and Imo rivers in south-eastern Nigeria. Pugh (1954b) is of the opinion that the entire Nigerian coastline represents a submerged lowland as evidenced by the wide estuaries of the Yewa and Owo rivers west of Lagos, the absence of deltas at the mouth of the Ogun and Osun rivers in the Lagos Lagoon, the extensive mudflat behind the sand bars in the Niger Delta and the wide estuary of the Cross river.

Emerged coastlines

Emerged coastline occurs where the land has risen relative to the sea or the sea level has fallen. Such coastlines are relatively rare because of the recent post-glacial world-wide rise in sea level. The chief features of emerged uplands are raised beaches, whereas those of raised lowlands are coastal plains.

Raised beaches consist of raised platforms backed by old cliffs. Platforms can occur at various levels. In fact Twidale (1976, p. 382) remarked that 'in many tropical regions particularly in the Pacific and Australasia regions, a multiplicity of narrow platforms is displayed within and relatively close to the tidal zone'.

Examples of such raised platforms covered by beach material are at Pumpone, west of Takoradi: and at Moree near Cape Coast. At Senya Beraku, 35 km west of Accra, three distinct fluvio-marine terraces comprised of pebbles and gravels can be recognized. The highest, at 18–20 m a.s.l., can be traced eastwards to Fete. Two lower beaches exist at 3 m and 1 m respectively above the level of Senya Beraku. A 12 m terrace has been traced in Accra. Two sets of raised beaches, at between 40–50 m and 10–12 m a.s.l. have been recognized around the Sierra Leone peninsula, especially around Hastings and Waterloo, where the beaches are developed on Tertiary and Quaternary sediments. The beaches are backed by relatively undissected cliffs. The towns of Freetown, Hastings and Waterloo are located on the upper beach. Other examples include those along the coasts of Kenya and Tanzania, such as near Bagamoyo, where three beaches have been identified at 2 m, 8 m and 12 m a.s.l. (Buckle, 1978), and those along the coast of Cameroon between the Sanaga river and Kribi at 2.5 m and 5.5 m a.s.l. (Kadomura, 1977).

Emerged lowlands are produced by the uplift of the neighbouring continental shelf, and can be very extensive where the continental shelf is relatively very wide. The lowland plains are often underlain by sand, gravel, clay and calcareous materials. The seaward side of the plain is marked by offshore lagoons, tidal flats, sandspits and sandy beaches. The classical example of such a plain is the coastal plain of south-east USA. However, examples in the tropics include the emerged northern shore of the Gulf of Mexico (Monkhouse, 1972), and the flat, low-lying plain of Mozambique extending for about 1 500 km from near Maputo to Nacala.

Rocky coasts

Rocky coasts are uncommon in the humid tropics. As observed by Tricart (1972), most high coasts rarely continue to the sea as they are usually fronted by beaches which preclude the formation of cliffs. Most rocky coasts are characterized by cliffs and other erosional features. Cliffs have been observed to develop on different types of rocks at several locations. The cliffs along the coast of western Ivory Coast, and south-west Ghana in West Africa, and along the eastern coat of Brazil are developed on gneisses and granites, whereas those at Madeleine Bay, Cape Verde Island, are developed on Eocene sandstones. Cliffs are also common on basalts, as evidenced at Cape Manuel, Fann Point, and the island of Goree along the coast of Senegal. They are also common on limestone, as shown by the limestone cliffs in south-west Malagasy, and the West Indies, as in Barbados. They have also been observed on bedded clays and sandstone in Ivory Coast.

Tricart (1972) recognized two major types of cliffs in the humid tropics:
1 vegetated sea cliffs;
2 bare sea cliffs.

Vegetated sea cliffs

The vegetated types are common where heavy, all-year-round rainfalls wash away the salt preventing it from playing a harmful ecological role. Examples of such vegetated cliffs can be found between Monrovia and Sassandra in West Africa, in El Salvador, and between Rio de Janeiro and Santos in Brazil. The steep cliffs are covered by rainforest except at the lowest few metres where they are bare, being exposed to sea sprays and waves. Above the bare part is a row of shrubs and spiny rattan palms.

The cliffs are in many places subjected to intense chemical weathering, leading to the formation of a thick regolith covered by dense vegetation. Undercutting of the weathered cliffs by waves and the formation of deep notches can undermine the regolith, leading to landslides and slumping that can extend upward to expose bare rocks across the surface of the cliff (Tricart, 1972). Apart from these catastrophic movements, there is also a continuous slow movement of masses of the weathered regolith towards the sea. As the regolith accumulates at the base of the cliffs, the sand and clay fractions are removed by waves to leave piles of boulders whose rate of disintegration depends on the nature of the rocks.

Bare sea cliffs

These are far more common than the vegetated types. They are especially common in areas of alternating wet and dry seasons where salt sprays prevent vegetal growth, especially where onshore winds prevail in the dry season, as at Cabo Frio, Brazil, or Cape Coast, Ghana. Limestones also tend to form bare cliffs in the humid tropics.

The morphology of bare cliffs varies with the nature of the rocks, especially their stratification, jointing, resistance to erosion, homogeneity, presence of bands of weakness (for example, fault or shatter beds), and the dominant geomorphic processes. Thus blocky or bouldery cliffs characterize exposures of igneous rocks, although sheet structure may be developed depending on the nature of jointing (Figure 12.8). Limestone cliffs are usually nearly vertical. It is generally held that the morphology of cliffs developed on bedded sedimentary rocks is affected to a large extent by the dip of the rocks. Where the dip is seawards, this facilitates the movement of rock masses down the bedding plane so that the cliff is rarely vertical. On the other hand, where the rock strata are horizontally bedded or dip inland, very steep, near-vertical-to-overhanging cliffs are common. Such cliffs may also display distinct benches at different levels. However, as observed by Small (1970),

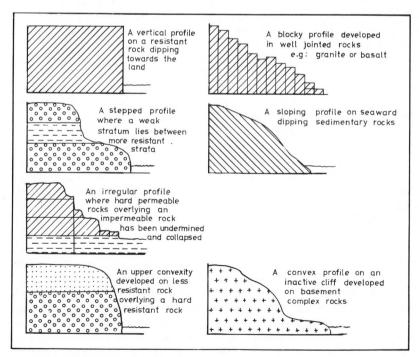

Fig. 12.8 Sea cliff morphology

exceptions can always be found to these simple rules, as cliff profiles and morphology are influenced by several other factors, especially sub-aerial geomorphological processes.

Small (1970) thus proposed that in studying cliffs, distinction should be made between 'active' and 'inactive' cliffs. The former, experiencing wave erosion at their base and sub-aerial erosion such as wash, gullying and slumping, are retreating backwards whereas the latter are isolated by sand or marsh deposits or by raised platforms. Active cliffs are thus characterized by free faces whilst inactive types, subject mainly to sub-aerial weathering and erosion, become comparatively gentle or even convex (Savigear, 1952).

Most rocky bare cliffs are characterized by several features, among which are corrosion etches or honeycomb pittings, lapiés in limestones, notches, pools, structural benches, and others like geos, caves, arches, stacks and so on.

Honeycombs consisting of micro-pits 3 cm wide and 3 cm deep have been observed on different types of rocks – for example, gneisses and arkose in Australia (Twidale, 1976, pp. 376–77), and mica schists in Rio de Janeiro, Brazil. They occur at a height of 5–6 m above the high water mark. Tricart (1972) attributes the formation of this feature on mica schists to the weathering of mica and the oxidation of other minerals in the rocks.

Coastal lapiés are common on bare limestones especially where the cliffs are not too high so that sea sprays can reach them. The features have also been observed on aeolianites. Examples of coastal lapiés are found on Neogene and Quaternary limestone at College Savannah, Barbados, in the West Indies, Bermuda, and on the western coast of the Malagasy Republic.

Several explanations have been proffered for the formation of this feature. As regards the formation of lapiés on aeolianites, Twidale (1976, p. 375) states that 'many rocks contain minerals which are vulnerable to solutional attack by sea water. Aeolianite, for instance, which is widely exposed on the coasts of southern Australia, consists of lime plus quartz, as sea water is not saturated in silica, solution is possible causing the rocks to crumble.'

Where the lapiés develop on limestones, the early stages consist of a profusion of small depressions and the formation of irregular hollows which gradually enlarge and join to form wide pits with sharp, steep edges.

Notches are common at the bases of cliffs especially where unprotected by beaches. In basement rocks like gneisses, granites, mica schists and so on, they result from the joint operation of hydraulic action and abrasion concentrated at the points of weakness at the base of the cliff – that is, along horizontal joint planes. As hydraulic action is especially effective in dislodging rock fragments from shattered zones, notches can originate as individual hollows or cavities located at the same level along the cliff base (Zenkovich, 1967). These hollows may reach a depth of 1 m or more. Eventually they may link up to produce a continuous notch. The process may be relatively fast where the rocks are highly jointed. However, where the rocks show no planes of weakness or exhibit any

366

heterogeneity, and the sea offshore is deep, notches may be completely absent.

Notches are common at the base of limestone cliffs. The base of the 100-m limestone cliff in south-west Malagasy is marked by an almost continuous notch up to 5 m deep and about 4 m high with a marked overhang. In Barbados, the notch at the base of the limestone cliff is 1–5 m deep, 0.5–3 m high. As wave energy is low in all these places, the notches have resulted from concentrated sea-water solution (Buckle, 1978). As notches become deeper, the upper parts become unstable and collapse. The fallen debris is broken down and removed by waves. This leads to a gradual retreat of the cliff.

Pools or *gnammas* a few metres wide and less than a metre deep often characterize rocky coasts, developing most especially on limestones at inter-tidal levels. Such pools, located at different levels on the rocky coasts, are invariably backed landward by steep scarps. Examples abound along the coast of Cape Verde Island, and at Cabo Frio, Brazil, where the pools have developed on mica schists. They have also been observed on the Hawaiian coasts at 1 m and 3–4 m a.s.l. (Bloom, 1978, p. 448).

The origin of pools is not clearly known. With regard to those developed on limestone, Twidale (1976, p. 375) attributing the formation to pool weathering writes as follows:

> In humid tropics, for instance, where rain falls almost daily (or in areas where there are many wet days), rain water is probably important in the solution of limestone. Because it is not saturated with lime, it is capable of taking the bicarbonate into solution. In addition it tends to remain as a distinct layer near the surface of the pool because of its lower density. Most weathering therefore occurs at a layer near the pool surface: hence what is called here *pool weathering*.

Bloom (1978) also emphasizes the importance of *water level weathering* in the formation of pools, even on non-carbonate rocks. Since the pools are inter-tidally located, they are subjected to alternations of wetting and drying – wetting at high tides, drying at low tides – a phenomenon which promotes chemical weathering. According to Tricart (1972), as a result of wetting and drying, minerals like biotite swell and cleave in tiny sheets causing granular disintegration, so that rocks rich in mica, such as mica schists, readily form pools.

Pools can coalesce to form solution benches or shore platforms.

Coastal platforms which occur between low and high tide levels consist of a wide, irregular surface of bare rocks sloping gently seawards with an overall concave upward profile. However, in detail, the surface of the platform is not smooth, as it may contain irregularities in the form of potholes or minor ridges resulting from differential erosion across steeply dipping strata from which the weaker rocks are etched out. On well-jointed rocks like granite, sheet structure may be exploited and the platform strewn with blocks and boulders.

According to Russell (1963), such platforms are not common in the humid tropics as the coasts exhibit negligible cliffing and as most of the cliffs descend almost vertically into the sea. However, such platforms have been observed around the coast of Oahu in the Hawaiian islands and on the Devonian sandstone seawards from the Castle Headland, and at several other locations around Accra, Ghana.

Coastal platforms include solution benches and wave-cut platforms. Wentworth (1938) distinguished four processes leading to the formation of coastal platforms, namely: (1) water level (water layer weathering), (2) solution bench, (3) ramp abrasion, and (4) wave quarrying. Opinions vary as to which of these processes or a combination of them are most vital to the formation of the platform.

Water layer weathering and solution have been shown to be important in the formation of pools which can grow and coalesce to form solution benches. However, most authors (King, 1972, p. 435–54) are of the opinion that most platforms result from cliff recession, which is heralded by the development of basal notches through the processes of abrasion and hydraulic action, followed by slumping of the undermined rocks and the removal of the rock debris by waves. Such developing platforms are covered by water at high tide but exposed at low tide. The platform may be covered by a layer of sand 15–30 cm thick which is moved to and fro across the platform by waves, and thus help to abrade the platform. The cliff behind the platform is subject to strong wave attack, mainly during the periods of storm when the sea level is raised by onshore winds. Such storms help to clear accumulating debris from the cliff base and expose it to further erosion.

Some rocky coasts in the humid tropics are characterized by erosional features such as geos, caves, blow holes, arches and stacks. However, all of these may not be present at any given location.

A *geo* is a narrow inlet walled in by steep cliffs developed by wave erosion, especially hydraulic action, along shatter zones or where the rocks are severely fissured.

Caves are also formed by wave action on the cliff base along zones of weakness, such as severely jointed or faulted zones, or along solution-widened bedding planes in limestones. They are found more frequently on headlands due to wave refraction. The most important process of cave formation involves hydraulic action and cavitation. As is well known, the impact of breaking waves exerts enormous pressure on the rocks. As shown by King (1972, p. 451), for a storm wave 3.05 m high and 45.8 m long, the observed pressure on the rocks is 0.593 kg/cm^2. The pressures generated by a succession of such waves are quite sufficient to cause extensive damage to cliffs, especially where air is compressed into the fissures in the rocks. When the water recedes, the air expands rapidly. Repeated expansion and compression loosens rock blocks and produce hollows or caves in the fissured zones. Examples of caves formed in hard basement rocks are common near Senya Beraku west of Accra, and those formed in limestones are common around Barbados.

The force of waves surging into some caves can weaken the joint blocks in the cave roof, leading to the loosening of the blocks and roof collapse. The resulting feature is a vertical shaft known as a *blow hole*. A good example is the one developed in basalt on the south coast of Réunion Island – *Le Souffleur* – located 3.5 km south-east of Savannah.

If caves develop on either side of a projecting narrow headland, the caves can be eroded back until they coalesce to form a feature referred to as an *arch*. Such arches are also common near Senya Beraku, where caves have developed along bedding planes in massive sandstone. Another example is the *Bogenfels*, a natural arch developed in Palaeozoic dolomite on the Namibia coast, south of Lüderitz (Buckle, 1978).

When the arch collapses, the island separated from the truncated headland is referred to as a *stack*. Buckle (1978, p. 205) describes the evolution of a stack in the Nkontompo headland located between Sekondi and Takoradi in Ghana between 1960 and 1967. The main process of cliff erosion started in 1963, and by 1965 the Nkontompo stack was created from the Nkontompo headland. When stacks are eroded and considerably reduced in height, barely protruding above the high water mark, it is referred to as a *stump*.

Coral coasts

Corals are bottom-dwelling, sessile, marine organisms. Some are solitary individuals, but the majority, especially the anthozoan corals, live and grow in colonies. The principal contemporary reef builders are the Madreporian (or Scleractinian) colonial corals, of the class Zoantharia, phylum Coelenterata. They secrete external skeletons of $CaCO_3$ which, together with the shells of Porifera, Echinoderms, Mollusca and Crustacea, constitute the major reef material. The spaces within and between these are filled by the accumulation of the skeletons of foraminifera. However, the most important binding and reinforcing agents are the coralline algae, which secrete an external calcareous coating or skeleton that bind the reefs together, especially on the outer margins of reefs and on reef crests.

Coral formations appear restricted to tropical seas, as the organisms thrive best in waters whose temperature is never below 18°C. Here, they are not found on western sides of tropical continents owing to the effects of cold currents. As they require light, they thrive at shallow depths, usually less than 25 m. They also require well-oxygenated water free from mud, so that the most favourable environment for coral growth is the outer margin where the water is more oxygenated and nutritious, but where waves break off several fragments which are cemented together to form a constructional slope.

Coral reefs can be classified according to their morphology. The main types include the fringing reefs, the platform or patch reefs, the barrier reefs and the reef islands (Figure 12.9). Other types include the apron reef (an embrayo fringing reef), and the table reef (small, open-ocean reef without central lagoons, King, 1972, p. 394). The fringing reefs develop just off the shoreline, and may be separated from it by a very narrow, shallow lagoon. Examples of fringing reefs are common along the coast of Kenya, Tanzania, the western side of the Republic of Malagasy, and along the coast of Queensland from Cape York to Cairns and Cape Flattery to Cape Melville.

The platform of patch reefs, which are generally ovoid in ground plan, are found on continental shelves as simple, low, reef patches about

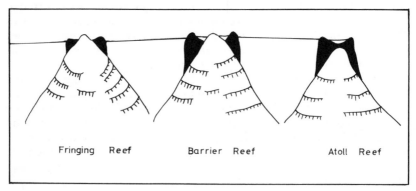

Fig. 12.9 Main types of coral reef

100 m across. Where exposed to strong winds, sand trains may partially cover the reef. In some cases, the sand trains may grow seawards to form twin horns which form a horseshoe-shaped structure, the interior of which then contains a shallow, small lagoon which is gradually filled with sediment. Examples of patch reefs abound between the coastline of Queensland and the outer barrier reef.

Barrier reefs consist of discontinuous, long ribbon reefs developed on the continental shelf but separated from the land by a wide lagoon. They are thus referred to as epicontinental reefs. The greatest barrier reef in the world is the Great Barrier Reef off the coast of Queensland, which extends for a distance of about 1 920 km from the Gulf of Papua to near Sandy Cape. The reefs comprise of a multitude of discontinuous ribbon or linear reefs of varying sizes, separated by passages, together with a vast array of other reefs – for instance, patch reefs, pseudo atolls and reef islands (Steers, 1929, 1937). The longest of the ribbon reefs can be found betwen Murray Islands (latitude 9° 58′S) and Trinity Opening (latitude 16° 20′S). It is about 24 km long and 1–1.5 km wide. According to Fairbridge (1950), the passages between the ribbon reefs represent Pleistocene stream valleys that crossed the continental shelf before post-glacial eustatic rise. As such passages now experience strong tidal currents, they are subjected to tidal scour and are thus kept open. Barrier reefs are strongly asymmetric as the inner part is usually gently sloping with the sediment wedge dotted by small reef patches, pinnacles and coral heads, while the outer edge drops abruptly seawards to depth of more than 1 000 m. The exterior zone of wave action is often marked by the growth of calcareous algae to several metres above the normal reef surface to form an algal ridge highly resistant to wave attack, which thus protects the reef from wave erosion (King, 1972). However, this outer face is diversified by numerous grooves, chutes and surge channels. Other examples of barrier reefs include those off New Caledonia, Borneo, Palau Islands, Mayotte, one of the four Comoro Islands, and Mikea in the Republic of Malagasy.

The reef islands are of three basic types: coral islands, continental islands and atolls.

Coral islands abound between the coastline of Queensland and the outer barrier reef. Fairbridge (1950) identified five main types: (1) simple sand cays (unvegetated); (2) vegetated sand cays; (3) shingle cays (with or without vegetation); (4) sand cays with shingle ramparts; and (5) enlarged reef islands.

Type (1) consist of a simple accumulation of loose coral sand and beach rock, and are generally located on the lee side of a coral platform. Examples include Sudbury, Undine, Mackay, Pickersgill and Chapman reefs, all off the coast of Queensland. Type (2) are similar to (1) but larger, more matured and covered by well-established flora. The island is fringed by beach rocks, which protect it from erosion during tropical cyclones. Examples include the Capricorn group of islands off Rockhampton in Queensland. Type (3) are similar to type (2), except that it is composed of large coral debris and is usually situated in the windward side of platform reefs. The debris are often cemented to form coral breccias. Examples abound in the Capricorn and Bunker group of islands, expecially Lady Musgrave and One Tree Islands of Rockhampton, Queensland. Type (4) are like type (3), except that they are restricted to the lee of barrier reefs where they are readily colonized by mangroves. Organic matter from the trees aid in the intense decay of the coral shingles and reef limestone to form swampy depressions. Examples include Bee Reef, Low Isles, Ripon, Berwick and Hope Islands off Queensland. Type (5) consist of older emerged coral reef with or without fringe of recent sand or shingle beach ridges. They are vegetated and stable. The best example is Raine Island (on latitude 125°S) off Queensland.

Continental islands are island reefs with cores of mainland rocks. Such rocks may be intrusive igneous types or occasionally gneisses. The islands vary a great deal in size. An example is Hinchinbrook Island, 36 km long, 8–15 km wide with a core of granite off the coastal stretch between the towns of Tully and Townsville, Queensland, Australia.

Atolls are horseshoe-shaped reefs enclosing lagoons (Figure 12.10). Three main types have been recognized (Fairbridge, 1968a), namely: shelf atolls, compound atolls and oceanic atolls. Shelf atolls, which are usually perfectly circular and smoothly rounded and found

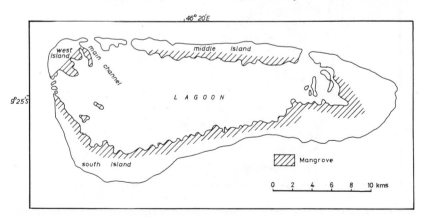

Fig. 12.10 Aldabra atolls (after Buckle, 1978)

371

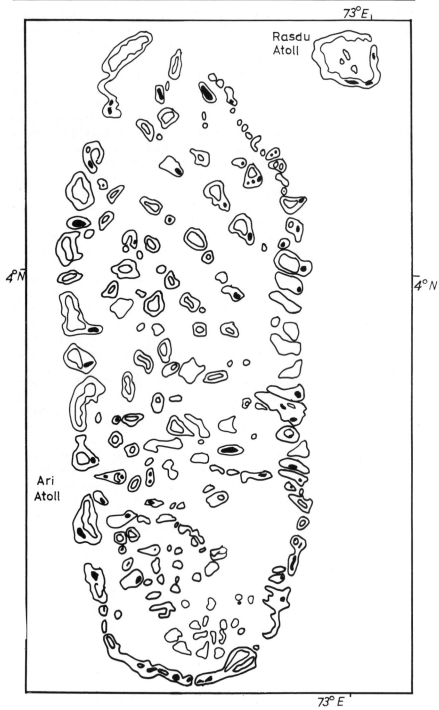

Fig. 12.11 A compound atoll, Ari atoll, Maldive Islands (after Bloom, 1978). Land areas, shaded in black, are very small

mainly off the north-eastern Australian continental shelf and off Borneo, have no volcanic foundations, having developed as open platforms or from earlier platform reefs. Several rise from depths of 400–500 m off the coast of Queensland. Compound atolls (Figure 12.11) are reefs with complex or compound tops consisting of several ring reefs (called *faros* in the Maldives), which grow upward from an initial annular foundation. These types of atolls are often referred to as the quasicratonic reefs, as their foundation resulted from *en bloc* subsidence of former continental crust (Fairbridge, 1968c). Examples include those in the South China Sea, especially the Tiger Islands off Indonesia, the Maldives and Laccadives groups of islands west of Sri Lanka, and the Bahamian block. Oceanic atolls are those in the mid-Pacific which rise from well-defined series of seamounts or guyots with up to 2 000 m of accumulated reef growth. These are often regarded as the true atolls, as they are believed to have resulted from subsidence (Fairbridge, 1968c). Examples include Fiji, Samoa, Gilbert, Ellice, Marshall, Fanning, Christmas, Phoenix, Cook and Tuamotu islands.

Various theories have been advanced to explain the formation of the different types of reefs. There is no problem in explaining the growth of fringing reefs. They simply grow seaward from the land, and in a low-energy environment such as the humid tropics, seaward reef construction can occur at a relatively fast rate (Tricart, 1972).

Two main theories have been advanced to explain the origin of barrier reefs and atolls: the subsidence theory, first advanced by Darwin (1837); and the non-subsidence theory espoused by Murray (1880). However, King (1972) suggests that in discussing the origin of these features distinctions should be made between structural coral reefs in which corals are actively contributing to the topographic development of the reef and coral communities which are an assemblage of reef organisms growing on a substrata that is other than their own.

The Darwin theory of subsidence with regard to the formation of barrier reefs and atolls is so well known as to need no reiteration. The main points of the theory are that (1) corals grow on volcanic islands as fringing reefs; (2) the islands subside gradually; (3) the corals grow upward and outward at the same rate as the islands are subsiding; (4) when the central parts of the islands are still visible, barrier reefs are prominent, turning to atolls when the islands completely subside.

Geophysical investigations and deep boring in the Marshall Islands and Eniwetok Atoll, where reef limestone was penetrated to depths of 1 450 m; in Tuamotu Islands (Ranson, 1955), where the basement rocks were encountered at more than 1 000 m; and in the Bahamas, where 4 488 m of reef limestone was penetrated on Andros Island, seem to confirm Darwin's theory. Based on the results of these investigations, King (1972, p. 393) suggests that since reef limestone is shallow-water material, those drilled date to Eocene period. She suggests rates of subsidence of 51.9 m, 39.6 m and 15.2 m/million years for the Eocene, Miocene and post-Miocene respectively.

However, the subsidence theory seems to apply only to the open Pacific Ocean, because elsewhere borings show that the reefs are very thin. For instance, at Bermuda, lava was encountered at depths varying from 171 m to about 20 m, while the thickness of the reef limestone in the Great Barrier Reef has been recorded at 115 m, 154 m, 121 m, 547 m and 225 m at various sites (King, 1972, p. 394). All these tend to confirm Murray's theory that in some cases, barrier reefs and stolls are formed on submarine hills or plateaux rising within a few metres from the surface.

A serious problem in studying the development of reefs is the determination of how much of the reef is due to contemporary reef-building activities and how much is inherited from earlier reef structures. There is no doubt that most modern reefs date to the last major interglacial period, and these have been affected by the various changes in sea level. During periods of low sea levels, older reefs were eroded, but with subsequent rise in sea level, reef growth began on the old foundation, so that features both recent and of older origin may be seen in coral reefs. As shown by Tricart (1972), Quaternary sea-level fluctuations have affected the development of fringing reefs in most places. He notes that during the early Holocene (Dunkirkian – 3 000–1 200 B.P.) post-glacial rise of sea level when the sea level was 1–2 m higher than the present level, reefs were built to a level slightly above that of the present low tide above which the corals cannot live, so that the corals have since died. The dead reefs are now broken into pebbles and sand which in places are deposited as spits and sand bars referred to as 'keys' or 'cays' in the West Indies. However, the corals grow strongly on the seaward side and build out farther and farther on banks of their own debris. The rate of development is still not well known, but according to King (1972), in some places, colony growth may be 1–10 cm/yr while reef growth is 1–2.5 cm/yr.

Conclusion

Most of the features described in this chapter can be observed in other morpho-climatic zones, as they are formed by distinctive geomorphological processes which cut across climatic boundaries, so that to a great extent, coastal processes are azonal. Also, as emphasized in this chapter, both the regional and the detailed form of coastal morphology are to a great extent controlled by the nature of the local rocks. However, there is no doubt that certain climatic parameters such as high temperature and rainfall, prevailing in the humid tropics, are directly or indirectly responsible for the formation of certain features restricted mainly to the tropics. Examples include the vast mangrove swamps, coastal lappies and coral reefs. Thus, to some extent tropical coasts can be examined within the context of climatic morphology.

As shown by Price (1955) and Davies (1964), the humid tropical coasts are low-energy coasts influenced by long, low, non-erosive swells. Thus, whereas the temperate coastal zones influenced by powerful waves generated by the high velocity westerlies are characterized to a great extent by erosional features such as cliffs, wave-cut platforms and so on, and by the deposition of coarse shingles on the beaches, humid tropical coasts under the dominant influence of gentle swells, and whose forelands and continental shelves have been subjected to deep chemical weathering, are characterized by the deposition of vast sand bars, spits and other depositional features composed mainly of sand, silt and mud, which are often colonized by mangroves. The low-energy coastal environment also encourages the proliferation of vast coral colonies which have formed different types of reefs. However, as our knowledge of the influences of climate on the coasts of the humid tropics is rather limited, more research is required into this vital aspect of coastal geomorphology.

References and further reading

Areola, O. (1977) 'Aerial photo-interpretation of land types and terrain conditions in the Lagos Coastal region', *Nig. Geog. J.*, **20** (1), 70–81.

Beaumont, E. de (1845) *Leçons de Géologie Pratique*, Paris.

Bird, E. C. F. (1961) 'The coastal barrier of east Gippslands, Australia', *Geog. J.*, **127**, 460–8.

Bird, E. C. F. (1969) *Coasts*, MIT Press, Cambridge, Mass.

Bloom, A. L. (1965) 'The explanatory description of coasts', *Zeits. für Geomorph.*, **9** (4), 422–36.

Bloom, A. L. (1978) *Geomorphology: a systematic analysis of Late Cenozoic Landforms*, Prentice-Hall, Englewood Cliffs, New Jersey, pp. 437–81.

Bruckner, W. D. and Morgan, H. J. (1964) 'Heavy mineral distribution on the continental shelf off Ghana', in *Developments in Sedimentology*, vol. 1, L.M.J.U. Van Straaten (ed.), Elsevier, Amsterdam.

Buchanan, J. B. (1958) 'The bottom fauna communities across the continental shelf, off Accra, Ghana', *Proc. Zool. Soc.* (London), **130**, 1–43.

Buckle, C. (1978) *Landforms in Africa: an introduction to geomorphology*, Longman, London.

Cotton, C. A. (1952) 'Criteria for the classification of coasts', *17th Intern. Geog. Cong., Abstract of Paper 15*.

Daly, R. A. (1924) 'The geology of American Samoa', *Carnegie Inst. Pub. No. 340*, Washington, D.C., 93–143.

Darwin, C. (1837) 'On certain areas of elevation and subsidence in the Pacific and Indian Ocean as deduced from the study of coral formations', *Geol. Soc. Proc.*, (London, **2**, 552–4.

Davies, J. L. (1964) 'A morphogenetic approach to world shorelines', *Zeits. für Geomorph.*, **8** (sp. ,no), 127–42.

Davies, W. M. (1902) *Elementary Physical Geography*, Ginn, New York.

Driscoll, E. M. and Hopley, D. (1968) 'Beach rock and conglomerate in the Townsville Area', *Br. Geomorph. Res. Group*, Occasional Paper No. 5, 89–96.

Ewing, M. *et al.* (1960) 'Revised estimate of ice volume and sea level lowering', *Bull. Geol. Soc. Amer.*, **71**, 1861.

Fairbridge, R. W. (1950) 'Recent and Pleistocene coral reefs of Australia', *J. Geol.*, **78** (6), 733–5.

Fairbridge, R. W. (1962) 'Eustatic changes in sea level', *Physics and Chemistry of the Earth*, **4**, 99–185.

Fairbridge, R. W. (1968a) 'Littoral processes – an introduction', in *Encyclopedia of Geomorphology*, R. W. Fairbridge (ed.), Reinhold, New York, pp. 658–62.

Fairbridge, R. W. (1968b) 'Limestone coastal weathering', in R. W. Fairbridge (ed.), *op. cit.*
Fairbridge, R. W. (1968c) 'Coral reefs', in R. W. Fairbridge, (ed.), *op. cit.*
Field, R. M. (1919) 'Remarks on beach rock of Dry Tortugas', *Carnegie Inst. Yearbook, No. 18*, 198.
Fournier, F. (1960) *Climat et Erosion: la relation entre l'érosion du sol par l'eau et les précipitations atmospheriques*, Paris.
Galliers, J. A. (1968) 'Barrier beaches and lagoons of the Ghana coasts', *Br. Geomorph. Res. Group*, Occasional Paper No. 5, 77–87.
Gardiner, J. S. (1903) *The Fauna and Geography of the Maldive and Laccadive Archipelagoes*, Cambridge, Univ. Press.
Gardiner, J. S. (1930) 'Studies in coral reefs', *Harvard College M. C. Zoological Bulletin*, **71**, 1–16.
Ginsberg, R. N. (1953) 'Beach rock in south Florida', *J. Sedimentary Petrology*, **23** (2), 82–92.
Guilcher, A. (1959) 'Coastal ridges and marshes and their continental environment near Grand Popo and Ouidah, Dahomey', *2nd Coastal Geogr. Conf. Baton-Rouge*, 189–212.
Guilcher, A. (1961) 'Le "beach-rock" ou gres de plage', *Ann. de Geog.*, **70**, 113–25.
Hodgkin, E. P. (1964) 'Rate of erosion of intertidal limestone', *Zeits. für Geomorph.*, N.P. (4), 385–92.
Inman, D. L. and Brush, B. M. (1973) 'The coastal challenge', *Science*, **181**, 20–32.
Jeje, L. K. (1978) 'Geomorphological features in the western half of the coastal plains of south-eastern Nigeria', *Nig. Geog. J.*, **21** (2), 153–60.
Johnson, D. W. (1919) *Shore Processes and Shoreline Development*, J. Wiley, New York.
Kadomura, H. (1977) *Geomorphological Studies in the Forest and Savanna Areas of Cameroon*, an interim report of the Tropical African Geomorphology Research Project 1975–76, Sapporo, Japan.
Ketchum, B. H. and Tripp, B. W. (eds), (1972) *The Water's Edge: critical problems of the coastal zone*, MIT Press, Cambridge, Mass.
King, C. A. M. (1972) *Beaches and Coasts*, Arnold, London.
Kuenen, P. H. (1933) 'Geology of coral reefs', *Snellius Exped.*, **5** (2), 1–125.
Langbein, W. B. and Schumm, S. A. (1958) 'Yield of sediment in relation to mean annual precipitation', *Trans. Amer. Geophysical Un.*, **39**, 1076–84.
Murray, J. (1880) 'On the structure and origin of coral reefs and islands', *Proc. R. Soc. Edinburgh*, 10, 505–18.
Monkhouse, F. J. (1972) *Principles of Physical Geography*, 7th edn., Univ. of London Press, pp. 280–321.
Neumann, A. C. (1966) 'Observations on coastal erosion in Bermuda and measurements of the boring sponge, *Cliona lampa*', *Limnology, Oceanography*, **11**, 92–108.
Neumann, A. C. (1968) 'Biological erosion of limestone coasts', in *Encyclopedia of Geomorphology*, R. W. Fairbridge (ed.), Reinhold, New York.
NEDECO (1954) *Report on the Western Niger Delta*, The Hague.
Nossin, J. J. (1964) 'Beach ridges on the east coast of Malaya', *J. Tropical Geog.*, **18**, 11–17.
Price, W. A. (1955) 'Correlation of shoreline types with offshore bottom conditions', A. and M. College of Texas, Dept. of Oceanogr. Proj. 63.
Pugh, J. C. (1953) 'The Porto Novo-Badagri sand ridge complex', *Research Notes 3*, Dept. of Geography, Univ. of Ibadan.
Pugh, J. C. (1954a) 'Sand movement in relation to wind direction as exemplified

on the Nigerian coastline', *Research Notes 5*, Dept. of Geography, Univ. of Ibadan.

Pugh, J. C. (1954*b*) 'A classification of the Nigerian coastline', *J. of West Afr. Sci. Assn.*, **1**, 3–11.

Ranson, G. (1955) 'Observations sur les îles coralliennes de l'archipelago des Tuamotu (Océanie française)', *C. R. Somm. Soc. Geol. de Fr.*, 47–9.

Russell, R. J. (1959) 'Caribbean beach-rock observations', *Zeits. für Geomorph.*, **3**, 227–36.

Russel, R. J. (1962) 'Origin of beach-rock', *Zeits. für Geomorph.*, **6**, 1–16.

Russell, R. J. (1963) 'Recent recession of tropical cliffy coasts', *Science*, **130** (3549), 9–15.

Savigear, R. A. G. (1952) 'Some observations on slope development in South Wales', *Trans. Inst. Br. Geog.*, **18**, 31–51.

Shepherd, F. P. (1963) *Submarine Geology*, 2nd edn., Harper & Row, New York.

Shepherd, F. P. (1968) 'Coastal classification', in *Encyclopedia of Geomorphology*, R. W. Fairbridge (ed.), Reinhold, New York.

Shepherd, F. P. (1973) *Submarine Geology*, 3rd edn., Harper & Row, New York.

Small, R. J. (1970) *The Study of Landforms*, 2nd edn., 1978, Cambridge Univ. Press.

Sparks, B. W. (1960) *Geomorphology*, 2nd edn. 1972, Longman, London.

Steers, J. A. (1929) 'The Queensland coast and the Great Barrier Reef', *Geog. J.*, **74**, 232–57, 341–70.

Steers, J. A. (1937) 'The Coral Islands and associated features of the Great Barrier Reefs', *Geog. J.*, **89**, 1–28.

Swift, D. J. P. (1968) 'Coastal erosion and transgressional stratigraphy', *J. Geol.*, **76** (4), 444–56.

Tanner, W. F. and Stapor, F. W. (1972) 'Precise control of wave run-up in beach ridge construction', *Ann. Geomorph.*, **16**, 4.

Tanner, W. F. (1960) 'Bases of coastal classification', *S. E. Geol.* **2** (1), 13–22.

Termier, H. and Termier, G. (1963) *Erosion and Sedimentation*, D. van Nostrand, London.

Tjia, H. D. (1970) 'Quaternary shorelines of the Sunda Land, south-east Asia', *Geol. Mijnbouw*, **49** (2), 135–44.

Tjia, H. A., Fujii, S., Kigoshi, K. (1977) 'Radio carbon dates of Holocene shorelines in Peninsular Malaysia', *Sains Malaysia*, **6** (1), 85–91.

Tricart, J. (1972) *Landforms of the Humid Tropics, Forests and Savannas*, Longman, London, pp. 86–123.

Tschang Hsi Lin (1962) 'Beach rock observations on Ping Chau Island, Hong Kong', *Chung Chi J.*, **1** (2), 117–22.

Twidale, C. R. (1976) *Analysis of Landforms*, J. Wiley, New York, pp. 366–400.

Usoroh, E. J. (1971) 'Recent rates of shoreline retreat at Victoria Beach, Lagos', *Nig. Geog. J.* **14** (1), 49–58.

Usoroh, E. J. (1977) 'Coastal development in Lagos area', Ph.D thesis, Univ. of Ibadan, 1977.

Valentin, H. (1952) *Die Kusten der Erde, Petermanns Geog. Mitt. Erganzungsheft*, **46**, Justus Perthes, Gotha.

Van-Straaten, L. M. U. U. (1965) 'Coastal barrier deposits in South and North Holland, in particular in areas around Scheveningen and Ijmunden', *Mededelingen van de Geologische Stithting*, N.S., **17**, 41–75.

Webb, J. E. (1958) 'The ecology of Lagos Lagoon – the lagoons of the Guinea Coast', *Philosophical Trans. Roy. Soc. of London*, **683**, 211, 307–19.

Wentworth, C. K. (1938) 'Marine beach-forming processes: water level weathering', *J. Geomorph.*, **1**, 6–32.

Zenkovich, V. P. (1967) *Process of Coastal Development*, Oliver & Boyd, Edinburgh and London.

13 Humid tropical geomorphology and development

In the preceding chapters, efforts have been made to describe the forms, materials, processes and the conditioning factors of the earth's surface in the tropical regions of the world. Although these factors are similar in other climatic regions, some of them have been shown to be sufficiently 'unique' or area-specific to have created an identity for a geomorphology of the humid tropics. In this chapter, effort will be concentrated on showing how these factors have influenced development in the region, and vice versa. This is done within the context of applied geomorphology or that area of science which seeks to apply in practical terms the findings of geomorphological research to the national, social, economic and strategic needs of man. This is in line with current trends in scientific investigations, including geography, with emphasis on the application of research to human problems.

The applied role of geomorphology relates mainly to the problem of environmental resource management. This role has been greatly enhanced recently by emphasis on the quantitative study of process–form relations through the application of various forms of process–response models. The area of most successful application is obviously fluvial geomorphology, although recent researches into tropical weathering have yielded tremendous benefits to development efforts as well. Another significant development in applied geomorphology is in the area of land analysis, classification, evaluation and mapping, a phenomenon which has spread to several countries of the humid tropics and so merits emphasis in this chapter. Finally, geomorphological systems usually embrace influences that extend well beyond the traditional confines of the subject and so have a way of surfacing in virtually any development project. With man getting into greater and greater prominence in these traditionally natural systems, especially in urban and other densely populated areas, the degree of relevance of geomorphology is bound to vary not only from place to place but with the cultural setting of the people who occupy an area and/or are involved with its development (Cooke and Doornkamp, 1974).

The humid tropical region is among the least developed in the world. It is also a region eager to develop, as shown by the several usually ambitious and over-enthusiastic periodic development plans, based at best on guesswork and estimates of the ingredients (including natural resources) needed for development. Geomorphology, with emphasis on detailed scientific study of forms, materials and processes, promises to provide the much-needed information for the planner in developing countries and so to make the exercise much less hazardous, but much more rewarding than hitherto.

Accordingly, this chapter will review some major themes in the study of geomorphology in the humid tropics with a view to highlighting not only their realized but also their potential uses in the development process. The themes discussed include land analysis and evaluation, the resource aspects of humid tropical forms, materials and processes; the negative role of humid tropical forms materials and processes; and man and land resources in humid tropical environments.

Land analysis and evaluation for planning purposes

'Land evaluation', as the phrase is used in the literature, currently describes the process of assessing the productivity and potentialities of land for purposes of landuse in agriculture, including forestry, water resources development, settlement and resettlement schemes, wildlife conservation, recreation and other socio-economic projects. The word 'land' here embraces the total natural environment of rocks, landforms, soils, climate, vegetation, water and so on. Although geomorphology covers a number of these, either as part of its subject matter (core topics), such as rocks, topography (landform) or soils, or as affective factors, such as climate, vegetation or water, emphasis in this work is on topography and its immediate determinants of materials (rocks, including soils and other superficial deposits), forms (land of various types), and processes (weathering, erosion and deposition). It is the aim of landform analysis, therefore, to measure and record the forms, materials and processes of the physical landscape for purposes of classification, evaluation and mapping.

This is similar to but not identical with Osunade's (1977, pp. 21–22) definition of a three-stage process of land evaluation, consisting of the following:

1 Analysis, or the simplification of the complexity which characterizes the natural environment. Here the significant attributes of a geographic area are described, identified and measured, where possible.
2 Classification, or the organization and arrangements of the information collected in (1) above (both qualitative and quantitative) for purposes of grouping and characterizing the different components of the geographic area.
3 Evaluation, or the manipulation, interpretation and assessment of the groups or classes arrived at in (2) above, for practical ends. In the final analysis, however, it is the final map which determines the extent of the

benefits derivable from effort put in stages (1)–(2). Landform analysis therefore places a high premium on mapping.

Landform mapping represents a significant change in the attitude of geomorphologists. Starting about the same time as the quantitative revolution, in the post-1950 era, this development represents a conscious effort by a few practitioners to enhance the application and applicability of geomorphological findings, through an improved system of landform representation on maps. Today, we have different types of landform maps, depending upon the complexity of the attributes included. Perhaps the most comprehensive is the land-system map (Cooke and Doornkamp, 1974), which shows a recurring pattern of topography, soil and vegetation, usually as shown on aerial photographs and other forms of remote sensing imagery.

By contrast, geomorphological mapping, the sense in which the term 'landform mapping' is used here, can be conceived in at least two ways: it may be designed as a source of basic geomorphological information on the mapped area; or it may be a purpose-oriented map and so contain information relevant for specific applications (Table 13.1). Basic geomorphological maps may also be converted into special geomorphological maps for use by other scientists (cf. Faniran and Ojo, 1980, p. 216).

It is pertinent here to comment further on the two latest developments in geomorphology – quantification and mapping. The point has been made earlier that they both came into vogue around 1950, but it seems that the comparison ends there. Whereas the quantitative revolution engulfed the discipline in almost every country, geomorphological mapping hardly leaves its European cradle. Effort to spread the technique seems to have met with effective resistance almost everywhere. Even the very influential and international groups like UNESCO did not record much success; it seems to have given up the idea after the 1970 seminar in the University of Dakar, Senegal, organized jointly by Professors J. Tricart and H. Vestappen.

A number of explanations have been offered for the relative lack of success of geomorphological mapping. First, the process (see Faniran and Ojo, 1980, pp. 216–29) is cumbersome and time- as well as energy- and resource-consuming, mainly because of the nature of the fieldwork and the cost of printing effective maps in colour. Secondly, the value of these maps, after they have been produced, has not gone unquestioned. Apart from the producer-bias factor and the associated problem of satisfying the multifarious users, there is as yet no globally derived mapping system which succeeds in showing local forms, materials and processes by a universally adopted set of symbols and conventional signs.

Extensive studies of the terrain with a view to its evaluation for resource exploitation have been attempted in many tropical countries, notably by the Australian CSIRO Division of Land Research and Regional Survey in Australia, Papua and New Guinea; and the British Directorate of Overseas Surveys Land Resources Division in Nigeria, Malawi, British Solomon Islands, Kenya, Fiji, Botswana and the New Hebrides. These surveys, adopting the landscape approach of terrain classification, were based on different sizes of land units, namely, the land

TABLE 13.1 Applications of geomorphological mapping in planning and economic development

1. National planning
   Territorial planning
   Regional area planning
   Conservation of the natural and cultural landscape

2. (a) *Agriculture and forestry*

Sectoral planning	Soil conservation
Forest conservation	Soil erosion control
Forest utilization	Reclamation of destroyed or
Crop cultivation and plantation	new areas
agriculture	Soil reclamation
	Drainage and irrigation

   (b) *Environmental design*
   Reconstruction and replanning of settlements, especially of towns
   Designing of industrial buildings
   Communications (roads, railways, canals, harbours)

   (c) *Water resources*
   Reservoirs and dams
   Regulation of rivers
   Natural and artificial waterways
   Irrigation canals
   Harbour construction
   Shore protection
   Fishing projects

   (d) Geological survey Mining
   Exploitation Potential and actual damage
   done by mining
   Reclamation of abandoned open-cast
   mines and waste dumps

   (e) Tourism and recreation

Source: Modified from Cooke and Doornkamp, 1974

zone, the land division, the land province, the land region, the land systems, the land facets and the land elements. The most frequently used are the land systems and the land facets.

The land system is defined as a recurrent pattern of genetically linked landforms, soils and vegetation recognizable on medium-scale aerial photographs and mappable at 1:250 000 to 1:1 000 000. Essentially the land-system approach views the entire terrain with regard to the whole range of possible land use interests. Thus Bawden and Tuley (1966) recognized and mapped thirty land systems in the Sarduana Province of Nigeria, and accompanied the maps by block diagrams which show the interrelations of the land elements. They also described the geology, physiography, soils, climate, vegetation and the present use in each land system and then assessed the lands' suitability for various crops, grazing and forestry. In the same manner, working in north-eastern Nigeria, Aichtison *et al.* (1969), based on the major rock suites, recognized five land provinces, each of which was further sub-classified into land regions and land systems. Each of the latter was broadly described in terms of geology, physiography and soils, and the resource potentiality evaluated with regard to agricultural development. However, as observed by Mitchell (1973, p. 65), this system of land resource analysis and evaluation is useful only 'where land is little developed or of relatively low economic value'. Thus most of the studies are essentially reconnaissance.

Many scholars believe that the land facet is more valuable for terrain evaluation than the land systems (Areola, 1977). The land facet is defined as 'one or more land elements grouped for practical purposes; part of a landscape which is reasonably homogeneous and fairly distinct from surrounding terrain' (Christian and Stewart, 1964). It is mappable at 1:10 000 to 1:80 000. It is generally recognized that land facets have distinctive patterns of slope, vegetation types and ground tones on aerial photographs, which is an indication of homogeneity of internal soil-drainage conditions. As a result of its uniformity, it is preferred as a basis for soil mapping, as Murdoch *et al.* (1976) did in mapping soils in the savanna area of western Nigeria.

With particular reference to humid tropical environments, Thomas (1969, p. 104) has observed among other things that land-system maps, usually at small scales (1:500 000) do not necessarily represent the most appropriate approach to African environmental conditions and development problems. Accordingly, even when landform systems are being mapped by the traditional land-system approach, the problem of the basis for recognition of these units needs to be verified from the background of the environment being mapped.

This system of land classification, based on recognizable attributes from aerial photographs, has been modified in the light of experience in the deeply weathered shields, a significant phenomenon of humid tropical lands. The pioneer work in this respect includes that by Mabbutt (1962), who listed four land systems for the Alice Springs area of the Northern Territory of Australia as:
1 erosional weathered land surface;
2 partially dissected erosional weathered land surface;
3 erosional surfaces formed below the weathered land surface; and
4 depositional surfaces.

The system adopted shows a complete departure from the traditional system of hill lands and plains, which, even when indicative of the constituent material, fails to incorporate the very important factor of process – namely, the humid tropical process of deep weathering. Chapter 3 underscores the significance of deep weathering in humid tropical environments, which process will be further discussed later in this chapter. In the meantime, it is recommended that Mabbutt's (1962) system of landform mapping as modified by Thomas (1969) be adopted for use in humid tropical regions, especially the shield regions. The maps so produced are likely to be more useful, at least to the extent that they indicate in addition the location of such valuable resources as duricrusts (laterites, ferricretes, bauxites, silcretes, calcretes and so on), rock outcrops (inselbergs of all types, including ruwarses, tors, rock pavements and so on) alluvial valleys, and slope forms (Table 13.2).

TABLE 13.2 Land types in Upper Ofiki Basin, Oyo State, Nigeria

Facet	Landform systems	Landforms	Soils and drainage	Vegetation and land use
Ha₁ Inselberg	Hill land	The facet is generally domes or forms a series of rugged broken sections of topography. Slopes are concave, and vary between 10 and 60° and are nearly vertical in some places.	The inselbergs have bare rock surfaces with soils in the joints and cracks.	Vegetation is absent, along the joints where these layers of soils have accumulated. Practically of no use for cultivation.
Ha₂ Hills and ridges	Hill land	Crests of elongated and rugged ridges. Some ridges and hills have flat tops and some sharp, convex summits. Slopes are concave varying between 10 and 30°. Facets have concave foot-slope.	The facet is co-vered by thick weathered soils. Corestones occur at great depths (observed in wells at Oke-Iho). The facet is freely drained.	Generally well-covered with vegeta-tion. Short bushes and trees predominate. Some parts are put into cultivation where slopes permit.
Pa₁ Ruware and turtleback topography	Plain	Rock outcrops; generally flat, or at the same level with flat to sloping land; some have round convex shape. Slopes are generally below 4°.	Covered by thin veneer of soil with an exposed small part of base granitic rock. Soil depth usually increases farther away from the exposed parts. Rock pits sometimes contain rainwater or partially weathered soil.	Poor vegetation consisting of grasses and small shrubs; practically useless for cultivation purposes but useful for for drying cassava, cassava flour, yam flour and pepper. The rock pits are also used for grinding where close to settlements or farmhouses.

383

Pe Broad valley	Plain	Wider and deeper channel incisions occur here than in facet Pd. Valley slopes are generally below 3°.	Very dark loamy sand soils. Freely drained except the stretches close to the stream channels	Dense wooded vegetation with sparse undergrowth. The facet is put under cultivation.
Pf₁ and Pf₂⁻ Very broad valley	Plain	Wider and deeper channel incisions than in facet Pe. Rocky floors exposed in some places. Streams are characterized by broad meanders and braids.	Very dark loamy soils and thick organic layer. Poorly drained near the river channels but improved farther upslope.	Dense wooded vegetation with sparse undergrowth. Some trees are festooned with climbers, which increased the density of the tree's canopy.
Hb₁ and Hb₂ Footslopes	Hill land	Concave slopes at the foot of inselbergs, hills and ridges. They connect the inselbergs, hills and ridges to the gentler long slopes below	Soils vary in depth from nothing to tens of metres. Drainage may be impeded during the rainy season.	Generally densely vegetated; thick layers of organic matter exist which make the area particularly fertile for cultivation.
Hc₁ and Hc₂ Long slopes	Hill land	Long slopes between the inselbergs, hills and ridges, sometimes very straight in outline between the concave footslopes and the valleys. Slopes vary between 5° and 30°. Gully and sheet erosion are dominant features in some places.	Covered by thick mantle of weathered soils. Generally, the top layer is gravelly. Top layers are organic. The facet is freely drained.	Well-vegetated except where the original vegetation has been interrupted by human activities.

Source: Modified from Osunade, 1977

TABLE 13.2 (continued)   Land types in Upper Ofiki Basin, Oyo State, Nigeria

Facet	Landform systems	Landforms	Soils and drainage	Vegetation and land use
Pb_1 and Pb_2	Plain	Gently sloping and smooth convex slope. Slopes are generally below 3°. Inter-crest distance may be more than 500 metres.	Facets are freely drained but drainage may be poor in facet Pb_1 during the rainy season.	Thick woodland vegetation with very sparse undergrowth on facet Pb_1 and abundant gasses dotted with short trees on facet Pb_2.
Pc_1 and Pc_2	Plain. Long connecting slopes	Long and gentle concave or convex slopes with slope angles varying between 3 and 6°.	Soils of the facets vary from dark coarse loamy sand to light brown sandy clay. Facet is freely drained except that drainage may be poor in facet Pc_1 at the peak of rainy season.	Facet Pc_1 is covered by thick wooded forest while facet Pc_2 is covered by abundant grasses dotted with few short trees. This is a well-cultivated facet in the area.
Pd Narrow	Plain	Broad and incised stream channel with valley slopes below 3°.	Dark loamy sand soils. Incised channels contain rain floods. Well-drained valley sides for most of the year.	Wooded vegetation cover on the valley sides. Vegetation usually distinct from these farther away from the valley.

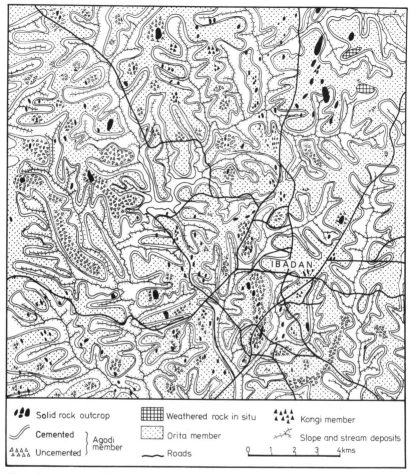

Fig. 13.1 Superficial deposits in Ibadan area

The above is the precise system adopted by the Tropical Section of the Road Research Laboratory of England in its attempt at terrain evaluation and mapping in order to aid the development of transport networks in the tropics, especially in Nigeria and Malaysia. The organization employed the land-system approach to study the location of roads, and the occurrence of road construction materials (rock outcrops, laterite pavements and laterite gravel) with respect to the geology, terrain characteristics, vegetation, soils and climate (Dowling, 1968).

Thus, in spite of the low reception rate of the idea of landform mapping and of the inherent problems involved, it is recommended that geomorphologists should not relent in their effort to map landforms. This is particularly necessary in the humid tropical world where periodic development plans are being drawn up which require vital information on landform attributes. In the words of Thomas (1969, p. 141), there is clearly a need to bring together the achievements within the fields of theoretical geomorphology, morphometric analysis and the designation of land systems for development purposes.

Humid tropical landforms, materials and processes and the development process

Landforms, their constituent materials and formative processes affect development in two main ways. First, they constitute a resource base and so serve as indicators, if not of actual, but of potential development. Secondly, they serve as constraints or hindrances to the development process, depending on the level of technological development attained by the people involved in the development, and the capital resources available to them. In fact, this idea is basic to the latest definition of the term 'resource' 'as an abstract concept which hinges upon man's perception of the means of attaining certain desired goals, within the limits of the biophysical environment' (Faniran and Ojo, 1980). Thus whether a landform, earth material or landform process will be regarded as a resource for, or a constraint to, development will depend partly on the extent to which man understands, or is aware of, the benefits derivable from the environment and partly on man's ability to realize the envisaged benefits therefrom. Table 13.3 provides a summary of landform systems for the purpose of aiding developers in making judgements involved.

TABLE 13.3 Descriptive notation for landform systems

A. **Dissected**   **(lateritized) etchplain**

1 **Landform complex:** 'laterite' implies association with two-phase stream valley
    unit landforms:
    a Laterite mesa with pediment;
    b Inner stream valley with active channel;
    c Rock domes (bornhardts).

2 **Morphometry:** drainage density 0.6; mean valley width 2.6 km; exposed
    laterite at least 15% of area, and occurring on 45% of
    transverse valley profiles; rock outcrops less than 0.1% of area.

3 **Detailed morphology unit and forms:**   **Facets**

a Laterite mesa   i Crest slope of primary vesicular
                                        laterite; slope generally less than 2%
                                        cambered at edges; mean area 1 000
                                        square metres;
                                    ii Breakaway: hardened laterite cliff or
                                        steep rubble;
                                    iii Deeply weathered pediment with
                                        veneer of fine laterite rubble and
                                        concretionary pisolitic gravel slope
                                        2–7%, length 200–400 m;

b Inner stream valley	iv Laterite bench of concretionary secondary laterite; slope 2–4%; v Minor laterite breakaway: hardened laterite forming break of slope; vi Pediment slope; often deeply weathered but with occasional outcrops, 100–200 m; slope 2–7%; vii Alluvial valley floor;
c Rock domes (bornhardts)	viii Convex rocky slopes up to 70% or more; relief 10–50 m.

4 **Soil catena types:** Most soil catenas have massive laterite cuirasse as summit feature (over 50%). Many catenas with 2 laterites – facets (i) and (iv) perhaps 25%. Less than 1% of catenas with rock outcrops.

B. Dominantly stripped etchplain

1 **Dominant rock type**: migmatites, acid, gneiss.
Planation surface remnants: none important.
Unit landforms: a Bornhardt domes;
  b Rock pavemented;
  c Tors;
  d Open stream valleys without clear evidence of multiphase incision.

2 **Morphometry**: Drainage density 0.99; mean valley, 1.08 miles (1.73 km); exposed rock surfaces at least 7% of area, and occurring on 25.66% of transverse valley profiles.

3 **Detailed morphology**:

Unit landforms	Facets
a Bornhardt domes	i Many contiguous convex rock surfaces carrying superficial debris; relief exceeds 10 m; bounding slopes generally over 30%;
b Rock pavements	ii Simple convex rock surfaces; relief less than 5 m, slope less than 15%;
c Tors (sometimes as landform complexes)	iii Complexes of contiguous joint blocks;
d Stream valleys	iv Deeply weathered crest slopes; less than 5%; v Valley flank (pediment); generally shallow weathering with quartz gravel surface layer; slopes 5–10%; length 0.4 miles (350 m). Interrupted by facets (i), (ii) and (iii); vi Alluvial valley floor.

4 **Soil and types** Many soil catenas with rock outcrops (over 25%); outcrops occur in all locations (only 44% on interfluves); some catenas with multiple rock outcrops. Very few catenas with laterite (none as massive duricrust on summits) – less than 1%.

388

Landforms, materials and processes as a resource base

The literature devoted to landforms and their associated materials and processes as a resource base is quite vast. Indeed, the most common definitions and classifications of earth materials, forms and processes, of which landforms are part, emphasize their positive (resource) rather than the negative (constraint) roles. Thus Howe (1979) defines natural resources as all the living and non-living endowment of the earth, while Faniran (1981a) restricts them to the naturally occurring resources and systems useful to man or that could be useful under any plausible technological, economic and socio-cultural milieu. Similarly, in connection with nomenclatures or classification, emphasis has always been placed on their positive roles in direct consumption, inputs into intermediate processing, consumptive uses in intermediate processing and *in-situ* uses (Faniran, 1981). These positive roles are here highlighted for the humid tropics under the three major headings of forms, materials and processes.

Landforms

The history of man the world over shows that landforms have always been viewed as a resource. The most important uses in the developing countries of the humid tropics relate to settlement, recreation, and religion. Faniran and Ojo (1980) describe the situation thus:

> Man has derived satisfaction in different forms from landform features . . . people worshipping mountains and other landform units; mountaineers deriving satisfaction from climbing either the highest peaks or taking the most dangerous approaches to climbing a mountain; mountains providing relatively safe sites for settlements; . . . tourists frequenting unique and spectacular landforms.

Reference in the above is limited to mountains and mountain sceneries, which comprise only a part of the landforms of an area. Others include plains of all categories; rivers and river valleys; and a whole range of landform examples which do not fall neatly into any of these major categories, such as lakes, weather pits, caves, and the coastline and its associated features.

The aesthetic role of landforms is universally acknowledged, and although humid tropical landforms have not many unique attributes which bring them out for special consideration, in conjunction with other attributes of the humid tropical landscape, they have attracted large numbers of foreign tourists, particularly from Europe and North America. By contrast, religious observance is unique, especially in its latest form. The role of landforms in land occupance and settlement has been widely discussed in the literature (cf. White and Gleave, 1971), and so need not be repeated here.

Aesthetics and landforms: One notable feature of present-day society is the marked increase in leisure time, especially in Western society, resulting in increased demand for recreational facilities. In an attempt to meet this demand, environmental planners have to grapple with a number of problems, including that of ascertaining the tastes of the people demanding recreational facilities (Obateru 1981), and developing such sites to afford maximum satisfaction for the visitors to the sites.

It is in this connection that geomorphology comes in handy, since the form of the ground, made up of unit landforms and landform complexes, constitutes significant inputs into a viable recreation site. The humid tropical regions have an additional advantage here, tied to psychological attractions of brilliant sunshine, perpetual summer, jungles and wildlife. After all, a very important aspect of choice in outdoor recreation sites is the personal emotional factor, which in this case has been strongly influenced by propaganda over the years of the tourism-based economies of tropical islands and countries.

The propagandist language is not limited to the tourist companies alone; it 'trespasses' into scientific literature, as shown in the following extract from Ouma's (1970) book. The piece relates to the 'touristic scenery and landscape' in East Africa, but the language is typical of other sections of the book. The extract is long in order to highlight as many of the relevant landforms as possible:

> The geological history, palaeoclimatic changes and concurrent geomorphological evolution of the East African landscape have ensured a great variety of scenaries now available to the visitor. Within a week, . . . not only is it possible . . . to relax on the *white-sand beaches* * off the *coral-fringed coastline*, or to plunge repeatedly into the *warm ocean waters*, of multiple hues, but also . . . [to] observe *perennial snow* gleaming in the equatorial sunshine at the *roof of Africa*. This is to say nothing of the nearby *multi-coloured soda lakes*, hemmed in . . . by the *world's largest crustal crack, the Gregory Rift Valley*. He can then drive through *sun-baked landscapes* around *Lake Baringo* . . . climb up the *rift wall* through *tropical rainforest* . . . make a quick short descent to Kericho, through *pine forests* . . . see Kisumu . . . to cruise in *The Victoria* . . . bid farewell to the *mesas of Mongo* and head northeastward for *Mount Masabao* or westward for Ptolemy's *'Mountains of the Moon'* . . . criss-crossed with a maze *of swamps, rivers threading lakes high and mountainous*, the *terraced terrain* of Kigezi and its *hot springs* and blue evening atmosphere rank the District among the most picturesque in creation. . . . on the approaches to Congo-Kinshasa (Zaire), one sees *Sabinio, Muhavura* and *Mgahinga*, Uganda's only share of the *Birunga Volcanoes* [pp. 58–59].

However, tourism in East Africa in particular and the humid tropical lands in general, is not simply sightseeing. Where the terrain permits, in mountainous regions, mountaineering is a very popular activity. Again, East Africa offers a very good illustration with its high mountains, inselbergs, granitic tors and small volcanic hills. Short excursions are arranged to the smaller landforms and expeditions to the

---

*Italics have been inserted by the authors, in order to highlight the variety of landscapes in the area.

tall mountains. The excursions and expeditions are usually arranged by clubs, based in different countries. Examples include the Uganda Mountain Club, the Mountain Club of Kenya, and the Mountain Club of East Africa. These clubs provide several facilities for tourists, with a view to making them comfortable so as to be able to enjoy their trip, come back again, and attract others. The facilities provided include huts for stops at convenient locations during ascent, trail guides, provision stores (at modest prices), rescue crews and so on.

Hills and religious sects: Although the practice whereby mountains and other landforms are worshipped as gods is fast disappearing as a result of conversion to Christianity and Islam, a new dimension has developed in the last few years. This has come with the advent of 'indigenized' Christian churches, including the Apostolic, Kerubim and Serafim and similar (Aladura) religious sects. These groups, especially the Kerubim and Serafim, have a special love for rock hills, where they either build a church (or churches), or pitch a tent. Around Ibadan, in south-western Nigeria, virtually all the known rock hills – inselbergs, tors and so on – and their foothills are either temporarily or permanently settled by the 'faithful' (such as prophets and their 'disciples'). One permanently settled site is an inselberg near Akinyele, on the Ibadan-Oyo road. The settlement, called in Yoruba '*Ori Oke Olorunkole*', meaning 'the hill on which God built a house', is led by a church leader, helped by a number of 'the faithful' or prophets. Their assignment is mainly to pray for the sick and other attendants, who may be 'hospitalized' or treated as out-patients. The busiest periods are weekends and public holidays, when people come from as far away as Lagos, and even outside the Nigerian borders, and when the scene resembles a typical picnic camp on a national holiday.

The popularity of this particular hill – there are several others like it spread all over the inselberg country of West Africa – derives partly from the belief that God and His angels live on hills a hangover from biblical days, but mainly from the story, popular among the adherents, that an angel once visited the hill and left his/her footprint. This print has since been walled round, and only devout believers are allowed to go near for special prayers.

Other uses of landforms: Emphasis so far in this section has been on mountains or hills, which are the most conspicuous landforms and so easily attract attention. There are, however, a number of other landform types which serve as resources. Some of them – for example, lakes and rivers – have been mentioned in relation to tourism in East Africa. Besides tourism, these resources are otherwise utilized by man, as in the generation of electricity from rivers (waterfalls), the use of rock surfaces for drying, threshing and grinding food items. Some weather pits provided abode for early men, such as the caves in Olumo Rock in Abeokuta, and water for various purposes, as the rock lakes of Ado Awaye, both in south-western Nigeria (cf. Faniran, 1981b).

It is perhaps stating the obvious that rivers have figured prominently in the affairs of man. Among the most important uses are in irrigation schemes, as political boundaries, as sources of domestic and industrial water supply, for fresh-water fisheries, and as a means of wate disposal. It is possibly this realization that influenced Chorley (1969) to devote seven of the chapters of the book *Water, Earth and Man* to this aspect of rivers. The drainage basin, a geomorphic unit, is also being increasingly used as planning units, and for similar reasons (cf. Faniran, 1972, 1980). Finally, the crucial function of rivers in both the environment and the development process has meant that several disciplines, including engineering (sanitary, hydrological, civil), administration and planning, study them. Several conferences, workshops and seminars have been held by interdisciplinary groups and books have been published, making rivers among the best-studied land units on earth.

Apart from the traditional uses in domestic life, humid tropical rivers are particularly important in at least two respects: in navigation and in the generation of electricity. These two uses specially relate to the environment in different ways. Because of the initial difficulties of access through jungles, tropical rivers provided the first routeways into the interior, thus featuring in the exploration of the hinterland. Examples are the Niger and Zambezi in Africa, the Amazon in South America and the Indus and the Ganges in the Indian sub-continent. Even today with marked improvements in the other transport media, tropical rivers continue to play significant roles, as they are improved through dredging and the construction of multi-purpose dams. They continue to supplement road and railway transport in many places (Figure 13.2).

The use of rivers in the humid tropics to generate electricity implies, of course, the construction of dams. These have attracted scholars from several disciplines, not only because they provide a convenient and meaningful rendezvous for the various academic disciplines whose activities relate to the manifold problems of development, but also because they provide a multitude of benefits for the people in their immediate environments.

Humid tropical rivers have certain advantages which make them suitable sites for dam construction for the generation of electricity. Large parts consist of the world's most ancient rocks underlying the most extensive shields or cratons. Africa, the most tropical continent, perhaps suits this description best, having been described as a continent of plateaux, each usually with well-marked edges. The rivers, in consequence, fall over these plateau edges, thus providing ideal sites for power dams, as in western Angola, the Kouilou gorge of the Niavi river in Congo (Brazzaville); the Kariba and other gorges on the Zambezi, and the cataracts and rapids on the River Nile. Indeed it has been estimated that Africa alone has power potential equal to over 40 per cent of the world's total. By inference the humid tropics would have well over 50 per cent – perhaps about 80 per cent – of the world's total power potential (Harrison-Church, 1968). This is most fortunate, because these areas lack in relative terms the traditional sources of fuel – coal and petroleum – for which rivers have compensated.

Unlike most other landforms, rivers are always in some sort of delicate balance or equilibrium: they are dynamic systems in which the form–process relations existing at a point are governed not only by the

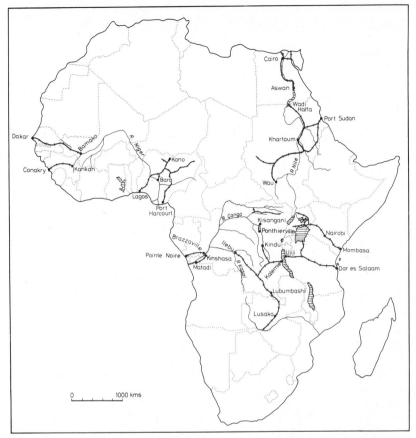

Fig. 13.2 Interconnections among railways, roads and waterways, tropical Africa

external factors (for example, precipitation and other inputs), but also by
such internal conditions as channel morphology and the contained water
and load (sediment). Although all uses affect the river, none has such
overwhelming effects as dams. These consequences are not limited to the
dam site alone but extend in virtually all directions, and particularly
upstream and downstream. Thus this is one of the major reasons for
preaching an integrated, if not a comprehensive approach to the
utilization of river-basin resources (cf. Faniran, 1972, 1980).

There are three major divisions of the landscape recognizable in
many humid tropical regions. These, according to Thorp (1970), are the
residuals and hills, the plains and the valley bottoms (see Chapter 6).
Burke (1967) called them hills, pediments and valleys. Two of these, the
hills, especially the rock hills, and valleys (rivers) have been covered,
leaving the plains and/or pediments, which, by definition, cover about 80
per cent of the land area in many countries, especially West Africa. Here
they include the interfluvial surfaces of dissected rolling topographies.

Both pediments – the low-angle terrains at the foot-slopes of hills –
and plains are not only widespread in humid tropical lands but are also

very intensively utilized for several purposes. Morgan and Pugh (1969) observed the following about the pediments of West Africa:

> [They] provide the obvious site for agricultural activity, being well-watered and well-drained with a soil cover often not present on the hillsides and lacking the heavy vegetation of the floodplains, which required much greater effort in clearing [p. 229].

Pediments are also favoured sites for settlements, especially in the savanna country with associated inselberg landscape (cf. Faniran, 1981).

As for plains, their most important role is in agriculture, especially in the drier (savanna–semi-humid) parts. Examples are the *fadama* or lowland fields of northern Nigeria, which produce virtually all the perennial crops, such as sugar cane. Recently, these plains have been 'extended' through the operation of irrigation agriculture. Examples are the Richard-Toll project in Senegal, the Gezira scheme in the Sudan, the Tiga area of Kano, Nigeria, the ricefields of coastal Sierra Leone, Guinea and Liberia as well as other schemes in tropical Asia, the Americas and Australia. Nigeria recently established eleven river-basin authorities, whose major function is irrigation of crops along the major river courses.

Materials

The materials relevant to our discussion here are the rocks, soils and mineral resources, especially those closely connected with the operation of certain humid tropical processes. These materials are not by any means restricted to humid tropical regions but are so common as to be 'typical' of them. They include the fresh rock outcrops (such as inselbergs) and the duricrusts (laterites, bauxites, iron-oxide ores, manganese-oxide ores and so on), as well as soils. Other materials of note are the sand and related deposits (including silt and clay) of floodplains, alluvial fans and deltas and the deeply weathered materials (mottled and pallid zones) of typical deep-weathering profiles. They are here discussed under two major headings of constructional materials and exploitable ore deposits. The soils and their uses, especially in agriculture, well merit inclusion here.

Constructional materials: These are by far the most widespread, occurring as they do nearly everywhere, whether as fresh rock outcrops; as partially-weathered hill rock (as in quartzite hills); as superficial deposits on slopes; or as channel (bed) deposits. Faniran (1973), in a study based on the Nigerian situation describes them thus:

> The earth materials which are of particular interest to the construction industry are fresh crystalline rocks such as granite and gneisses; partially weathered quartzites. . . . deeply-weathered and duricrusted materials such as laterites, silcretes, ferricretes, etc. loose sand; and clay. . . . For these materials to be used . . . two things need to happen; their locations have to be mapped and their engineering properties adequately established [pp. 153–54].

Figure 13.1 shows an attempt to record some of these materials for the Ibadan area: similar maps exist for several other parts of Nigeria, which provide useful information for would-be users, while others show particular features such as inselbergs (Figure 13.4). Currently these materials are being exploited at an accelerated rate, such as in quarries, gravel pits, soil pits and clay pits. The towns and their environs are obviously the areas of greatest activity since the towns constitute the major consumers of such materials.

Very few studies exist, and those available are far from definitive, especially as far as volumes, rates and other useful measures are concerned. The situation is exemplifed by a study of river-sand mining in the Lagos area of Nigeria (cf. Faniran *et al*., 1976). The investigation was carried out along the rivers and creeks between Ikorodu and Lagos, and involves the nature of the material being mined, the locations of the operations, the methods of operation, the quantity involved, and an appraisal of the relative contribution of sand mining in the building industry. It was found that both the physical (humid tropical riverine) and the human (urbanizing) environments favoured the operation. The physical environment of deep tropical weathering, fluvial transportation, sorting and deposition, and deltaic-lagoonal situations favour the preponderance of fine (sharp), quartz sand, which is ideal for the sandcrete industry. Table 13.4 indicates growth in this industry. For its

TABLE 13.4   Sand lorryloads counted along Lagos–Ikorodu road for 10 days in the dry and wet seasons 1974–5

	Dry season		Wet season	
	Lorryloads	Tonnage	Lorryloads	Tonnage
1	721	3 605	150	750
2	789	3 945	142	710
3	543	2 715	123	615
4	747	3 735	154	770
5	544	2 720	137	685
6	758	3 790	147	735
7	705	3 525	112	550
8	821	4 105	133	665
9	784	3 920	180	900
10	900	4 545	140	700
Total	7 321	36 605	1 418	7 090

Sources: Fieldwork Oct. 1974–Sept. 1975

part, the insatiable demand of residential, commercial, industrial building of Greater Lagos added immensely to the high value of the material, which varied between 5 and 22 per cent of the cost of the building in the area. The income accruing to the operators is another indicator of the value of the river sand (Table 13.5). Very similar situations apply to the other sources of material for the building industry – namely, sandy soils and gravel, including weathered duricrusts, quartzite rubble, quarried granites, gneisses, blocks and slabs of varying shapes and sizes.

TABLE 13.5 Sand-mining industry: estimate of income and expenditure per month by a sand mining contractor

Income per month (N)		Expenditure per month (N)	
		Rental	
3 boat trips daily = 6 lorry loads of sand		Boat rental per day	15.00
30 days = 90 boat trips		30 days rental	450.00
90 boat trips = 180 lorryloads of sand		Labour	
1 lorryload of sand	26.00	Sand digging per boat load	8.00
180 lorry loads of sand	4680.00	3 boat loads daily	24.00
		90 boat loads in a month	720.00
		Other equipment	
		Rope, basket, mat, etc. per month	12.00
		Total expenditure	450.00
			720.00
			12.00
			1 182.00
Total income	4 680.00		4 680.00
			1 182.00
		Credit balance =	3 498.00

Source: Modified from Faniran *et al.*, 1976

Duricrusts and duricrust profiles as exploitable minerals: Perhaps the most significant contribution of the humid tropical process is the formation of earth materials with 'added' values. These materials (described in Chapter 3) vary with environmental factors, especially of geology and topography. The most important materials are:
1 iron-ore deposits associated with environments on quartz-banded iron ore and ultra-basic to basic rocks;
2 bauxite-ore deposits associated with basic and ultra-basic rocks, shales, clays and nepheline syenites;
3 manganese associated with gondites; and
4 nickel ore on basic to ultra-basic rocks.
Others include opals associated with some silcretic duricrusts; calcium carbonates with calcretes; and laterites (Faniran, 1971, pp. 60–61). Iron ore and bauxites, which, with their intermediate variant – laterite –

constitute the most widespread occurrences in this respect, are discussed further here for exemplification.

There are three major sources of the world's iron deposits. They are deposits associated with igneous rocks; sedimentary deposits and residual deposits. We are interested here in the residual deposits which have formed from deep tropical weathering. Faniran and Ojo (1980, p. 357) described the material thus:

> Residual deposits of iron minerals consist of the brown, yellow, black and red oxidized ores commonly associated with deep weathering. Depending upon the formative factors, these deposits can be extensive and of high quality. However, because of their topographical location (on hill tops and plateaux), they have been largely destroyed, following millions of years . . . of erosion.

Consequently, the most important occurrences of these, and some other, iron-ore deposits are associated with hill and mountain terrains, as in the Kukusan mountains of Indonesia, the Kaloun Peninsula of Guinea (Conakry), the Sula mountains of Sierra Leone, and the Bomi and Bong hills of Liberia. In Nigeria, the Agbaja plateau deposits are among those studied in connection with the country's iron and steel complex at Ajaokuta. Where deposits occur at the foot of hills and mountains or in river channels and gorge bottoms in dissected terrains, transportation and redeposition from adjacent hilltops have been suspected.

The situation with bauxites or aluminium ores is not too dissimilar. Unlike iron ores, the most extensive deposits of bauxites are the residual types located in topographically well-drained terrains, as in the Arkansas region of the United States, Ghana, Guinea, Cuba, Jamaica, Guyana and Surinam.

Laterites contain relatively high proportions of both iron and aluminium oxides and are often mined for one or both substances. In Nigeria, the iron-ore deposits are mostly the lateritic type, with relatively higher iron- than aluminium-ore contents. They are, however, hardly distinguishable from one another physically, being mostly concretionary. At the moment, laterites are used mainly in civil engineering works (cf. Faniran, 1974). For the future, perhaps when most of the rich bauxite and iron-ore deposits have been exhausted, and when extraction methods get more sophisticated, laterites and related materials, widespread as they are in both tropical and extra-tropical (fossil) regions, especially the former, may be the main sources of these valuable materials.

Deep weathering and water supply: The deep-weathering process has another beneficial effect worth mentioning here – namely, the provision of ground-water in basement complex areas. Basement complex rocks are intrinsically poor in aquifers and have accordingly been written off in several countries as possible sources of groundwater following disappointment created by low ratios of success recorded in well-sinking operations in these areas (cf. Asseez, 1972; Faniran, 1975). However, recent investigations in these areas, especially in Nigeria (cf. Faniran 1975, 1980; Faniran and Omorinbola 1980 and Omorinbola 1979) have

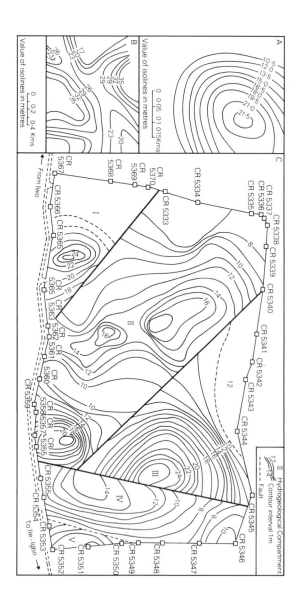

Fig. 13.3 Depth of weathering. A. Dan Mongu area; B. Harwell Paddock, Jos Plateau; C. Ile-Ogbo Housing Estate, Iwo area, Oyo State

shown that the application of geophysical techniques, among other methods, promises to improve on the success rates and so render the groundwater reserves in these areas accessible for exploitation for the benefit of local inhabitants. These reserves occur in basins of decomposition (Enslin, 1943), joints and fissures, especially the former, created by deep chemical weathering, and consisting of mainly elastic rocks with high rates of water ingress during rains. Examples of these basins of different shapes and sizes are shown in Figure 13.3, which has been prepared from geophysical data. Unfortunately, the preparation of such maps requires much expertise, time and capital and are not easy to come by, especially in the developing countries of the humid tropical world. It is, however, hoped that research will soon provide alternative methods which are both faster and cheaper. Then the valuable water resources in basement regoliths can be expected to serve the people who need it most – the isolated, small rural communities, especially those remote from the traditional urban-based water schemes (Faniran, 1974; Faniran *et al.*, 1980).

Processes

There are fewer direct uses specific to humid tropical processes than in the case of forms and materials. This is to be expected since process produces forms and materials which in turn affect process. Moreover, previous discussions of forms and materials have included process – for example, in running water and the generation of hydroelectricity as well as in the discussion of materials produced through the operation of humid tropical processes. Finally, process is universal and, although varying in magnitude in different places, does not constitute a sufficiently recognizable and identifiable theme in the applied sense, outside of the forms and materials it produces. Running water, whether in rivers or along the coast, behaves alike over large parts of the earth's surface under the influence of local controlling factors; it is therefore not surprising that there is little spatial or regional variation in the application of the processes.

Landforms, materials and processes as constraints to development

There are usually two sides, positive and negative, to any issue, and this is true of landforms, materials and processes in the humid tropics, which have both beneficial and harmful or limiting aspects. For example, an impressive escarpment or a rugged, broken mountain terrain, although providing an ideal site for tourists, also constitutes a great impediment to other economic activities such as road, railway or other civil engineering constructions. The extent to which an impediment or limitation can be overcome depends upon technology and capital availability. Where technology is advanced and capital is available, the limitations are easily overcome. Conversely, where, as in large parts of the humid tropical regions, the level of technological development is low if not primitive and capital a very scarce resource, the effectiveness of the limitations is absolute or at best at a maximum. A few examples will suffice.

Landforms

As observed above, the best-known countries in humid tropical regions to overseas people are those with scenic landscapes consisting of spectacular landforms such as mountains, hills, plains and rivers, to mention only a few. Examples are the East and Central African countries, each with a well-developed tourist industry. Now, these same features, especially the mountains, constitute major impediments to infrastructural developments, among others. In fact, it appears that in many countries, the areas designated as national parks, nature reserves and so on, for tourists have been so labelled deliberately, because the areas could not otherwise be utilized to greater advantage. Consequently, there is often a correlation between landforms and certain indices of development such as length of good-quality roads and urban centres.

Landforms affect the pattern of development at both regional and local scales. In general terms, ruggedness limits the opportunities for cultivation among other activities, as in, say, the Aberdare mountains or the Ruwenzori region in East Africa; nevertheless, there are local variations, which the inhabitants utilize to wrest a living from the generally unfavourable sites. Similarly, within an area of generally low relief – such as in Sukuma-land, Tanzania – there are isolated hills with slopes too steep for normal cultivation. Even plains constitute problems for development. This is particularly true of many coastal plains in the equatorial region where the thick vegetation, insect infestation and generally poor drainage combine with the low (primitive) technology and abject poverty of the inhabitants to preclude effective colonization. Many plains are also underlain by poor soils or are constantly under water.

The nature of the coastline also directly affects development. The African continent again provides a good illustration of the adverse effects of coastlines on the pace and cost of economic development. Not only did such coastlines delay penetration by external modernizing influences, but they still constitute a major limitation to both internal and, more importantly, external trade. Except in very few cases, ports along the African coasts are artificial, constructed and maintained at very high costs.

Finally, it is pertinent to note that every development represents a mitigation of the constraining influence of landforms. In fact, it is the major objective of technological advance to minimize the limiting aspects of man's environment. Often these activities of man create other environmental problems, especially where planning or execution, or both, are defective. It is for this reason that caution is called for in man's attempts to overcome the 'obstacles' to development. It is not only absurd, but also difficult, if not impossible, to attempt to pull down the Himalayas or fill up the African rift valley in order to facilitate agriculture or road construction. Each landform unit has several possible benefits or uses as well as negative aspects. It should be the major task of rational planners to identify the most suitable use for each landform unit so that it will be subject to the minimum of constraining influences.

Materials

Of all the materials found in humid tropical regions, those with the most obvious limitations are the soils, especially those which have resulted either from the operation of deep weathering or have formed on weathered duricrusts. The deep-weathering process is characterized, among other things, by:

1 deep and intense weathering of the country rock under high temperature, humidity and organic-acid conditions;
2 rapid decay of organic matter;
3 mildly acid to mildly alkaline (pH 5–8) reaction;
4 decomposition of clays into kaolinite; and
5 the formation of a hard, sometimes impermeable, layer (duricrust) near the soil surface.

The soils themselves belong to different classes which have been variously called lateritic soils, latosols, ferruginous tropical soils, ferralitic soils, oxisols and so forth. The following are some of the features of these soils which limit their agricultural capability:

1 low cation exchange capacity (CEC);
2 low base saturation in the colloids, owing to excessive hydration;
3 rapid removal of bases and low absorption and adsorption capacities of the clays;
4 accumulation of the sesquioxides in the upper part of the profile;
5 a predominance of the kaolins, or a tendency towards these conditions.

Consequently, these soils are not the best for agricultural purposes. The initial fertility, where this exists, soon disappears following clearing and cultivation for a few years, thus forcing, under traditional practices, a system involving constant movement to new locations. This system is referred to as shifting cultivation or bush fallowing.

Processes

The negative aspects of humid tropical processes take different forms; the most serious ones relate to flooding and related drainage problems; deep chemical weathering, eluviation, humification and other depletive soil processes; rapid growth of plant and animal life and the associated problems of land management and health; and the problem associated with the phenomenon of landslides and related processes of mass wasting. We discuss some of these briefly in what follows.

Flooding: Floods are among the most dramatic forms of interaction between man and his environment, and they emphasize the limitations of man in his attempt to control the sheer force of nature. When they occur, whether in the developed or the developing world, they are always associated with heavy losses of life and property, misery, hardship, disease and, at times, famine. This is because although man has been responding to floods from time immemorial, the phenomenon remains one of the least understood by hydrologists (including fluvial geomorphologists), engineers, planners and politicians alike. This is even more so in humid tropical regions where floods occur most commonly and with the greatest intensity.

The reasons for the frequency and magnitude of floods in humid tropical regions are connected with the basic causes. The most important cause is rainfall: excessively heavy and/or prolonged rainfall that occurs in several humid tropical regions constitutes the most universal cause. Other floods are associated with snowmelt, coastal storm surges, landslides and dam failure (Figure 13.4). The Ojerami dam in Bendel State, Nigeria, collapsed in 1980 and gave rise to the worst flood in the area for centuries.

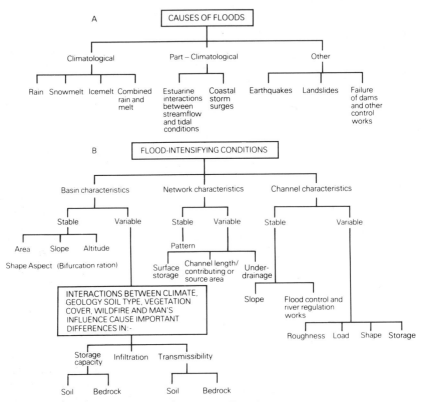

Fig. 13.4 Floods: causes and intensifying conditions

That the humid tropics are the areas most liable to the biggest floods is perhaps best illustrated by Ward's (1978) account:

In the early 1970s alone there were at least ten major flood disasters around the world. A tropical cyclone struck the shores of the Bay of Bengal in November 1970 and devastated the emerging nation of Bangladesh. . . . 6 000 people died on just one island. . . . In the following year flooding in India caused damage exceeding £300 million. June 1972 saw Hurricane Agnes cause the single largest natural disaster in terms of damage, in the history of the United States with estimated losses approaching 3 000 million dollars. In August

1973, the worst floods for a century hit Mexico. . . . In the same month 1 500 people died and damage worth £400 million occurred as floods devastated huge areas of Punjab and Sind Provinces of Pakistan [p. 1].

Other cases reported by Ward (1978) include those of eastern Australia, Brazil and Honduras, all in the humid tropics. However, it is the situation in Ibadan, south-western Nigeria, that is more familiar to the authors. Here, third-order rivers (the Ogunpa and Kudeti) break their banks so frequently that flooding in the area has been described as the Ibadan annual festival (Akintola, 1974). The latest of these events occurred on 31 August 1980, following 275 mm of rain in nine hours, and caused large-scale damage to property and life, the effects of which were still being felt in such vital services as water, electricity, telephones and road transport more than six months after the event, in February 1981.

Tropical weathering: All natural materials are subject to weathering processes, espcially if they were formed under conditions different from those of the present environment. The weathering process, especially chemical weathering (see Chapter 3) is most intense in humid tropical regions and so has both the most beneficial as well as the most deleterious effects. Examples of such destructive effects, which unfortunately have not been widely studied, include deep weathering of solid rocks and the problems created for civil-engineering construction; destruction of fabric through the process of hydration, hydrolysis and so on, the rapid decomposition and mineralization of organic matter, leading to humus-deficient soils; and eluviation, caused by the constant downward movement of soil water, again leading to impoverished soils. In connection with civil engineereing works, it is on record that additional expenditure has been incurred in the attempt to locate bedrock for bridge, dam and other structures, either by boreholes or through geophysical surveys.

Mass wasting: As with floods, the areas of maximum effects of mass wasting include the humid tropical regions (cf. Chapter 4), where the high rainfall, deeply weathered regolith, which is rich in clays, and the constant flow of water in basal channels (through flow, including the lateral flow on basal surfaces) induce the movement downslope of the overburden. Subsidence, or the vertical downward movement of surface rocks, is also prominent. Not all forms of mass movement on slopes are of geomorphological significance, and only the major, more dramatic types, such as landslides, mudflows, subsidence and some clumps, are relevant here. Again, as with several other humid tropical phenomena, detailed studies of examples are lacking. Nevertheless, there are numbers of reports of landslides which have blocked roads and drainage ways, of buried buildings – sometimes whole settlements – and of 'annexed' agricultural lands. Two examples were described recently from the forest region of south-western Nigeria (cf. Burke and Durotoye, 1970; Jeje, 1979). In the first, a granite boulder about 200–300 tonnes in weight was reported to have slid over a distance of about 80 m following heavy rainfall, damaging all the trees, including cocoa trees, and other crops in its path. In the second example, Jeje (1979) described the movement of about 250 m³ of earth over a distance of 355 m, uprooting the cocoa trees, among other crops, along its way. Slumps of widespread distribution have also led to crop damage, and structures such as buildings and telegraph posts on unstable slopes have either been pulled down completely or otherwise damaged. Finally, many houses in Gombe, Bauchi State, Nigeria, have developed cracks under the influence of a slowly subsiding foundation, caused by the removal of the underlying limestone sub-structure.

Man and land resources in humid tropical regions

As in other regions of the world, man in the humid tropical regions has discovered several of the resources mentioned in this chapter and exploited them for his various needs. The rate of exploitation has been accelerating of late, especially following the establishment of colonial administration and of foreign trade. In particular, the Western world has found many of the agricultural and mineral resources indispensable and so has done everything possible to encourage their exploitation and shipment to the industrialized (metropolitan) countries. Consequently, the impact of man is being felt in various humid tropical countries so much as to warrant mention here.

One way in which man has made his impact felt in humid tropical regions is in what may be described as man-made landforms. Man's geomorphic activities have, unfortunately, not been given adequate attention in this region, except perhaps in connection with accelerated soil erosion and the development of gullies. However, man-made and related landforms result from several other activities of man, including road building and other civil-engineering constructions, mining and refuse disposal.

The rapid rate of infrastructural development in many developing countries of late has created new artificial landforms, including road and other cuttings, dumps, embankments and jetties, many of which have induced geomorphological processes such as slumping, sliding and vertical collapse of surface material and severe erosion. Mineral extraction is another activity, especially the open-cast and adit systems, which directly create landforms.

Another outcome of resource utilization is the physical pollution of the environment. Opinion is divided on the extent and seriousness of this process in tropical regions generally, but anybody who has visited humid tropical cities or travelled extensively in the countryside would certainly be amazed at the extent to which our environment has been spoiled. Examples are the refuse dumps, and the blocked and/or washed drains (Faniran, 1980b) in the cities and the extensive deforestation and soil erosion in rural areas. The relationship between man and land resources needs to be addressed if the resources are to continue to serve man, not only in the humid tropics but also elsewhere.

Conclusion

Much of the material that has been covered in this chapter is yet to be studied in detail. The tropics constitute the major field of operation of applied geomorphology generally and in the application of geomorphic research to environmental management in particular. Examples of what needs to be done in each area were given by Cooke and Doornkamp in their book *Geomorphology in environmental management*, published in 1974. It is a matter for regret that since the publication of that text, not much progress has been recorded in any of the recognized fields, especially in humid tropical regions. An example is the issue of geomorphological mapping, where little or nothing has been done apart from writing about it. Even in the much easier matter of tourism, countries in the humid tropical world are lagging behind others, even though they have much to publicize in the form of varied landforms, materials and processes.

The picture is not all gloom, though. Large tracts in several countries have been mapped, albeit at reconnaissance scales for land capability purposes. Examples are Papua and New Guinea, the Northern Territory of Australia mapped by the Australian Commonwealth Scientific Industrial Research Organization, and the large tracts of Africa mapped by both the British and the French government agencies and private companies. Although biased generally towards agricultural development projects, these reports highlight specific problems and possible solutions to them. What is called for is co-ordinated and integrated scientific research into the various dimensions of the forms, materials and processes (including their utilization), not only in the humid tropics but also other regions.

References and further reading

Aitchison, P. J. *et al.* (1969) *The land resources of north-eastern Nigeria*, LRD/Overseas Devt. Administration/Foreign and Commonwealth Office, Publ. No. 9.
Akintola, F. O. (1974) 'The parameters of infiltration equations of urban land use surfaces', Ph.D thesis. Univ. of Ibadan.
Areola, O. O. (1977) 'A review of landfacets as soil mapping units with particular reference to the L.R.D. survey of the soils of the Western State, Nigeria', *Savanna*, **6** (1), 85–9.
Asseez, L. O. (1972) 'Rural water supply in the basement complex of Western State, Nigeria', *Bull. Intem. Assoc. Hydrol. Sci.*, **175**, 97–110.
Bawden, M. G. and Tuley, P. (1966) 'The land resources of Southern Sardauna and Southern Adamawa Provinces, Northern Nigeria', LRD/DOS No. 2.
Burke, K. C. (1967) 'The scenery of Ibadan', mimeo, Dept. of Geology, Univ. of Ibadan.
Burke, K. C. and Durotype, B. (1970) 'Boulder slide on an inselberg in the Nigerian rainforest', *J. Geol.*, **78** (6) 733–5.
Chorley, R. J. (ed.) (1969) *Water, Earth and Man*, Methuen, London.
Christian, C. S. and Stewart, G. A. (1964) *Methodology of Integrated Surveys*,

UNESCO, Conference on Principles and Methods of Integrating Aerial Survey Studies of Natural Resources for Potential Development, Toulouse.

Christian, C. S. and Stewart, G. A. (1968) 'A methodology of integrated studies', *Proc. Toulouse Conference, 1964, UNESCO, Paris*, 135–280.

Cooke, R. U. and Doornkamp, J. C. (1974) *Geomorphology in Environmental Management*, Clarendon Press, London.

Dowling, J. W. F. (1968) 'Land evaluation for engineering purposes in Northern Nigeria', in *Land Evaluation*, G. A. Stwart (ed.), Macmillan, Victoria, Australia, pp. 147–59.

Enslin, J. F. (1943) 'Basins of decomposition in rocks', *Trans. Geol. Soc. S. Africa*, **46**, 1–12.

Faniran, A. (1971) 'Implications of deep weathering for the location of natural resources', *Nig. Geog. J.*, **14** (1), 59–69.

Faniran, A. (1972) 'River basins as planting units', in *Planning for Nigeria: a geographical approach*, K. M. Barbour (ed.), Univ. of Ibadan Press.

Faniran, A. (1973) 'Geography and the construction industry in present day Nigeria', *Nig. Geog. J.*, **16** (2), 151–60.

Faniran, A. (1974) 'The extent, profile and significance of deep weathering in Nigeria', *J. Tropical Geog.*, **38** (1), 19–30.

Faniran, A. (1975) 'Rural water supply in Nigeria's basement complex, a study in alternatives', *Proc. 2nd World Congress on Water Resources*, **3**, New Delhi, 89–100.

Faniran, A. (1980*a*) 'On the definition of planning regions: the case for river basins in developing countries', *Singapore J. Tropical Geog.*, **1** (1), 9–15.

Faniran, A. (1980*b*) 'The shallow water reserve in basement complex areas of south-western Nigeria', *24th I.G.U. Congress, Tokyo, Japan, Abstract*, **1**, 234–5.

Faniran, A. (1970*c*) 'Economic and ecological aspects of solid waste disposal in developing agricultural economies: a case study of Ibadan', *Proc. Assn. Adv. Agric. Sciences in Africa*, Douala, Cameroon.

Faniran, A. (1981*a*) 'On the outstanding issues in natural resources inventory and utilization', paper presented to *24th Nigerian Geog. Assoc. Conf., Kano.*

Faniran, A. (1981*b*) *African Landforms*, Heinemann, Ibadan (in press).

Faniran, A. and Ojo, O. (1980) *Man's Physical Environment*, Heinemann, London.

Faniran, A. and Omorinbola, E. O. (1980) 'Evaluating the shallow groundwater reserves in the basement complex areas', *J. Min. Geo..*, **17** (1), 62–76.

Faniran, A., Sada, P. O. and Areola, O. (1976) 'River and mining, its organization and building industry in the Lagos area', *Quart. J. Admin.* (Univ. of Ife), **10** (4), 423–35.

Faniran, A., Filani, M. O., Akintola, F. O. and Acho, C. C. (1980) 'Improvement of water supplies for small nucleated settlements in rural Nigeria', Dept. of Geography, Univ. of Ibadan; URC, Ottawa.

Harrison-Church, R. J. (1968) 'Dams in Africa, a geographical view', in *Dams in Africa, an interdisciplinary study of man-made lakes in Africa*, W. M. Warren and N. Rubin (eds), Frank Cass, London.

Howe, C. W. (1979) *Natural Resource Economics: issues, analysis and policy*, J. Wiley, New York.

Jeje, L. K. (1979) 'Debris flow on a hillside along Ife-Ondo road, western Nigeria', *J. Mining and Geol.*, **16** (1), 9–16.

Mabbutt, J. A. (1962) 'Geomorphology of Alice Springs area' in *Lands of the Alice Springs Area, Northern Territory, Australia, 1956–57*, R. A. Perry *et al.* (eds), Land Res. Sev. CS. RO. Aust. **6**, 163–78.

Mitchell, C. W. (1973) *Terrain Evaluation*, Longman, London.

Morgan, W. B. and Pugh, J. C. (1969) *West Africa*, Methuen, London.

Murdoch, G. *et al.* (1976) *Soils of the Western State Savanna in Nigeria*, LRD, Ministry of Overseas Development, Tolworth.

Obateru, O. I. (1981) 'Outdoor recreational behaviour of Ibadan residents: a geographical analysis', Ph.D thesis, Univ. of Ibadan.

Omorinbola, E. O. (1979) 'A quantative evaluation of groundwater storage in basement complex regoliths in south-western Nigeria', Ph.D. thesis, Univ. of Ibadan.

Osunade, M. A. (1977) 'An analysis of the physiographic and the parametric methods of land evaluation', Ph.D thesis, Univ. of Ibadan.

Ouma Joseph, P. M. B. (1970) *Evolution of Tourism in East Africa*, East African Literature Bureau, Nairobi.

Stewart, G. A. (ed.) (1968) *Land Evaluation*, Macmillan, Australia.

Thomas, M. F. (1969) in *Environment and Landuse in Africa*, M. F. Thomas and G. W. Whittington (eds), Methuen, London.

Thorp, M. B. (1970) 'Landforms', in *Zaria and its Region*, M. J. Mortimer (ed.), Ahmadu Bello University, Dept. of Geography, Zaria, Occasional Paper No. 4, 13–32.

Ward, Roy (1978) *Floods, A Geographical Perspective*, Macmillan, London.

White, H. P. and Gleave, M. B. (1971) *An Economic Geography of West Africa*, Bell, London.

Appendix glossary of slope terms

The following definitions are based on the work of Young (1963); Savigear (1965); Waters (1965); and Doornkamp and King (1971).

A *slope* is a surface area which is inclined at about 2° to 40°.

A *flat* is a surface area which is horizontal or is inclined at an angle of less than 2°.

A *cliff* is a surface area which is vertical or inclined at an angle of more than 40°.

A *slope unit* consists of either a slope segment or a slope element.

A *slope segment* is a rectilinear portion of a slope profile.

A *rectilinear segment* is one within which the internal variation in angle is less than 2° or 3° (Ahnert 1970). Doornkamp and King (1971) described it thus: 'where two successive measured lengths differ by less than 1° or where three successive lengths differ by less than 2°, they may be considered to form a continuous rectilinear unit'.

A *slope element* is a curved portion of a slope profile.

A *maximum slope segment* is steeper in angle than the slope units above and below, but may also form the lowest unit on a profile having a gentler unit above it.

A *minimum slope segment* is gentler in angle than the slope units above and below it.

A *crest segment* is one bounded by downward slopes in opposite directions.

A *basal segment* is a segment bounded by upward slopes in opposite directions.

A *convexity* is a part of a slope profile marked by a continuous increase in angle downslope and may be terminated by a change of curvature to concavity or by a rectilinear segment.

A *concavity* shows a gradational progressive decrease in angle downslope and may be terminated either by a change of curvature to convexity or by a rectilinear segment.

A *break of slope* is a discontinuity on the ground surface between two slope segments.

An *inflexion* is the point, a line or zone of maximum slope angle between two adjacent concave and convex slope units.

An *irregular segment or element* is a slope element or segment possessing distinct surface irregularities in form of micro segments or elements that are too small to be represented at the scale of the field map.

A *profile sequence* is a portion of a slope profile consisting successively of a convexity, a maximum segment and a concavity.

A *slope profile* is a line orthogonal to the direction of the drainage line, from the latter to the hill or interfluvial crest. A simple profile comprises a profile sequence while a compound type comprises two or more slope sequences.

Index

411